AF323432

About the Biggest, the Smallest, and Everything Else

Travelling Through the Universe with a Physicist Guide

About the Biggest, the Smallest, and Everything Else

Travelling Through the Universe with a Physicist Guide

Serge Parnovsky

Taras Shevchenko National University of Kyiv, Ukraine

World Scientific

NEW JERSEY · LONDON · SINGAPORE · BEIJING · SHANGHAI · HONG KONG · TAIPEI · CHENNAI · TOKYO

Published by

World Scientific Publishing Co. Pte. Ltd.
5 Toh Tuck Link, Singapore 596224
USA office: 27 Warren Street, Suite 401-402, Hackensack, NJ 07601
UK office: 57 Shelton Street, Covent Garden, London WC2H 9HE

British Library Cataloguing-in-Publication Data
A catalogue record for this book is available from the British Library.

Every effort has been made to trace copyright holders and to obtain their permission for the use of copyright material. The publisher apologizes for any errors or omissions in the above list and would be grateful if notified of any corrections that should be incorporated in future reprints or editions of this book.

ABOUT THE BIGGEST, THE SMALLEST, AND EVERYTHING ELSE:
Travelling Through the Universe with a Physicist Guide

ISBN 978-981-125-603-5 (hardcover)
ISBN 978-981-125-604-2 (ebook for institutions)
ISBN 978-981-125-605-9 (ebook for individuals)

For any available supplementary material, please visit
https://www.worldscientific.com/worldscibooks/10.1142/12828#t=suppl

Desk Editor: Rhaimie Wahap

Typeset by Stallion Press
Email: enquiries@stallionpress.com

DEDICATION

I dedicate this book to my teachers, colleagues, and collaborators who have helped me enjoy my life in science throughout my career as researcher and scholar.

PREFACE

What is this book about? About almost everything! The birth of our Universe by the Big Bang; its stormy youth; the appearance of stars; why the Sun shines; how scientists have learned that the Solar System was formed from the remnants of an exploded star; diamond rains on Uranus and clouds of sulphur acids on Venus; our Earth, its interior, oceans, and the atmosphere.

It is also about the reasons why the humankind is unlikely to conquer the Galaxy; about black holes, neutron stars, atoms, their nuclei, elementary particles and quarks; about nuclear reactors, including those created by nature; about rainbows, mirages and lightning. It is about the scientific method and how it appeared; about the differences between science and philosophy.

This is a book about nature and science; about nature, as seen by science, and about science which studies nature. There are many such sciences, but the main one is physics. It, like a master key or a lock pick, opens almost all doors and locks which obstruct the path to the place that contains answers to all possible questions in our world. Science has not arrived at this place yet, but it is on its way.

The great scientist Ernest Rutherford said that all natural sciences are divided into physics and stamp collecting. I do not share Rutherford's extreme assertion and believe that nature is one and the same. As a result, in this book you can find knowledge from different branches of science: astronomy, cosmology, chemistry, geology, meteorology and others, but, most of all, from physics.

If you try to describe in one word what it is about, then the word "Universe" comes to mind. Although the book does not describe everything in the world, the most important known things are probably mentioned here. With one exception: I tried to avoid everything related to life, which deserves a separate big book.

In this book, I take you on a journey that no travel agency can offer you. And I will be your guide on this tour along the ladder of scales from the Universe as a whole to the microcosm and back.

Why this route, specifically? Physics completely permeates our world and many different tracks and footpaths can be laid in it. I suggest going down the scale because I believe this is the most natural choice. Indeed, it would be strange to arrange unusual physical processes and effects alphabetically.

So, welcome to the field trip! On the way, we will make several stops where we will get acquainted with the life and customs of local inhabitants, among which we will single out dangerous creatures. We start with the largest, gigantic object perceived by the human mind — the Universe,[a] and then we descend down the scale of objects, phenomena and processes, ending our journey in the depths of the microworld — the kingdom of fundamental particles and fields. And we will be amazed that we have returned to the very magical place, from which we set off.

Surprisingly, the opposite ends of the ladder of scales — the smallest scale of the microworld and the huge cosmological scale that describes our world as a whole — are often coupled, influencing each other. Physicists now study the Universe together with the properties of elementary particles, that is, the smallest objects together with the largest. Therefore, scales in space behave like the ancient alchemical symbol of the *ouroboros*, depicting a snake biting its tail. Opposites unite to the delight of dialectical philosophers.

---

[a]Sometimes it was also called the Metagalaxy. I remember this term because in some old Soviet encyclopaedias the article "Universe" was commissioned to a Marxist philosopher, but scientific ideas, primarily cosmological and astronomical, were presented in the article "Metagalaxy".

The book is organized in the following manner: we walk through various scales of phenomena, from the Universe to elementary particles, discussing in detail each step of this ladder. In the first chapter this is the entire Universe as a whole, including its formation 13.8 billion years ago, expansion, the appearance of atoms and molecules, and the formation of structures, from galaxies to stars. In the second chapter, we descend to the scale of the Solar System, occasionally recalling our Galaxy. I will tell you about the structure of the Sun and processes inside it, planets, and cosmic rays. In the third chapter, our attention will be focused on the Earth and a little on the Moon. The fourth chapter tells about some of the familiar phenomena of the world around us.

Then we will go to the microcosm. In the fifth chapter, I will introduce you to the mysterious world of quanta and the structure of atoms. In the sixth chapter, we will look at the atomic nucleus and elementary particles. In the seventh chapter, the story will focus on phenomena and objects that do not belong to any one scale or combine the properties of the microcosm and the macrocosm. These are neutron stars, black holes, white dwarfs. And I will also tell you about superfluidity and superconductivity.

The last, eighth chapter does not apply to our journey. It can be considered as a kind of "bonus miles", a gift to travellers who have completed this entire fascinating journey together with me. It will tell about what the scientific method is, how science works and what is definitely not science. And also about the history of natural sciences, about their first steps and about what makes science and philosophy related and different.

Along the way, while travelling, I will describe the most fundamental basis of physics, talk about particles, waves and fields, about forces and interactions, about gravity, electricity and magnetism, about nuclear forces and radioactive transformations, dark matter and dark energy. However, I expect that the reader already knows the most basic concepts of physics and does not need to explain what energy, momentum, angular momentum or frequency of light is.

You will hardly find any formulas here. This is both good and bad. Good, because it makes the book accessible to people who fear or hate formulas. And it is bad, because the language of physics is mathematics, equations and their solutions. Without formulas, it is impossible to adequately describe many concepts and phenomena, which, due to their importance, could not be thrown away. Fortunately, there are many books with formulae available to help fill this gap, from the simplest textbooks to complex scientific books.

I want to thank those who helped write it and improve it, first of all my friend Andrey Varlamov, my son Aleksei Parnowski and my nephew Leonid Parnovski, who read the first version of it and advised how to improve it.

Serge Parnovsky

CONTENTS

Chapter 2.   The Sun and Surroundings. Scale: Solar System   89

LIST OF FIGURES

1   IT COULDN'T GET ANY BIGGER. SCALE: UNIVERSE

"Big is beautiful." This is what fashion magazines, real estate agencies and yacht owners assure us, not to mention those who send out specific spam. But this is best suited to our Universe, which we will talk about. In this part I will tell you about gravity, cosmology, atoms, radiation, dark matter, and dark energy. Some of the concepts mentioned here will be explained in detail in the parts that follow. You will learn about phase transitions in Section 3.2, about the concept of field in Section 2.3 and about true and false vacuum as well as about quantum tunnelling transition in Section 5.3. But for this chapter, you just need a basic knowledge about them.

1.1. How It All Began or Forgotten Birthday

We usually celebrate our birthday: the birthdays of our loved ones, friends, and sometimes historical and state figures, besides New Year or Christmas. But no one celebrates a birthday that should be vital for each of us — the birthday of the Universe. After all, if it did not exist, none of us would live, as well as our planet, star system, galaxy; in general, there would be nothing.

Let us try to bring justice and celebrate this day just by remembering it. First, let us find out when this event happened. I will not name the exact date, since about 13.8 billion years have passed since then.[a] This is a huge amount of time. There is a concept "astronomical time scale," which deals in millions and billions of years. When something changes on such a large scale, it implies that it is practically invariable on the scale of our lives. Therefore, intuitively, it seems that such a mature Universe like ours can change only at time scales comparable to its current age.

But it turns out that this is not at all the case! When the Universe was born, it immediately began to develop and the process of this evolution went very quickly at the beginning. Both the astronomical millions of years and the biblical 7 days of creation seem like huge time intervals compared to the characteristic time scale of the initial stages of the evolution of the newborn Universe. Do you think it was an hour? A minute? A second? The latter assumption is exaggerated somewhere 10^{30} times (physicists, rather than writing such a large number, say "by 30 orders of magnitude").

1.1.1. *Big Bang*

So, the Universe had a beginning, which is called the Big Bang ever since 1949, thanks to the British astronomer and cosmologist Fred Hoyle. Hoyle himself did not believe in the formation of the Universe; he believed in a theory called the stationary model, which is practically forgotten today. Hoyle was one of its authors; therefore, the term Big Bang was chosen specifically. But the irony soon disappeared, and the term took root in both science and the mass consciousness.

When and where did the Big Bang take place? I have already given an estimate of the age of the Universe, but the place can be indicated exactly. This is the tip of your nose. However, any other place will do just as well. If we choose any two points in the Universe and trace their position during a mental journey into the past, then each of them forms what is called a world line in the theory of

---

[a]Although if we decide that this happened on Wednesday, we will certainly be mistaken by no more than 3 days. I agree that this is amazing accuracy for an event that is almost 14 billion years old!

relativity. And they all start from the point of the Big Bang. And no matter how we choose the points, the distance between two pieces of the world lines taken at the same time will decrease to zero as we approach the Big Bang. In other words, the world lines of all points available to us in the Universe started from the same place.

What happened before the Big Bang? No one knows. If you are asked about this and you want to look like a cool scientist, feel free to answer that there was no "before", because time itself appeared at that moment. And before it, time did not exist.

This, of course, is as meaningless as the bombastic answer, but no one can refute it. Indeed, from the point of view of mathematics and physics, the Big Bang is a singularity (from the Latin *singularitatem*), but nothing is known about this mysterious object. We don't even know if the laws of physics we are used to did work at the time of the Big Bang.

And how could we check it? The Universe did not provide us with a second Big Bang. And exploring something unique in a unique Universe, with almost no opportunities to observe an object and having no way to influence it, seems like a task beyond the capabilities of the scientific method, which we will talk about further, in Section 8.1.

In physics, experiments should give repeatable results and preferably carried out by different groups of researchers for verification. How is it done in the Big Bang theory? Very bad! Fortunately, the events that took place shortly after the Big Bang left behind some evidence that we will discuss.

But let us start with an unexpected question. I wrote that the Universe did not provide us with another Big Bang. And why did new universes stop being born? Or haven't they? I don't know for sure, but even if a couple of universes were born in the last five minutes in your room, you could have simply not noticed them.

From the outside, they may look like unusual elementary particles, but once you get inside such a particle, you will find a new Universe. But would you be able to return home to announce your discovery?

1.1.2. *Fluctuations*

Physicists believe that the most likely cause of the Big Bang is the so-called quantum fluctuation. For readers who are not well familiar with this concept, it is worth explaining that fluctuations (from the Latin word *fluctuatio*) are random deviations from the average value of physical quantities characterizing a system of a large number of particles; they occur due to the thermal or quantum mechanical movement of particles, which has an element of randomness.

Each of us has observed phenomena associated with the manifestation of these very fluctuations. Gas molecules, due to their thermal motion, can move a little closer to or move away from each other, forming a density fluctuation. A small area with increased or decreased density appears, but in both cases this is the result of thermal fluctuations.

These areas appear all the time in different places and then quickly disappear. When light passes through them in the atmosphere, an area of increased air density acts as a weak converging lens and a low-density area acts as a diverging lens. As a result, light from distant stars begins to twinkle at night, increasing and decreasing their apparent brightness.

During the day, fluctuations appear in a different way. It is known that blue light with a higher frequency is scattered by the smallest particles with sizes smaller than the wavelength of a light wave much stronger than red with a lower frequency. Therefore, the sky appears blue, and the red traffic signal in fog is visible from a larger distance than the green one.

However, after the rain has washed away all the dust from the air, the sky becomes even bluer. In this case, light scattering occurs not on dust grains, but on fluctuations in the density of air; this is called molecular scattering.

So, most readers have seen fluctuations in air density with their own eyes. Physicists are accustomed to fluctuations on both the micro and macro scales. But how can they work on a cosmological scale and give birth to a huge Universe?

1.1.3. *Mass defect*

There is a well-known phenomenon that is called a mass defect. If we take two bodies that are attracted to each other and allow them to combine into one composite object (for example, a proton and a neutron combined into a deuterium nucleus or two such nuclei became an alpha particle), then the potential binding energy transforms into kinetic energy, and then into another type of energy, such as radiation, nuclear excitation, or heat in the case of macroscopic bodies. In any case, energy will be released. According to Einstein's formula $E = mc^2$, this means that the mass of the composite system in less than the sum of the masses of its parts.

This mass defect is well known in nuclear reactions and is the cause for the release of energy in both nuclear decay reactions and thermonuclear fusion reactions; it is the source of energy for atomic and hydrogen bomb explosions, as well as for the energy that consumers get from nuclear power plants. Gravitational defect of masses when they are attracted is less known and less advertised in popular literature. But it ultimately provides power to the hydro-electric power plant.

There are innumerable bodies that are mutually attracted in the huge Universe; therefore, the gravitational mass defect can be significant. If the density of the Universe is greater than a certain critical value, it is said to be closed;[b] its volume is finite, but it has no boundaries, just as there are no boundaries at the surface of the globe with its finite area. In addition, the total mass of such a universe is very easy to remember.

It is 0 (zero). The gravitational defect fully compensates for the huge total mass of individual galaxies and other contents that fill it. And giving birth to something weightless with the help of quantum fluctuations does not look as hopeless as it seems at first glance.

---

[b]When it comes to the Universe around us, this word is written with a capital letter, but the universe as a mathematical model is written with a lowercase letter. Similarly, when we mention the Galaxy, we mean exactly the Milky Way; all other galaxies are written with a lowercase letter.

It is possible that we will never know exactly how the Big Bang happened, but we now have more information about what happened shortly after. Well, we know the most important thing about the Big Bang: it happened.

1.2. Expansion of the Universe

1.2.1. *Inflation of the Universe*

Let us return to the newborn Metagalaxy. Scientists are discussing many important milestones in the history of the early Universe (in cosmology it is customary to talk about the early stages and the early Universe, but they do not use the term "young Universe"), marked by the appearance of new objects or phase transitions (even though water did not turn into ice, the adolescent Universe had enough phase transitions).

We skip the earliest, hypothetical stages, about which we can only speculate, and move on to the one that influenced what we are seeing now, and has had a significant impact. This is the stage of inflationary expansion of the Universe, which ended when it was about 10^{-32} seconds old.

Cosmologists believe that during this short moment, each piece of the Universe managed to expand at least 10^{26} times, and over almost 14 billion subsequent years, approximately 10^{26} times more — a convincing illustration of the fact that any job can be done quickly or slowly.

Cosmological inflation is a rapid, almost exponential expansion of space in the early Universe that lasted for a tiny fraction of a second. It has nothing to do with inflation, which is more familiar to us — an increase in the money supply, which usually leads to an increase in the prices of goods and the cost of services, and a depreciation of money.

The last type of inflation worries the common man much more than does cosmological inflation, but in fact, most likely, without the latter, there would be no money, no goods, or the common man himself. It is possible that the Universe itself would not have lived its 13.8 billion years.

Naturally, a question arises, or rather two questions. What caused inflation, and why did it end and not continue to this day? The first question is difficult to answer. But only because we have a lot of different theories of inflation, created by different scientists, for example, Alan Guth, Andrey Linde, Alexei Starobinsky, and others, and inflation is good at covering its tracks.

More precisely, different mechanisms of inflation provide very similar pictures of the observed Universe after its end. And these mechanisms can be quite dissimilar.

Let us start with the simplest options. The fact that space expanded so rapidly means that the energy density of the Universe was large during inflation and greatly decreased after inflation. One of the possible explanations is the introduction of a so far unknown field, which has existed in the Universe since its birth, but quickly faded away. However, for a brief moment of its existence, it was the driving force behind the inflation of the Universe.

A more complex mechanism of inflation is associated with the concepts of false and true vacuum, which I will discuss in more detail in Section 5.3. It is caused by the decay of the false vacuum, which occurred in a short but not zero time. There are two main possible mechanisms for this decay, which I will illustrate with an example.

Imagine two lakes. One is in the mountains; the other is below them in the valley. The water in each of them is at a local minimum of the potential energy of the gravity field. In other words, it is as low as it can get. The bottom and the coast do not allow the water to go down further below.

But the potential energy levels of water are clearly different. If the lakes are connected with a pipe, water from the upper lake will flow to the lower one. What could be done if there is no pipe? Then the water can fall down either by slowly seeping through the bottom or by making a gully for itself on the shore.

Let us go back to the vacuum and refine the analogy. The lower lake corresponds to a true vacuum with the lowest possible energy; this is its global minimum. The upper lake corresponds to a false vacuum with a local energy minimum that is higher than the global one. While the Universe is filled with such a false vacuum, its increased energy density provides inflationary expansion.

When the vacuum somehow goes into a state with minimum energy, that is, into a true vacuum, inflation stops. The energy released during the transition can heat up the Universe to enormous temperatures. The fact that the Universe was hot is known from observations, such as cosmic microwave background radiation, which I will talk about below.

How did the Universe end up in a false vacuum? It could have just been born like this. Finally, it could be born in a state of "true" vacuum, but in a tiny fraction of an instant some processes occurred that changed the energy densities of the two states. The former vacuum has become energetically less beneficial and has lost the right to be called the proud word "true".

As a result, the Universe suddenly found itself in a state of false vacuum. Imagine that the same two lakes were side by side on a plain. Suddenly, the one that was in a lower place (then it was a true vacuum) was thrown up by some tectonic processes. Let's say a piece of land has risen and turned into a mountain. True and false vacuums were reversed.

But how does the vacuum in the Universe go from one state to another? It can simply transition in the usual way (in our analogy, seeping or even flowing through a pipe); this is called a slow roll-down of the Universe. Recall that this slow slide lasted about 10^{-32} seconds, so it is slow only in the sense that the Universe managed to expand by a massive number of times during this time.

If the false vacuum cannot transform into an energetically favourable true vacuum in the classical way, then it will do it with the help of a quantum tunnelling transition. In Section 5.3, I'll discuss how this capability is implemented in an atomic nucleus that emits an alpha particle. The Coulomb repulsion of two positive charges makes this process energetically beneficial, but atomic forces provide a potential barrier that an alpha particle cannot overcome without quantum effects.

You have to tunnel through this barrier, going practically through the wall. Alpha decay has long been known to physicists, and every act of alpha particle emission from a nucleus illustrates the possibility of quantum tunnelling. It remains to be imagined that the entire Universe is capable of such a trick.

1.2.2. *Multidimensional worlds*

But there are a lot of much more complex and exotic options. Among them, I mention only the one related to the so-called multidimensional theories. We know very well that the world is three-dimensional. Stand in the middle of the room. You can move north or south, west, or east (if the room is not at the geographic pole, of course), you can jump up or crouch down. That is, you can shift along any one of three mutually perpendicular spatial coordinates. Time is often added to them while talking about a four-dimensional space-time.

Time has one important difference from spatial dimensions. We move along the time coordinate from the past to the future, unable to stop or change the direction of movement. And the rest of the world does the same.

But physicists consider theories in which the world has a greater number of dimensions, for example 11. And even though they are not yet supported by anything other than the imagination of their authors, nothing prevents us from considering this assumption. Naturally, the question immediately arises as to why we do not see these additional dimensions and cannot move along them.

The answers are generally divided into two groups, not counting the more exotic options. The first group includes the so-called brane theories (*brane* comes from the truncated word *membrane*). They assume that we exist, for example, on a four-dimensional hypersurface in a five-dimensional space-time.

What do these words mean? A surface (a continuous set of points that has length and breadth but no thickness) is a two-dimensional section of a three-dimensional space. A hypersurface is a section of a multidimensional space which has more than three dimensions. The dimension of a hypersurface is less than the dimension of the space, and it has at least three dimensions. We can consider the three-dimensional world around us as a hypersurface of a four-dimensional space-time.

Is it unclear? Let's give an analogy. We have a film, cloth, or some other two-dimensional object that is hung in the usual three-dimensional space. Insects, unable to fly or jump, move along it; they are tied to this surface inside space. Add one more temporal and one additional spatial coordinate and get the basis for one of the brane theories.

They differ in some aspects, for example, whether gravity propagates there along a hypersurface or straight through five-dimensional space, the laws of this interaction, and a dozen other details. But it is important to note that in these theories, we do not see additional dimensions because this is how the world works. Matter or light cannot go beyond the brane; this property is originally incorporated into the model.

To some extent, we all are also tied to the surface of the Earth. However, we have an idea of the height. One can imagine a world in which its intelligent inhabitants would live on a two-dimensional surface, unaware of the existence of a third coordinate. And this surface is somehow placed or even moved in the usual three-dimensional world. According to brane theories, we are two-dimensional creatures with a single unobservable and inaccessible hidden coordinate. The difference between us and these imaginary creatures is that they exist in a two-dimensional world, and we exist in a three-dimensional one.

The second group of theories suggests that extra dimensions exist, but they are compactified. That is, the sizes of our world in these dimensions are very small, for example, comparable to a value which is called the Planck length. It is equal to approximately $1.6 \cdot 10^{-35}$ m and often considered the minimum size that modern physics can deal with.

Naturally, it is difficult to move at such a small distance, so the existence of additional dimensions is almost imperceptible. This can be demonstrated with a very simple example. Take a piece of paper, that is, a two-dimensional object, and roll it up into a thin paper tube. If you manage to get a tube with a radius of $1.6 \cdot 10^{-35}$ metres, it will be practically indistinguishable from a one-dimensional line segment or a bar of zero thickness. One of the two coordinates becomes compactified and unobservable.

Imagine that our Universe was born with a larger number of dimensions; say, the same 11, as in the so-called M-theory, popular in certain circles of theoretical physicists. Immediately after the moment of their birth, 7 out of 10 spatial dimensions began to compactify, and simultaneously, the Universe grew rapidly along the other three dimensions. In so doing, it was a single process, and inflation ended after the completion of compactification.

As you can see, there are many opportunities for inflation. Is it possible to choose a real one from them, discarding beautiful, but incorrect hypotheses? It is very difficult. Basically, cosmologists are guided by the period of the very end of inflation, its completion.

All mechanisms provide some perturbations that arise at this stage and their marks can be found by analysing the data of astronomical observations. Different mechanisms generate different disturbances, both qualitatively and quantitatively. But for now, we can discard the obviously incorrect mechanisms, rather than singling out the only correct one. We are not even talking about identifying a certain group of possible mechanisms.

Where did the idea that the very early Universe went through an inflationary stage come from? The answer to this question is not so simple. We have no direct evidence of inflation. That is why the Nobel Prize in Physics was not awarded for this theory, despite repeated tattle on this topic.

However, the idea of inflation made it possible to brilliantly solve several different problems that puzzled cosmologists and astronomers. These include the problem of the flatness of the Universe, the constancy of the cosmic microwave background temperature, the large number of photons per baryon, and the absence of observed exotic particles and objects that could form with the rest of the Universe in the Big Bang. The details of these problems require information, which I will provide later, so we will return to the question of why the idea of inflation is so attractive in Section 1.6.

So, we are convinced by some indirect evidence that the inflationary stage existed shortly after the Big Bang. Each of them is not persuading us by itself, but together they look quite impressive. All of them are associated with predictions by the theory of inflation and have astronomical confirmation. I believe that this stage has earned its own separate section as the oldest event in the Universe, with at least some observable confirmation. But we do not know of any direct evidence of inflation.

The concept of the inflationary stage underlies the popular nowadays hypothesis of the Multiverse, which is described in Section 8.3. The reason it is described separately rather than in this section is quite simple. This beautiful theory, unfortunately, fails the falsification test. It is impossible to refute it. But it is interesting as a complete logical construction based on both science and philosophical ideas.

1.2.3. *What does it mean that the Universe is expanding?*

So, the Universe successfully passed the inflationary stage and continued its usual business. And what is the Universe constantly doing? It is expanding! Just a few decades ago, when our knowledge was less accurate and reliable, cosmologists assumed that it is possible that in the distant future it will stop expanding and begin to shrink, ending this process by universal compression into another singularity, which is called the Big Crunch. But as of now, we know that this option is not possible and the Universe is likely to expand eternally. As a result, I have to answer two natural questions: What does it mean to say that the Universe is expanding and how is this manifested?

Let's start with the first one. Imagine a sheet of metal with something painted on it. The sheet is evenly heated and it expands. The sheet remains flat, but all parts of the drawing become larger, retaining their shape, but increasing in size. If you move along the surface in fireproof shoes, measuring the distance between the marks drawn on the sheet, it increases. Moreover, the increase depends only on the initial distance between these marks and the initial and final temperatures, but not on the position or mutual orientation of the marks.

If we draw a measuring ruler next to the drawing on the sheet, the distance between strokes will increase and any part of the drawing measured by an exact copy of this enlarged ruler will contain the same number of strokes as before heating.

You can draw something on the surface of a balloon and then inflate it. All distances between any two elements of the picture increase in proportion to a certain scale. You can use the radius of the balloon as the natural unit of measurement for this scale. In other words, when the ball is inflated, the elements move away from each other, but the ratio of distance to radius is preserved.

1.2.4. *Hubble's expansion and redshift*

We can observe all this in our Universe. We see that distant galaxies are moving away from us. This is called Hubble's expansion. More precisely, they move away from any arbitrary point that is motionless against the background of general expansion. If astronomers exist in one of such distant galaxies, they will find that all other galaxies are moving away from them, and their velocities are proportional to the distances between the alien astronomer and the galaxy observed.

But not only distances, but all the lengths increase in proportion to the change in the scale of the Universe (it is usually called the scale factor). Electromagnetic waves, including light, are characterized by their length. Wavelength is the distance between two adjacent nodes. Therefore, it also changes in proportion to the scale factor. As the Universe expands, the radiation wavelength increases. The frequency of an electromagnetic wave is inversely proportional to its wavelength, so it decreases.

This is called a redshift. Having received the radiation spectrum of a distant object, an astronomer can very accurately find out how many times the scale factor has increased since the moment the light was emitted by this object. After all, the light was emitted a long time ago and could reach us millions and billions of years later. So, astronomers can estimate the distance to the object quite accurately.

What happens to the contents of the Universe as it expands? Let us consider this using the example of a gas in a vessel. If you increase the volume of the gas, for example, by allowing it to press on the piston and expand, both the density and the temperature of the gas will drop. Density decreases because the volume is increasing, and the temperature decreases because the gas does some work by pushing the piston and spends its internal energy on it.

The same happens when the expanding Universe is filled with matter, (but not the mysterious dark energy, which will be discussed in Section 1.9). The matter cools down, and its density and pressure drop. If we use a cylinder with mirrored walls and place in it the electromagnetic radiation instead of gas, it can be considered the "gas of photons". Then, with the same initial parameters, the density and temperature of the radiation will drop even faster due to the greater pressure of light with the same energy density compared to ordinary gas. The "more rarefied and colder" principle works here as well. The density and temperature of the matter that fills the Universe, as well as radiation, decrease as the Universe expands.

1.2.5. *Formation of structures: Stars, galaxies, clusters, superclusters*

Note that I did not mention another important quantity — the degree of homogeneity of the Universe. Not everything is obvious with this degree. Lots of people are familiar with the concept of "heat death of the Universe", which appeared in the middle of the 19th century. It was believed that due to the increase in entropy[c] the Universe tends to a homogeneous state with uniform temperature and density. Although the concept of heat death has survived in the public consciousness, science has long abandoned it.

The reason is simple. When discussing an entropy increase, scientists of the past neglected the gravitational field of the matter they were considering. Initially, this was clearly stated in the articles, but over time, the reservation was forgotten, especially in lectures. And when physicists began to take into account the gravitational field, they realized that the state to which everything tends is not matter evenly distributed over space, but matter collected in clumps by their mutual gravitational attraction.

---

[c]Entropy is a measure of disorder. In closed systems, it cannot decrease.

A homogeneous universe is unstable. This was shown by the Soviet scientist Yevgeny Lifshitz. The growth of small fluctuations of density and velocity begins after the Big Bang and the inflationary epoch. As these fluctuations grow, they gather matter into stars, galaxies, clusters, and superclusters of galaxies and form the currently observed large-scale structure.

In recent years, these processes have been actively simulated on supercomputers. The result of this simulation can be seen in Fig. 1.1.

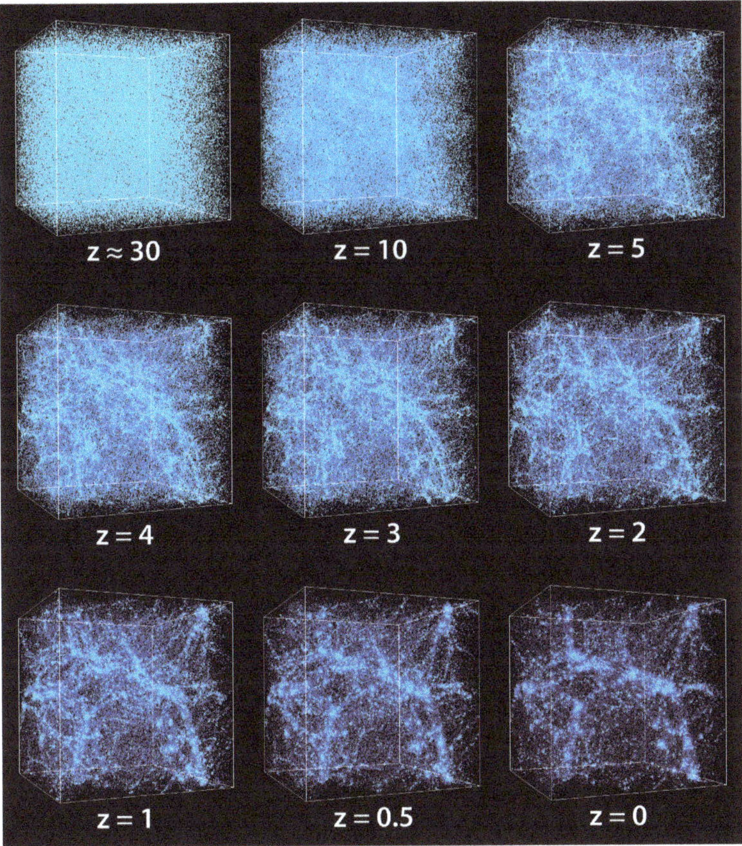

Figure 1.1.   Formation of a large-scale structure in the Universe. Each point in the picture is, for example, a galaxy. It can be seen how an almost uniform distribution of matter turns into lines or filaments, forming a cellular structure over time. Simulations were performed at the National Center for Supercomputer Applications by Andrey Kravtsov (The University of Chicago) and Anatoly Klypin (New Mexico State University). Visualizations by Andrey Kravtsov.

It was carried out at the National Center for Supercomputing Applications by Andrey Kravtsov (University of Chicago) and Anatoly Klypin (University of New Mexico).

The value marked with the letter z is called the redshift parameter. It is related to the scale factor or the size of the Universe. The scale factor of the Universe was $1 + z$ times smaller than it is today at a certain moment in the past, characterized by the redshift parameter z. In the course of evolution, z has fallen from an infinitely large value at the time of the Big Bang to zero right now.

To continue my story about what the Universe was doing in infancy, I need to discuss atoms and molecules.

1.3. Atoms and Molecules

In the 19th century, physicists and chemists jointly established that all bodies are composed of molecules, and molecules are composed of atoms. But the very word *atom* had appeared more than two thousand years before. It was proposed around 440 BC by the ancient Greek philosophers Leucippus and Democritus from the city of Abdera in Thrace. The Greek word ἄτομος means uncuttable or indivisible.

By atoms, they understood something quite different from what modern science means by this word. An atom was something that cannot be split, cut, or broken into pieces. Although other ancient philosophers like Plato, Epicurus, and even poets like Lucretius were fond of speculation about atoms, ancient Greeks and Romans essentially did not contribute anything to the modern concept of atoms, but its name.

Leucippus atoms could have different shapes, sizes, and other characteristics that determine their properties. According to the philosophers of the Abdera school, the atoms of burning fire or salt were sharp, the atoms of solids were rough and hooked, the atoms of water were smooth and slippery, and the atoms of air were light and whirling. The sweet taste was caused by large, round, and smooth atoms, and the soul also consisted of round atoms of various types.

The number of types of atoms is infinite according to Leucippus. There must be a void between them. This idea was strongly disliked by Aristotle, who, through the reasoning described in Section 8.2, came to the conclusion that there can be no emptiness or vacuum, and therefore no atoms.

However, the disputes of the ancient sages have long since ceased to be relevant. Let us talk about what modern science understands by the word "atom". To begin with, we have no doubt about the existence of atoms and the molecules formed from them. We can photograph some atoms with an electron microscope. Optical devices are not suitable for this, because atoms are much smaller than the wavelength of visible light. Physicists from Cornell University in 2018 obtained photographs of individual atoms with a record resolution — less than half an angstrom, more precisely $0.39\,\text{Å} = 3.9 \cdot 10^{-11}\,\text{m}$.

1.3.1. *Mendeleev's table*

In nature, there are no thousands and millions of kinds of atoms. Only atoms of specific chemical elements exist. At the time of the discovery of atoms, the number of known elements did not exceed a hundred, but it has now risen to 118. All information about them is collected in a table known as the periodic table or Mendeleev's table, after its creator, Dmitri Mendeleev.

All chemical elements are located in it in the ascending order of a parameter called atomic number. At first, it was just the number of each element in the table, but then it turned out that the atomic number is equal to the number of electrons in the electron shell of the atom. Electrons are the first type of elementary particles, i.e. objects smaller than the atom, that physicists learned about.

They form something like a cloud of electrons surrounding the compact massive nucleus of an atom, consisting of protons (p) and neutrons (n). These two elementary particles form the nuclei of atoms, so they are also called nucleons. The name comes from the Latin word *nucleus* or "core". So, nucleons are just "pieces of a nucleus" or "nuclear particles". How scientists figured out the structure of the atom is described in detail in Section 5.2.

The masses of the proton and the neutron are very close to each other. When talking about atoms and nuclei, scientists are accustomed to using special units of measurement for the usual quantities. They do not express mass in kilograms, but in conventional atomic mass units, abbreviated as amu, or daltons, abbreviated as Da or u; this is the same unit, but named after the British scientist John Dalton. $1\,\mathrm{Da} = 1.66 \cdot 10^{-27}\,\mathrm{kg}$. It is very close to the mass of a proton or neutron, so the mass of the nucleus, expressed in Da, almost coincides with the total number of nucleons in it.

1.3.2. *Atoms and ions*

The atomic nucleus is surrounded by electrons. Their negative electric charge compensates the positive charge of the nucleus provided by protons. More precisely, if it does not compensate and we are dealing with a charged object, then it should be called an ion. It can be called an atom only when it is electrically neutral. Ions can be either positive, meaning they have fewer electrons than the atom of this chemical element, or negative, meaning they have more electrons than the atom.

The name "ion", derived from the Greek word ἰόν or "being", was coined in 1834 by the British physicist Michael Faraday. He also referred to cations as positively charged ions moving in a solution to a negatively charged cathode and anions as negative ions moving to a positively charged anode. Chemists also consider complex ions containing several atomic nuclei, but in this book I write exclusively about simple ions with one nucleus.

The process of transformation of atoms into ions is called ionization. The decay of neutral molecules into ions in electrolytes is electrolytic dissociation. The creator of the theory of this process, Svante August Arrhenius, received the Nobel Prize in Chemistry for it in 1903.

We say that bodies are composed of atoms, although we have known for over a century that sometimes these are still ions of different signs. The so-called ionic crystals are made up of ions bound together by electrostatic attraction. An example is a crystal of common salt or sodium chloride (NaCl). Its solution or melt conducts electric current well because of the presence of positive sodium ions and negative chlorine ions.

1.3.3. *Lifetime and half-life*

I mentioned three particles that are required to create an atom. Electrons form a shell, and nucleons form atomic nuclei. Only the nucleus of the lightest hydrogen isotope (sometimes it is called protium) consists of a single elementary particle called a proton. It has a positive electric charge equal in absolute value to the charge of an electron. The proton does not decay.

All other nuclei also contain electrically neutral particles called neutrons. They decay in a free state (weak interaction, which I will tell you about, is responsible for this decay), but can exist in the form of non-radioactive nuclei for an arbitrarily long time. The lifetime of a free neutron is 880 seconds, or nearly 15 minutes, which is almost an eternity for the microcosm.

Since I mentioned a new concept called the lifetime of an elementary particle, I feel obligated to explain what I mean in accordance with the precepts of the philosopher Francis Bacon. It is clear that this does not mean that neutrons exist for 880 seconds before decaying all at once. Each of them will decay after a different time interval, ranging from zero to infinity. The lifetime is the average time it takes for one particle to decay, or the time it takes for the number of non-decayed particles to decrease by a factor of e, where the number of $e \approx 2.72$ is the base of natural logarithms.

Let's take a billion neutrons, find out the decay time of each, and calculate the average time of their existence. For a period equal to this lifetime, out of a huge number of particles, a share will decay, equal to $1 - 1/e \approx 1 - 1/2.7182818 \approx 63.2\%$. Over the next time interval equal to the lifetime, a share of 63.2% of the remaining will decay, and so on.

If, at an arbitrary moment, we consider a set of non-decayed particles, then its properties will be the same. This means that the lifetime of the particles does not change with time. This condition unambiguously determines the law of radioactive decay, which states that the number of non-decayed particles decreases exponentially over time.

The radioactive decay of nuclei is also described by the same law. People usually talk about the half-life, during which 50% of the nuclei of some isotope decays. It is clear that the half-life is shorter than the lifetime, during which 63.2% of the nuclei decay.

Nothing prevents us from linking the half-life of particles or nuclei with their lifetime. It is easy to calculate that the half-life will be equal to the lifetime multiplied by a coefficient equal to $\ln(2) \approx 0.693$. Therefore, the half-life of a neutron, or rather a large group of free neutrons, is 610 seconds, i.e. more than 10 minutes.

1.3.4. *Isotopes*

Back to atomic nuclei. I have already mentioned that the number of protons in the nucleus determines the ordinal number of a chemical element in the periodic table. The number of neutrons in the nucleus determines which isotope of this chemical element we are dealing with.

The point is that there are atomic nuclei with the same number of protons, but different numbers of neutrons. These are isotopes (from the Greek words ισος — "equal", and τόπος — "place") of one chemical element. They differ in atomic mass, which is indicated in isotope notation as a superscript before the chemical symbol. The mass is given in daltons, so it is close to the number of nucleons in the nucleus of a given isotope. For the discovery of isotopes of chemical elements, the English radiochemist Frederick Soddy was awarded the Nobel Prize in Chemistry for 1921.

Different isotopes of the same element have the same chemical properties, but different physical ones. Therefore, they manifest themselves in the same way in chemical reactions. In addition to mass, the angular momentum or the magnetic moment of the nucleus may differ.

How to find out the number of protons and neutrons in the nucleus of an isotope by its designation, which includes letters with numbers? The letters correspond to the name of a chemical element and indicate its atomic number, equal to the number of protons in the nucleus. Sometimes, it is indicated directly at the bottom before the element symbol.

The number above corresponds to the atomic mass of the isotope in Da, that is, the total number of protons and neutrons in the nucleus. Isotopes are usually pronounced as the name of an element followed by a number indicating its weight. For example, the isotope ^{238}U (pronounced uranium-238) has 92 protons (the number is equal to the ordinal number of uranium in the periodic table) and 146 neutrons (calculated as 238 minus 92).

Consider stable oxygen isotopes as an example. There are three of them. They, like any isotopes of one element, have a different number of neutrons in the nucleus. The nucleus of an isotope ^{16}O contains 8 neutrons, while isotopes ^{17}O and ^{18}O have 9 and 10 neutrons, respectively. They each have 8 protons and their electron shells contain 8 electrons. The isotope ^{16}O is the most abundant in nature (99.762%). Another 10 oxygen isotopes (from ^{12}O to ^{15}O and from ^{19}O to ^{28}O) are unstable and transform into other elements by radioactive decay. Radioactive isotopes exist, but their lifetimes are limited. This confirms that the distinction in the physical properties of different isotopes can be very significant.

The standard naming convention is not usually used for hydrogen isotopes called by their own names: ^{1}H is sometimes called protium, but this name is rarely used; ^{2}H is called deuterium, and ^{3}H is called tritium. You will come across these names more than once in this book. The letters D and T are often used to denote deuterium and tritium, respectively, as if they were separate elements. The American scientist Harold Clayton Urey received the Nobel Prize in Chemistry for 1934, for the discovery of deuterium in 1931.

Ions are designated in the same way as the corresponding atom, but with the addition of a superscript plus or minus. If the modulus of the ion charge is greater than the electron charge, then the number in front of the sign corresponds to the degree of ionization, which is greater than 1.

Moreover, the designation ^{1}H^{+} is used extremely rarely in physics, because it is the same as a single proton. The deuterium nucleus or ^{2}H^{+} ion is also called a deuteron. There is only one ion of the isotope of another element, specifically $^4\text{He}^{2+}$, which has its own name. It is called an alpha particle, and even then, only when it is a by-product of some nuclear reaction or when one is referred to a stream of alpha particles; in all other cases it is still called the nucleus (or doubly ionized ion) of helium-4.

After I have specified about isotopes, I can go back to the atomic mass unit. There were times when 1/16 of the mass of natural oxygen was used as amu. Later, chemists decided to introduce a new definition of Da associated with the mass of a particular isotope. It is now defined as 1/12 of the mass of the ^{12}C carbon atom.

By the beginning of 2016, 3211 isotopes of all elements were discovered. Only 339 of them are stable or almost stable. Among them, 286 belong to the so-called primary elements, which have existed since the formation of the Solar System. In other words, these are isotopes that were present in the cloud from which the Sun and planets were formed and have survived to this day. Of these, 252 isotopes are considered stable, and 34 have half-lives longer than one hundred million years.

Tin has the largest number of stable isotopes; there are 10 of them, and 26 chemical elements have one stable isotope each. For example, gold is found in nature only in the form of the ^{197}Au isotope. But physicists are able to obtain 36 artificial radioactive isotopes of this element with atomic masses from 169 to 205.

Elements like uranium and thorium do not have a single stable isotope; only their radioactive isotopes exist in nature. However, some of them have long half-lives. All isotopes of transuranic elements are obtained artificially and do not live long, sometimes only a fraction of a second.

1.3.5. *Molecules*

Atoms combine to form molecules. The Latin word *molecula* is a diminutive of *moles* — mass. "Molecule" is the name of the smallest part of a substance that has all its chemical properties. All molecules are assembled from atoms — unique parts of the constructor, although not quite a Lego set.

Some of them, for example, the atoms of noble or inert gases — helium (He), neon (Ne), argon (Ar), krypton (Kr), xenon (Xe) and radon (Rn) — practically do not enter into chemical reactions and do not like to form couples and groups. They live alone like hermits. Therefore, it is believed that they are simultaneously the molecules of the so-called monatomic gases.

In chemical reactions, atoms of other, more active gases such as atomic oxygen or hydrogen are often released, but at the first opportunity they enter into chemical reactions with other atoms or form polyatomic molecules such as molecular oxygen O_2, hydrogen H_2, or ozone O_3.

Most molecules are composed of at least two atoms. And some complex molecules boast thousands or even millions of atoms in their composition. These include proteins, polymers, and many other high-molecular compounds.

However, in some substances, such as crystals, it is difficult to talk about individual molecules. At high temperatures, atoms are often ionized and turn into plasma — a mixture of neutral atoms, positively charged ions, and electrons. Under very exotic conditions, for example, inside a neutron star, instead of atoms, you will find a soup of neutrons, protons, and electrons.

You see how many "buts" have to be added to the simplest statement that matter consists of molecules, which in turn are made of atoms. In Section 1.9, I will tell you about the mysterious dark matter, which does not consist of atoms.

1.4. Formation of Chemical Elements and Other Contents of the Universe

As the Universe expanded, its temperature dropped continuously, passing through the temperatures of a variety of phase transitions. This process could have started even before inflation and continued after it. Now they are discussing the Planck era, the era of quantum gravity, the era of grand unification, the era of inflation, electroweak interaction and its violation, the quark era with its transitions in the quark–gluon plasma, the hadronic era, the lepton era, etc. Too many epochs, transitions, and changes of masks? But by the end of all this leapfrog, our Universe was only 10 seconds old!

You've been reading this paragraph for a longer time than all this happened. As a matter of fact, I wrote this stream of strange words only to achieve this effect. But for physicists studying processes in the early Universe, each of them has a very definite meaning and deserves study.

What about the evidence for these stages? Bad! There is no direct evidence of what was happening in the Universe in those first seconds, because there is nothing that would have formed then and would have been observed or recorded by our instruments now. Perhaps elementary particles, but from them it is impossible to understand when and how they were formed.

Of all the signals coming from matter in the inflationary universe, only relic gravitational waves could, in principle, reach us. Light and other electromagnetic waves would be immediately absorbed by the medium (not interstellar, because there were no stars then and could not be, no galaxies, just plasma that fills the Universe).

Gravitational waves could have existed from the end of inflation to the present day, but, like all other sizes, their length would increase by 10^{26} times during inflation and become obscenely large.

At the time of this writing, mankind has recorded the arrival of only a small number of gravitational waves generated by the merging of black holes and neutron stars. The equipment used in this — the LIGO detector — is tuned to detect waves with a much shorter wavelength. To catch relict gravitational waves (not as old as during inflation), we need a detector in space, and as of now, they are just getting ready to create it (information about LISA and eLISA projects can be found, say, on Wikipedia).

We are convinced of the existence of an inflationary stage by some circumstantial evidence. All the other listed stages are the result of the thinking and calculations of theoretical physicists. They proceed from the simple and confirmed (more on this later) fact that the Universe was either born very hot or became such at one of the many stages of its earliest youth.

Temperature is the average kinetic energy of particles making random movements; it is just that it is traditionally measured not in joules or ergs, but in degrees. In the early Universe, there were a lot of them: degrees, ergs, and even joules. The particles thrashed back and forth at speeds beyond velocities that the most powerful accelerator on Earth can provide, and most likely the most powerful one that will be ever built.

The temperature, that is, the energy of these movements, fell during the expansion of the Universe. Therefore, theorists are simply trying to imagine what processes took place with one or another giant energy of particles. With all my curiosity, I will not describe the details of theoretical predictions at all these stages, except for inflationary ones, since they are based on speculation and have a minimum of confirmation.

In addition to phase transitions, the course of the evolution of the Universe was also influenced by what it is mostly filled with. In the very early Universe, there was a period when electromagnetic radiation prevailed in it. But since its density during expansion fell faster than the density of matter, the latter soon became the main component that filled our world. This happened when the size of the Universe was about 3200 times smaller than it is now. Since then, the density of matter has exceeded the density of radiation.

Soon after the Big Bang, atomic nuclei formed. In the very early Universe, there were none, only a mixture of elementary particles, including protons and neutrons. So the protium nuclei ^{1}H, i.e. the protons, were always there from the very beginning. Nuclei of other elements or heavier isotopes of hydrogen could not yet exist, because they would be immediately broken down into protons and neutrons in collisions with very high-energy particles that fill the entire Universe and constantly collide with each other. Then, after a further drop in temperature, the intensity of collisions and the energies of bumping particles decreased and conditions arose for the formation of more massive atomic nuclei.

1.4.1. *Primary nucleosynthesis*

This process of nucleosynthesis, that is, the formation of the atomic nuclei of chemical elements, is the oldest process that has direct confirmation. It occurred in the early Universe, everywhere and simultaneously, and it did not go on for long.

Steven Weinberg's well-known scientific bestseller is called *The First Three Minutes: A Modern View of the Origin of the Universe*. It gives an idea of how long it took to make the first portion of nuclei. It is better to talk about the first 20 minutes, but the title of the book is so popular that it is the most cited assessment of this period among amateurs. It is interesting that this book has not lost its relevance now, although it was written a third of a century ago. And the theory of primary nucleosynthesis was developed even earlier, by Georgiy (George) Gamow and his colleagues.

Georgiy Gamow was born in Odessa and made a scientific career in the USSR, but wanted to move to the West with his family. He had no legal path and planned to simply escape. An attempt to cross the Black Sea to Turkey in a kayak with his wife was thwarted by a storm, but went unnoticed by the Soviet border guards. Gamow's family had to go back and prepare for a new attempt. He and his wife did not return to the USSR after a trip abroad in 1933. He was lucky that the Soviet authorities allowed his wife to go with him, because the wife was usually held as a hostage in the USSR.

A brilliant theoretical physicist, he was not mistaken in his calculations. In 1938, two of his closest friends and colleagues were arrested by the NKVD (predecessor of KGB). Matvey Bronstein was executed by a firing squad, and the future Nobel laureate Lev Landau spent a year in prison, from where he was released at the request of Niels Bohr and under the surety of another future Nobel laureate, Pyotr Kapitsa. It is possible that for Gamow, had he stayed in the USSR, everything would have ended badly.

In 1940, Gamow, who changed his name to George, received US citizenship. He lived in the USA until his death, in 1968. When, during World War II, the USSR deployed an extensive spy network in the United States to collect information about the details of the Manhattan Project to create the atomic bomb, the Gamow family was not spared either. According to the testimony of the Soviet intelligence officer Sudoplatov, who was in charge of this activity, he recruited Gamow's wife to spy for the USSR, threatening to repress the relatives who remained in the USSR. However, it is unlikely that she could provide the necessary information, which went through many channels. Gamow himself did not participate in the Manhattan Project, but he could know some rumours.

Gamow was the author of three Nobel Prize-worthy achievements: the theory of alpha decay, primary nucleosynthesis, and the idea of triplet coding of information in the DNA. He did not live to see the Nobel Prize for the discovery of cosmic microwave background radiation, which he could clearly count on.

But let's get back to nucleosynthesis. In the early Universe, numerous protons and neutrons interacted in collisions, flitting back and forth at tremendous speeds. It is clear that there were many more collisions of two particles than collisions of three or more particles; therefore, reactions requiring three-particle collisions proceeded at a significantly lower speed. Considering the very short period of time of primary nucleosynthesis, we can say that they practically did not go.

Nucleosynthesis went on until the temperature of the particles, that is, their speed fell below the threshold required for the course of synthesis reactions. At the end of this process in the Universe, in addition to protons — hydrogen nuclei — there were also nuclei of deuterium D and tritium T, as well as nuclei of helium-3 (^{3}He), helium-4 (^{4}He), and lithium-7 (^{7}Li). But there were no other, more massive nuclei that are formed when three particles collide or that require a whole cascade of slow-progressing reactions.

The radioactive isotopes tritium and ^{7}Be (it appeared for a moment) quickly decayed and can be ignored. So, after their decay, the total mass share of hydrogen was 75%, ^{4}He — 25%, and quite a few other isotopes: $3 \cdot 10^{-5}$ of deuterium, $2 \cdot 10^{-5}$ of ^{3}He, and 10^{-9} of ^{7}Li.

These elements are called primary. It is amazing that physicists have calculated the composition of the Universe at the time of the completion of these processes, that is, a dozen minutes after the Big Bang, and astronomers were able to verify this prediction. The result was in perfect agreement with the predicted theory of the content of primary elements. This is direct evidence of the correctness of the theory of primary nucleosynthesis.

Interestingly, the 1948 *Physical Review* paper that did the necessary calculations was written by Gamow and his graduate student Alpher. For the sake of beauty, they invited the famous physicist Bethe to be a co-author to make the set of authors look like Alpher–Bethe–Gamow, which resembles the beginning of the Greek alphabet. Naturally, this article is often referred to as the $\alpha\beta\gamma$ article. The fact that the article was published on April 1 adds to its allure.

1.4.2. *All the gold in the world: The formation of elements in stars and in supernova explosions*

The nuclei of elements heavier than lithium were formed inside stars; this process is called stellar nucleosynthesis. The gravitational field of a star, like the walls of an invisible vessel, prevents the nuclei of hydrogen and other elements heated to a high temperature by thermonuclear fusion reactions from flying away. The fusion reactions turn protons, neutrons, and lighter nuclei into the nuclei of heavier elements.

In stellar nucleosynthesis, the frequency of particle collisions was high because of the high density inside the star, and this process lasted for quite a long time, from millions to billions of years. In this case, not only helium nuclei but also heavier elements, such as carbon and iron, are formed from the hydrogen nuclei.

Okay, but there are deposits of uranium on Earth, which is obviously a heavier element than iron. How was it formed? For the answer, we need the last type of nucleosynthesis — the explosive one. If a star explodes like a supernova as a result of a loss of inner equilibrium, a huge amount of energy is released, including streams of particles that bombard the star's material.

In this case, some processes take place that are aimed not at reducing the energy of nuclei, such as normal processes inside stars, but processes of rapid nuclear transformations "under the radiation beam", in which energy is simply absorbed by atomic nuclei. This is how heavy elements arise that cannot be formed in any other way. The explosion scatters these nuclei across space and then a new star or planet can emerge from them.

Our Sun belongs to such "second-hand" stars; it was formed from the remnants of an ancient outburst of a nova or supernova star, enriched in heavy elements that were formed by explosive nucleosynthesis. Calculations show that the Sun with its planets has too many heavy elements, which requires at least two explosions of its predecessor stars. More scientifically speaking, the Sun is a third-generation star.

If you have a piece of gold jewellery — a ring, an earring, in general, everything that contains gold — then take it out and examine it carefully. Each atom of gold that went into it appeared long ago during a supernova explosion. So, your gold is not from a jeweller, but from this not very frequent event. Nature had to work hard to provide gold to those who are crazy about it. And even those who are indifferent to it.

1.5. Light and Other Electromagnetic Waves

We listen to the radio, watch TV, use cellular communications and Wi-Fi, and occasionally take X-rays. Therefore, we have some idea of electromagnetic waves. We can see thanks to them. And we understand that although they are also considered to be matter, they are somehow clearly different from a stool or a cobblestone.

I will tell you the most basic concepts of electromagnetic waves, including light. Over the centuries of the existence of physics, the opinion about their nature has repeatedly changed. Each of the changes marked a revolution in optics. These revolutions are described in Section 5.1, together with the concept of photons that is accepted by all physicists. But in it, we discuss the world of quanta, focusing on a quantum view of electromagnetic waves.

In the meantime, I will talk about the classical point of view. In most cases, it is sufficient to understand the optical phenomena, although not all of them. For example, the principle of operation of lasers cannot be described within the framework of classical physics. In Section 2.4, I will go into greater detail about electromagnetic waves, specifically the polarization of light.

How does a classical electromagnetic wave arise? It is not difficult to create an alternating electric field by swinging an electric charge in the air or by applying an alternating voltage to a capacitor. It is also not difficult to create an alternating magnetic field by swinging a magnet in the air or by passing an alternating current through a solenoid. As a result, a wave will arise that propagates in all directions from a source that creates an alternating electromagnetic field. This process is somewhat similar to the dispersal of ripples on the surface of a pond after a stone falls into it.

If we hang the charge like a pendulum or on a spring and allow it to oscillate, or if we rotate it in a circle, then it will perform periodic motion, returning to the same place after a certain interval of time, called the period of oscillation. It will create a diverging electromagnetic wave that will repeat itself[d] with the same period.

The reciprocal of the oscillation period is called frequency, and we will denote it by the Greek letter nu, which looks like this: ν. The frequency is usually measured in hertz, i.e. in inverse seconds: $1\,\mathrm{Hz} = 1\,\mathrm{s}^{-1}$. The distance between the crests of such a wave is called the wavelength and is often denoted by the Greek letter λ (lambda). The product of these two quantities is equal to the wave propagation speed. A wave that has a fixed frequency and length is called monochromatic. In it, oscillations occur along a sinusoid.

---

[d]Provided the oscillations continue forever. If they once began, then the wave occupies only a part of the space and expands at the speed of its propagation. Strictly speaking, such a diverging wave does not repeat itself over a period of oscillations. But it usually does this at every point within the propagation area.

But no one forces us to limit ourselves only to the periodic movement of the charge. We can do this in an arbitrary way, but the generated electromagnetic waves can always be viewed as a combination of monochromatic waves of different frequencies. Their distribution by frequency or by wavelength is called the emission spectrum.

1.5.1. *Scale of electromagnetic waves: From radio to gamma rays*

Electromagnetic radiation is named differently depending on its wavelength. The longest length is for radio waves; it starts from about 1 mm and can be arbitrarily long. Radio waves are conventionally divided into ranges: long (wavelength over a kilometre), medium (from 100 m to 1 km), and short, which in turn are divided into decametre, metre, decimetre, centimetre, and millimetre waves.

The existence of radio waves was proved by the German physicist Heinrich Rudolf Hertz in 1888. With this, he not only confirmed the predictions of the theory of electromagnetism, but also opened the way for radio communication. He also established the basic properties of radio waves, which completely coincided with the properties of light that were well known at that time. The last gaps in the comparison of light and radio waves were later eliminated in a series of precision experiments by the Russian physicist Pyotr Lebedev.

Let's move on to shorter wavelengths. Infrared (IR) radiation follows the region of microwaves and submillimetre or terahertz waves. This thermal radiation has wavelengths from 1 to 2 mm to the border of the visible light range, more precisely to 0.74 microns. It is conventionally divided into three regions: near (0.74–2.5 microns), middle (2.5–50 microns), and far (50–2000 microns) IR.

Hot bodies emit infrared waves, which is why they are also called thermal radiation. Infrared radiation was discovered in 1800 by Frederick William Herschel, an English astronomer of German origin.

Then, there is a small interval of wavelengths called visible light. Light detectors in the human eye work in the wavelength range from 380 nm (frequency 750 THz, violet light) to 780 nm (385 THz, red). The musician would say that all visible colours fall into one octave, i.e. in the interval from a certain frequency to doubled frequency. But how much information this single octave[e] brings us!

Some animals can also see ultraviolet (UV) radiation that is invisible to us. Its wavelengths range from 10 to 400 nm; astronomers divide this interval into near and far ultraviolet. Naturally, the eyes of animals can only see a small part of the near-UV range. The German physicist Johann Wilhelm Ritter discovered ultraviolet radiation in 1801 while trying to determine which light acts most effectively on silver chloride, which was later used in photography.

Shorter wavelengths of electromagnetic radiation are referred to as X-rays and gamma rays. They differ from each other only in the way they were emitted. X-rays arise when electrons pass from one atomic orbit to another, i.e. their source is the electron shell of the atom. They were discovered in 1895 by the German Wilhelm Conrad Röntgen. He received the first ever Nobel Prize in Physics in 1901 for this discovery.

Gamma rays are emitted by an atomic nucleus. They were discovered in 1900 by the French physicist Paul Villard while studying radioactive radiation from radium.

Naturally, X-rays can differ from gamma quanta only by their frequency and the energy determined by it. Conventionally, the boundary between them corresponds to a wavelength of about 10^{-10} m. Sometimes hard, i.e. high-energy, X-rays have a higher frequency and energy than soft gamma rays, but usually the energy of gamma rays is higher. The gamma rays end the scale of electromagnetic radiation used by physicists, which is shown in Fig. 1.2. At

---

[e]But the frequencies of sounds heard by a human vary by about 1000 times, from 20 Hz to 20 kHz, which is less than 10 octaves. However, we get a lot more information with our eyes than with our ears.

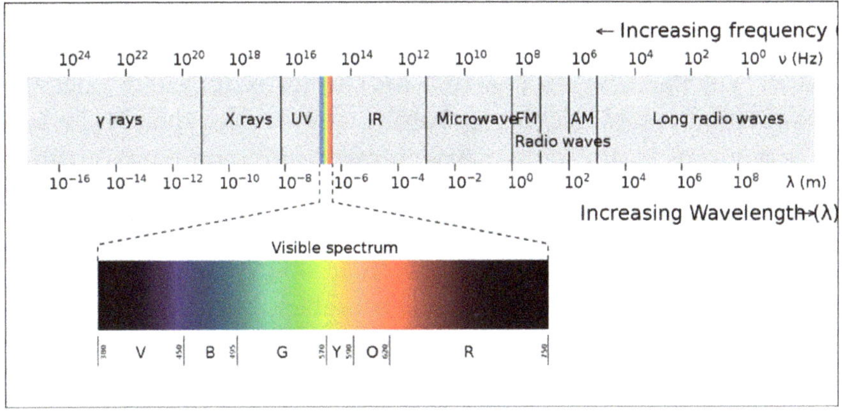

Figure 1.2.   The scale of electromagnetic waves shows the wavelengths and frequencies of radio waves, visible light, IR and UV ranges, and X-ray and gamma radiation. Illustration by Philip Ronan distributed under a CC-BY 3.0 license.

this edge of it, the quantum properties of light play an essential role, and gamma quanta are more similar in their properties to a stream of particles than to waves. I'll cover this in detail in Section 5.1.

Note that both electromagnetic and other waves carry energy, momentum, and angular momentum. Everything is clear with energy; we love to bask in the Sun and cook food in the microwave. The fact that light has an impulse is seen by the pressure of light, which deflects the tails of comets from the Sun and creates a propulsive force acting on the spaceships with the solar sail. The pressure of light was first measured by the Russian physicist Pyotr Lebedev in 1899–1907.

1.5.2. *How are electromagnetic waves emitted?*

It is quite obvious that different electromagnetic waves are emitted and received in different ways. Radio waves are associated with transmitters and receivers, i.e. devices with a set of radio components, with antennas, sometimes quite large.

Microwave radiation is generated by special devices, most often a magnetron, which was invented and patented back in 1910 by the German engineer Hans Gerdien. Naturally, this vacuum tube has been significantly improved over a century. It can be found inside every microwave oven.

IR, visible, UV, and X-ray radiations require significantly smaller emitters. Molecules and atoms are used for this. The vibrations of atoms in molecules cause infrared radiation. The transition of electrons in the shell of an atom from one orbit to another is accompanied by the emission of a quantum of light, usually visible, ultraviolet, or X-ray.

Heated bodies also radiate. The higher the temperature, the shorter wavelengths prevail in the spectrum of thermal radiation. If you heat a metal object, such as a poker or a sword, it will at first simply emit "heat waves" and then heat up and begin to glow with red light, which has the longest wavelength of the visible range.

The hotter molten iron glows with a white light. The high-melting tungsten filament in an incandescent light bulb has an even higher temperature.

The surface of the Sun is heated to 6000 degrees. More massive stars have an even higher temperature and their light is whiter, or rather less yellow, because the radiation spectrum is shifted to the violet side compared to the Sun. The spectrum of less massive stars called red dwarfs, on the contrary, is shifted to the red end. Even less massive stars, called brown dwarfs, have gone further. The record for coolness belongs to a brown dwarf with a surface temperature of 98°C, which is less than that of a boiling kettle. It shines like a kettle in the IR range, but emits no visible light.

The sources of gamma radiation are exclusively atomic nuclei. Here, we hit the natural limit of the electromagnetic wave scale.

Is there something interesting on the other side that I forgot to mention? These are super-long waves with a length of more than 10 km, which include extremely low frequency (ELF) waves with a length of more than 10,000 km. They bend around the curved surface of the Earth and are capable of penetrating water to a fairly great depth. They are used primarily for military purposes, for example, to communicate with submerged submarines.

Here, you cannot do without a particularly long radiating antenna. A version of the Tu-142MR "Eagle" aircraft with a special super-long-wave radio station R-826PL "Frigate" was developed in the USSR. It released an 8.6 km long cable antenna in flight, which followed the plane like a huge tail.

1.5.3. *Telescopes and satellites for astronomy*

Electromagnetic radiation — and in the last couple of years, also gravitational radiation — is the main source of our astronomical information. In addition to these, there are also cosmic rays, including neutrinos, and experiments to search for dark matter particles, the story of which is ahead.

For centuries, astronomers have looked at the sky, first with the naked eye and then with telescopes. Telescopes were getting better until further progress was hampered by the problem of atmospheric influence.

Anyone who has seen the starry sky knows that the stars twinkle. The reason is random changes in the density of air in the path of light, the very fluctuations I mentioned in Section 1.1. Because of them, the image on the photo detector constantly trembles, slightly deviating in different directions, and becomes less clear.

How to deal with the problem? Nowadays the largest telescopes are often located high in the mountains, where the layer of the atmosphere is thinner and its influence is less. At the same time, astronomers try to choose a place with a good astroclimate. This word refers to all the nuances of the weather affecting the quality of the atmosphere and the possibility to make observations. The blowing wind can spoil the image quality. Observations are impossible on cloudy or rainy days because the telescope cannot see through the clouds.

One of the most ideal places for astronomers is the Atacama Desert in Chile. It is not only located at a high altitude but there are also no strong winds. This is one of the driest places on Earth. For four centuries, there has been no significant precipitation here, so astronomical observations can be carried out every day, or more precisely, every night. In addition, the Chilean government is donating land in this desert for free to build observatories that will bring fame and new jobs to the country.

Another method of dealing with atmospheric influences is to try to continuously compensate for the effects of fluctuations in air density. For this, instead of the previously used fixed mirrors, telescopes began to use so-called adaptive optics.

Under the control of the computer, the mirror "moves" all the time, slightly changing its shape on the way to compensate for all random processes in the atmosphere. In recent years, new techniques have been developed in adaptive optics that has dramatically improved image quality.

You can easily find on the Internet images of the planet Neptune taken by a telescope called the Very Large Telescope, which is located in Chile. Perhaps some readers have already guessed that it is not just in Chile but also in the Atacama Desert at an altitude of 2635 m. Just compare the images taken with and without the latest adaptive optics techniques and feel the difference.

However, the problem of the atmosphere can be solved radically. To do this, you need to send a telescope beyond it, into space. This is exactly what NASA did in 1990 by launching the Hubble Space Telescope, which is still a source of invaluable astronomical information.[f] The much larger James Webb Space Telescope was launched on December 25, 2021, successfully flew to its destination, and is preparing to start its work at the end of June 2022.

Launching the telescope into space allows us to solve another problem related to the Earth's atmosphere. The fact is that it absorbs or reflects electromagnetic waves with wavelengths that do not fall into the so-called transparency windows. Visible light and infrared and UV radiation close to it, as well as radio waves, reach the Earth's surface. The harsher ultraviolet radiation is absorbed by atmospheric oxygen and ozone.

But the atmosphere is opaque to many other wavelengths. This is the basis of the well-known greenhouse effect. How to make astronomical observations in these ranges? This is currently possible from satellites only, but it may be possible in the future from

---

[f]NASA celebrated the 30th anniversary of this telescope in 2020. Use the hashtag #Hubble30 to find some articles and photography.

the surface of the Moon. Now, a lot of astronomical satellites are orbiting the Earth, launched by NASA, the European Space Agency (ESA), China, Russia, Japan, and some other countries. Observations are carried out in the IR, UV, X-ray, and gamma ranges, in particular.

1.5.4. *Doppler effect*

I would like to tell you about one important effect, named after its discoverer, the Austrian Christian Andreas Doppler. This is a change in the frequency of a wave emitted by a moving body.

The spectrum emitted by an approaching object is shifted towards the violet side, i.e. the frequency of the waves registered by the observer increases. The spectrum of light from receding objects undergoes a redshift, which means its frequency decreases. This applies not only to light but also to sound.

Moreover, the Doppler effect does not require any waves; any periodic processes or events are sufficient. I will use an analogy that I have encountered in popular science literature, expounding it in my own words.

D'Artagnan rides in a carriage from London to Paris, carrying diamond studs to Queen Anne of France. Every two hours, he writes a letter to Madame Bonacieux in Paris, telling how he loves her. And he writes another letter to Milady Winter in London. Every two hours, he takes two letters written during this time and sends them to the addressees with the help of two carrier pigeons that fly straight to Paris and London at a constant speed.

And what is the result? Constance Bonacieux is pleased because she receives letters at intervals of less than two hours. The reason is simple: each letter is sent two hours after the previous one, but the carriage has managed to get closer to Paris during this time and the pigeon needs to cover a shorter distance. But Milady Winter is offended because she receives letters at longer intervals. After all, the pigeons need to spend more and more time on the way to her.

For the same reason, not only the frequency of receiving the letters changes, but also the arrival of the wave maximums, i.e. observed radiation frequency. The magnitude of the effect depends on the ratio of the speed of movement of the emitting body to the speed of wave propagation.

By observing the frequency of the incoming light or sound, you can find out at what speed its source is approaching or receding. This is actively used by astronomers, doctors, traffic cops, and security alarm specialists.

1.6. Cosmic Background Microwave Radiation

So, the Universe spent the first 20 minutes of its existence on the processes described and not described in Section 1.4. Perhaps it decided that it had already become a genuine universe that survived the primary nucleosynthesis, and should not rush. Be that as it may, the Universe was in no hurry with the next important event. It is called recombination and had to wait for about 380 thousand years. Childhood impatience has passed, and respectable universes can afford some slowness and imposing.

Where have these millennia gone? It took time to cool down. It must be understood that during the period of primary nucleosynthesis, the temperature of the Universe was enormous. Therefore, it took so long to wait for the temperature to fall due to the expansion and reach about 3000 K.

Why are we interested in this particular temperature? At higher temperatures, protons and alpha particles, that is, the nuclei of hydrogen and helium, move at great speeds, and electrons do the same. They all form plasma. Plasma is opaque to light and other electromagnetic waves. Photons are emitted but absorbed almost immediately.

A photon emitted deep in the Sun can travel to its surface for hundreds of thousands of years, all the time being absorbed and emitting again. I talk about this in detail in Section 2.7. Therefore, while the Universe was filled with plasma, it was practically opaque. This is why astronomers cannot observe radiation emitted in the very early Universe.

1.6.1. *The oldest trace of the birth of the Universe*

But after cooling down to 3000 K, the plasma transforms into hydrogen atoms. This process is the reverse of ionization and is called recombination. From this moment on, the Universe becomes transparent to light. Moreover, almost all the light that was emitted at the time of recombination travels through the Universe to this day. Its wavelength has increased by about 1100 times along with the Universe itself. As a result, it is no longer visible or IR light, but microwaves with a frequency of about 160 GHz.

We can observe them using special receivers. These waves are called cosmic microwave background (CMB) or relic radiation. The area in which the relic radiation was emitted is called the surface of the last scattering. This is the oldest thing that humanity has ever observed. There appears to be no way to detect earlier electromagnetic radiation. So, if it is ever possible to detect earlier signals, they must be of a completely different nature, such as gravitational waves, or associated with currently unknown particles or fields.

I have already mentioned Gamow, who did not live to receive the Nobel Prize for the discovery of CMB radiation, which was awarded in 1978 to two American engineers — Arno Allan Penzias and Robert Woodrow Wilson. They got half of the prize; the second half was received by Pyotr Kapitsa for completely different discoveries. The relic radiation was discovered quite by accident while tuning the antenna for space communications. Penzias and Wilson did it before the team of cosmologists, who purposefully searched for microwave radiation from the Big Bang, in addition using the equipment developed by the same cosmologists. The cosmologists did not receive the prize at that time.

That changed in 2019, when half of the Nobel Prize in Physics was awarded to one of the leaders of this team, Phillip James Edwin Peebles, with the formulation "for theoretical discoveries in physical cosmology".

Naturally, the relic radiation is an irreplaceable source of information about the early Universe. It is studied both from the surface of the Earth and from satellites and balloons. The goal is to observe and measure the parameters of radiation coming from different parts of the sky, that is, from different parts of the surface of the last scattering.

This is the easiest to do in space, so data on the CMB was successively received by the spacecraft *COBE* (The Cosmic Background Explorer) and *WMAP* (Wilkinson Microwave Anisotropy Probe) from NASA, their missions culminated in the receipt of the 2006 Nobel Prize in Physics, and *Planck* of the European Space Agency.

Data from high-altitude balloons, including those obtained in Antarctica, supplement them, especially in the area of high angular resolution. The distribution of temperature and polarization of the relict radiation over the sky is investigated.

What temperature are we talking about? Plasma, immediately before recombination, emitted waves of different lengths according to the laws of radiation of a heated body. Moreover, I mean a body very similar to the ideal absolutely black body, so beloved by theoretical physicists. I will describe it in Section 5.1.

We observe CMB also as thermal blackbody radiation, but due to the expansion of the Universe and the radiation wavelengths, its temperature decreased by the same 1100 times as the wavelengths increased. This follows from the laws of blackbody radiation, which state that radiation is emitted most of all with a wavelength inversely proportional to its temperature.

Now the radiation temperature is only 2.725 K. It is difficult to call such a cold body heated. But at the moment of recombination, the temperature of the luminous medium was $2.725\,\mathrm{K} \times 1100 \approx 3000\,\mathrm{K}$.

The very fact of the existence of the CMB proves that the Universe at the beginning of its existence was hot, very hot. It is not so important whether it was born that way or warmed up when it exited from the inflationary stage. The thermal nature of the CMB is reliably confirmed by the fact that the distribution of its intensity over wavelengths is the same as that of a heated body.

1.6.2. *CMB tells about the early Universe*

However, the temperature of the radiation coming from different parts of the sky is insignificant, but different. This is called the CMB anisotropy. It is very small, and the standard deviation of the temperature is approximately $18\,\mu K$ ($1\,\mu K = 10^{-6}\,K$), that is, less than 0.01% of the average temperature.

The cause of the anisotropy is fluctuations in the parameters of individual parts of the surface of the last scattering, such as temperature and density. It can be seen that they were small at the time of recombination, and the Universe was practically homogeneous. On the Internet, you can find the map of the temperature distribution according to the data of the Planck space mission.

Scientists were able to obtain a lot of useful information when processing this data. For example, it turned out that our Universe is practically flat, that is, if its space is curved on a large scale, it is insignificant. This is exactly what the inflation theory predicts; without it, this result could be explained only by chance.

By the way, the CMB itself as a whole also confirms the theory of inflation. Without inflation, areas that could somehow exchange energy and have the same temperature would occupy areas on the map only a few angular degrees. Add a couple more problems that I kept silent about, but which also worried cosmologists, and you will understand why the theory of inflation was accepted by them with a bang.

Although we do not have direct evidence of the existence of an inflationary stage of the Universe in the past, there are many indirect ones. Some of them can be explained by other reasons, but none of them can clarify other observed manifestations of inflation. And then we must assume the simultaneous existence of many causes, one for each manifestation.

William of Ockham does not like this and threatens those who increase the number of entities beyond what is needed with his razor. Well, if this mediaeval monk and his razor are unknown to you, you can meet them in Section 8.3.

1.7. Forces in Nature

What do all processes in nature have in common? In them, some bodies or fields somehow affect others. Or, as physicists say, they interact with them. What is meant by this word? Sometimes they are a source of force that acts on others. This is the simplest example of interaction.

When we hear the word *force*, we usually think of brute mechanical force. Everyone who has lifted heavy objects or dragged a treasure chest full of silver piastres from a sunken pirate ship has an idea of it. In the boxing ring, one boxer's fists interact with the head and body of another. Naturally, interaction can be more complex in science, for example, causing the transformation of some objects into a set of others. Let's try to understand all these terms, starting with the simplest examples.

1.7.1. *May the force be with you*

As far as I mentioned boxing, why not use examples from sports, especially since the Olympic motto "Faster, higher, stronger!" or "*Citius, Altius, Fortius!*" in Latin directly mentions a force?

The athlete throws a hammer or puts a shot and the hammer flies into the distance. The force of its weight acts on the hammer. This is the result of its gravitational attraction to the Earth or, in other words, the impact of the Earth's gravitational field on him.

The hammer moves almost in a parabola — almost because, in addition to the gravity of the Earth, much weaker forces act on it due to the influence of the Sun, Moon, Jupiter, the Galaxy, other galaxies, etc. A little further, in Section 1.8, I will talk about the influence of these forces, called tidal ones. They got this name because they are responsible for the regular change of high and low tides on the coast. The most important thing is that tidal forces are weak compared to the influence of the Earth's gravity. Both a high jumper and a weightlifter lifting a heavy barbell fight against gravity.

The hammer is also influenced by forces associated with the Earth's atmosphere. Air resistance slows down a flight, especially if you don't throw a hammer, but a lump of fluff or an inflated balloon. If you throw a correctly folded paper airplane, then a lift appears, which partially compensates the Earth's gravity and allows the airplane to fly longer.

But after all, gravity also affects the thrower, and usually no one calls him weightless. Why doesn't he fall down to the Earth's centre? If he stands on the podium or even away from it, then the force of his attraction to the Earth is completely compensated by the force called the normal force. It is the component of a contact force that is perpendicular to the surface of the podium. Simply put, this is the force with which the podium presses on the soles of the thrower's feet, or the reaction of the floor to the attempts of our feet to deform it by the force of our weight. We, too, do not fly to the centre of the Earth, even if we are just spectators. Of course, if we are not skydivers.

There are situations where weight and reaction forces do not cancel each other out. This happens when a soccer ball is placed on a side of a mountain or an inclined plane and it rolls off from there. Or when the weight is hung on a string and it swings like a pendulum.

The tension force which occurs when you try to stretch the thread is directed along the thread, and the force of the weight is strictly vertical; therefore, their sum provides a force that returns the pendulum deviated from the position directly below the suspension point.

And then there is the friction force. Imagine a runner who is not on the usual running track or treadmill, but on very slippery ice. He is clearly not up to records. But reducing the friction force helps the ice-skater and curler.

Try pushing a sumo wrestler or a heavy cupboard slightly to the side.[8] For some reason they won't budge. This is because a horizontally directed friction force of the cupboard on the floor appears, fully compensating the force with which you push it. To move the cupboard from its place, it is necessary to apply a force exceeding the maximum possible frictional force of the floor.

---

[8]For safety reasons, I advise you to carry out this experiment on a cupboard.

In addition to those already mentioned, there are many other forces at work in nature. A curve ball thrown by a football or baseball player, or a rotating table tennis ball, does not move like a cannonball because it is additionally acted upon by a force caused by the combination of its translational movement and rotation and associated with the action of air.

1.7.2. *Inertial frames of reference*

So what is a force? It is defined in Wikipedia as any interaction that, when unopposed, will change the motion of an object. In mechanics, the effect is manifested by a change in the speed of the body on which the force acts. If there is no force, then the speed of this body does not change. This statement is called Newton's first law.

Note that if the speed is constant, then it does not follow that there are no forces acting on the body, but only that their sum, called the resultant force, is zero. For example, if we clench a sponge by a palm, then the forces will obviously act on it, otherwise what has squeezed it? But their sum is zero, so the sponge speed will not change. But some force is certainly involved if the speed changes.

A nonzero total resultant force changes the speed of a body. Its acceleration is equal to this force divided by the body's mass. This is Newton's second law, which includes not only a qualitative statement, like the first, but also a formula. It is clear that this formula implies that the acceleration of the body vanishes at zero total force, i.e. its speed is constant. So, Newton's first law is a special case of the second law. The problem is that Newton's first law does not always work straightforwardly, but only in special frames of reference, which are called inertial.

Imagine that you are sitting in a train at the platform. Through the window you see the immobile building of the railway station. The resultant force acting on it is clearly zero, so the speed of the station is constant. But the train starts to move and we see that the building begins to move faster and faster towards the back of the train. What force sets it in motion?

You were probably taught something similar to this reasoning in your school physics course. You are using a frame of reference in which you yourself are stationary. In this case, it coincides with the frame of reference attached to the train carriage. While standing at the platform, this system was inertial or almost inertial. But as soon as the train began to accelerate, the system became non-inertial, and Newton's first law does not apply in such a nasty system.

Everything is fine, but two problems arise. First: why do we need laws that do not always work? Second: if they work only in exemplary reference systems, called inertial, then how to find such a "chosen" system, at least one? All other inertial systems move uniformly and rectilinearly relative to it. Newton proposed to use a special system, fixed to distant stars. Now we understand that distant stars are moving relative to each other and they cannot be used to determine the inertial system as Newton assumed.

1.7.3. *Non-inertial frames of reference*

But physicists solved this problem back in the 19th century. Who said that only inertial reference frames can be used? In mechanics, non-inertial systems are also used, which are not so ideal. Additional forces act on the body in this case due to the imperfection of such a system. They are called inertial forces.

Let me clarify that although the forces of inertia are somehow related to motion by inertia, they are still different concepts. At one time, natural philosophers, following Aristotle, believed that movement requires a constant application of force. The movement stops instantly as soon as the force runs out. It was the great Italian scientist Galileo Galilei who proved in his experiments that this is not so.

I will talk about the ideas of Aristotle and the discoveries of Galileo in Section 8.2, but for now I will just remind you that an object on which the force has ceased to act continues to move until it is stopped by another force, such as friction or resistance.

You can take a run and slide quite far on slippery ice. If you're lucky, then you will finish on your feet. The Moon orbits around the Earth for billions of years. In both cases, nothing pushes the person or our satellite forward. This movement is called inertial motion.

But this phenomenon does not depend in any way on what frame of reference we use to describe it. It just exists.

Forces of inertia appear only if we describe motion using a non-inertial system. And the question immediately arises: why do we need this? Maybe you shouldn't contact dubious companies and give a vow to lead a healthy lifestyle and not use non-inertial systems? Alas, this is hardly possible (I'm not talking about a healthy lifestyle). There are two reasons for this.

I'll start with a more fundamental one. If we decide to abandon non-inertial systems, then we will have to use inertial ones. Where can I get them? Let me remind you that they do not accelerate or rotate. Newton thought a system of fixed stars would do. Now we know that the stars of our Galaxy revolve around its centre, the stars of other galaxies revolve around their centres, and the galaxies themselves move relative to each other. Where can you find these fixed stars?

In the most accurate system of astronomical coordinates in use today, astronomers took care of the practical absence of rotation. The orientation of its axes is tied to the position on the celestial sphere of very distant objects, such as pulsars (described in Section 7.1). It is called the International Celestial Reference System (ICRS) and is used for extremely accurate measurements, including the global positioning system (GPS). Its origin is at the barycentre, i.e. the centre of mass of the Solar System.

But even this system, introduced in 1997, does not guarantee the absence of rotation of its axes. It only guarantees that this rotation is negligible in comparison with the accuracy of all existing instruments. So, the issue is resolved practically, but not in principle.

The second reason why we use non-inertial frames of reference is more pragmatic. We do this because it is more convenient for us. No one could forbid us to do this if the use of non-inertial systems gives some advantages.

And these benefits usually don't come down to a mere whim. For example, we live on the planet that spins around its axis, and the frame of reference associated with the Earth is clearly non-inertial.

But it is the system in which the overwhelming majority of all scientific experiments are carried out. And a lot of the processes that we observe around us are associated with the fact that the Earth rotates. And this is not only a change of day and night, but also phenomena in which it is clearly possible to introduce inertial forces acting on bodies.

So let's take a closer look at these forces. We are thrown backward, against the direction of acceleration, while sharp braking throws us forward. In the general case, we say that if the frame of reference moves with the acceleration $\vec{a}$, then a body of mass m will be acted upon by an inertial force $\vec{F} = -m\vec{a}$ directed in the direction opposite to acceleration.

It is this force that causes an overload at the start of a spacecraft and flattens an object that hits a solid wall. It throws us forward in the event of sudden braking and only a car airbag will help us avoid injury. In this case, the car, to which the reference system of the passenger or driver is attached, is clearly not moving uniformly relative to distant stars. So, the forces of inertia rage.

1.7.4. *Centrifugal forces*

There is also centrifugal force, which squeezes out wet laundry in the washing machine and deflects seats hung on a rapidly rotating carousel. If the frame of reference rotates with an angular velocity ω, and a body of mass m is at a distance r from the axis of rotation and, because of this, rotates with a velocity $v = \omega r$, then in this system, it will be acted upon by a centrifugal force, $F = m\omega^2 r = mv^2/r$, directed away from the axis.

The same formula describes the so-called centripetal force, so they are often confused. What's the difference? Centrifugal force is a type of inertial force; it occurs when we use a rotating non-inertial frame of reference. Centripetal force is the resultant of real forces, necessary for the body to make a circular motion. It is directed towards the axis of rotation.

As Mark Twain told us, Johnny Miller bought a chance from Tom Sawyer to whitewash the fence in exchange for a dead rat and a string to swing it with. While Johnny was painting the fence, Tom was spinning the rat over the head.

Consider the physics of this exciting process. In the frame of reference of the spinning rat, which is undoubtedly non-inertial, two forces act on the rat, balancing each other. One is the pulling force of the rope, and the other is the centrifugal one.

Since the resultant force is zero and does not act on the rat, its speed is constant in the rotating system. More specifically, it is motionless. Look, we applied Newton's first law in a non-inertial system and it started working after the introduction of inertial forces!

Now, let's use the system associated with the fence that Johnny Miller whitewashes. It is not entirely inertial, but if you forget about the rotation and movement of the Earth, it will pass for this. Be that as it may, it is much more similar to inertial than the system of a rotated rat and we will consider it as "more or less inertial."

Only one force acts on the rat in this frame of reference: the tension force of the rope. It also acts as a centripetal force that makes the rat follow a curved path. In the fence system, the rat moves with the acceleration $a = v^2/r$ directed towards the axis of rotation, i.e. Tom Sawyer's hand. Note that the centripetal force also exists in a non-inertial frame of reference if a body moves in a circle in this system.

The rat is not often twisted on a rope; the laundry is squeezed out in a centrifuge occasionally, but not permanently. But there is a merry-go-round that we rarely jump off, almost never: our Earth with its rotation. It leads to the fact that, due to the centrifugal force, the surface of the planet and the ocean has acquired a shape close to an oblate spheroid. Simply put, the Earth is flattened in the direction of its axis. This conclusion was reached by the great physicists of the 17th century — Newton and Huygens.

But the famous French astronomer Cassini claimed that the Earth is prolate along its axis. Scientists prefer not to argue, but to find out and measure. It was decided to send expeditions and measure the length of the meridian arc, corresponding to one degree of latitude, near the equator and closer to the pole to answer the question about the shape of the Earth.

The latter was carried out by an expedition to Lapland in 1736–1737, led by a retired captain of musketeers, a member of many academies, the Frenchman Pierre-Louis Moreau de Maupertuis.

The expedition, which confirmed the correctness of Newton and Huygens, was reflected in literature, in particular in Voltaire's *Micromégas*. In letters to Maupertuis, the author referred to him as "my dear, who flattened the worlds and Cassini", which did not prevent him from making fun of Maupertuis himself in other letters.

Centrifugal force separates milk into cream and skim milk in a separator, or deflects us to the side when we are driving in a fast-moving car and it turns sharply. In school textbooks, inertial forces are sometimes called fictitious forces. Tell this to an astronaut who has been flattened by centrifugal force in a centrifuge or by acceleration at the launch of a rocket.

1.7.5. *Coriolis force*

But centrifugal force is by no means the only inertial force caused by rotation. I'll try to convince you of this with a simple example. Imagine you have at home a turntable and two house dwarfs — Sleepy and Dopey.

And so you stand and watch as the disc on the turntable makes its 33 1/3 revolutions per minute, and the dwarfs have fun. Sleepy falls asleep directly on the axis of the turntable and rotates in the centre of the record. And Dopey runs along the edge of the plate with the same angular velocity ω as it rotates.

Consider the forces acting on Dopey in Sleepy's rotating frame of reference associated with the plate. This is the centrifugal force $mv^2/r = m\omega^2 r$, where r and $v = \omega r$ are the radius and speed of rotation of the edge of the plate, and m is the mass of the dwarf. The force is directed from the centre. Dopey moves in a circle; this requires a centripetal force, also equal to $m\omega^2 r$, but directed towards the centre. So, the directions of the acting centrifugal force and the resulting centripetal force are opposite. The ends do not meet the ends.

Let us now consider the same process in the frame of reference of a stationary observer, i.e. you. When Dopey runs in the direction of rotation of the disc, i.e. clockwise, if you look from above, then its speed doubles, because to the speed of the plate, v, you need to add the same speed of the dwarf relative to the plate v and get the speed in the system of a stationary observer, equal to $v + v = 2v$. The centripetal force required for this is $m(2v)^2/r = 4mv^2/r$.

And if the dwarf runs in the opposite direction, then its speed relative to the stationary observer is equal to $v - v = 0$, i.e. he is motionless. For such a movement, the total force must be zero.

None of these values coincide with what we counted in Sleepy's frame of reference. But this is impossible, because in physics the result of a calculation cannot depend on which frame of reference we use. So, we missed something important.

This is the second of the inertial forces associated with rotation. It is called the Coriolis force in honour of the French scientist Gaspard-Gustave de Coriolis, who discovered it in 1835. It acts on bodies moving in a rotating frame of reference. In our case, on Dopey, but not on Sleepy, who is motionless in this system.

If the body moves with speed $\vec{v}$ in a frame of reference rotating with angular speed $\vec{\omega}$, then it will be acted upon by a Coriolis force equal to $\vec{F} = 2m\,[\vec{v}\vec{\omega}]$. Let me remind you that the direction of the angular velocity vector coincides with the axis of rotation, and where the beginning and end of it are usually determined according to the corkscrew rule. That is, if you take a right-handed corkscrew in your hand and turn it in the direction of the body's rotation, it moves in the direction of the vector $\vec{\omega}$. The same applies to a screw with a right-hand thread.

From this, it is easy to find out that the vector of the angular velocity of the Earth's rotation is directed from the South Pole to the North, and the same vector for the disc, spinning on the turntable, is down. Square brackets mean the vector product of two vectors, in our case, $\vec{v}$ and $\vec{\omega}$. It is always directed perpendicular to the vectors being multiplied.

If you are not very familiar with the vector product, then the so-called Zhukovsky rule may come in handy: the Coriolis acceleration can be obtained by projecting the body's velocity vector onto a plane perpendicular to the angular velocity vector (i.e. onto the "equator" plane) by increasing the resulting projection into 2ω times and rotating it 90 degrees in the direction of rotation. And the Coriolis force is obtained after multiplying this acceleration by the mass of the body. As you can see, this force, in contrast to the centrifugal force, does not depend on the position of the body, but depends on its speed.

Let's see if it can clarify the situation with the gnomes for us. The sum of the two centrifugal forces caused by the rotation of the plate and Dopey's running along the circumference is $mv^2/r + mv^2/r = 2mv^2/r$. It is easy to calculate that the Coriolis force associated with its motion in a rotating coordinate system is also equal to $2mv^2/r$. Moreover, if the dwarf runs against the movement of the plate, i.e. remains in place for a motionless observer, then it is directed to the centre and the sum of all inertial forces is zero, as it should be.

And if Dopey runs in the direction of rotation of the disc, then it is directed from the centre and the sum of all inertial forces is $4mv^2/r$. We counted the same force in a non-rotating frame of reference. It will act on the dwarf, but not alone.

There must be some kind of force that compensates for it and does not allow the dwarf to simply fly off the plate with screams and curses in the gnomish language. It can be just friction between the soles of his shoes and the plate, or the tension force of the string, by the ends of which Sleepy and Dopey are pulling. Or maybe the force of electric attraction if they have corresponding charges of different signs. It is important that this force, let us say the tension of the twine, can be measured by the device and the result must coincide with the calculated value. This, as we have seen, is the same for two frames of reference.

The Coriolis force can be felt on the fast-spinning carousel. If you try to move along its radius, then you will clearly be deflected to the side. You can throw a ball or another object and watch how it moves. Without considering the Coriolis force, this movement cannot be described.

1.7.6. *Coriolis force and the Earth's rotation*

Is there some effect associated with the Coriolis force because of the rotation of the Earth? There are a lot of them. It is clear that the small angular velocity of this rotation, equal to 2π radians per day, makes them not very strong. So, they are especially noticeable if something on the planet moves quickly or over long distances, which means extended time period of Coriolis force acting. In these cases some impacts can be observed. This applies to deviations of projectiles, winds, and sea currents.

Fast-moving projectiles deflect slightly to the side, especially when firing at long distances. They turn slightly to the right in the Northern Hemisphere and slightly to the left in the Southern Hemisphere. The Coriolis force manifested itself during the famous naval battle near the Falkland Islands on December 8, 1914, between the German cruising squadron of Vice Admiral Maximilian von Spee and the British Royal Navy of Vice Admiral Doveton Sturdy.

When analysing the actions of the parties, it turned out that the German artillery hit more often than the British. The shells of the latter constantly fell a hundred metres to the left of the German ships. But the ships of "the Mistress of the Seas" Britain were modern for the time and their cannons had recently been zeroed near the British coast.

However, the sailors did not take into account that the Coriolis force not only deflects shells but also does it in different directions in the two hemispheres. Zeroing was carried out in the north, and the battle in the south, at a latitude of about 50°.

Additionally, the Germans could not ignore the correction for the Coriolis force when shelling Paris from a distance of 110 km from the cannon nicknamed Kaiser Wilhelm Geschütz ("Emperor William Gun", not to be confused with the famous "Big Bertha" howitzer) during World War I. Without this, the shells would have deviated by 1.5 km. However, the effect of the shelling was mainly psychological.

Winds blow and sea currents flow over long distances. They are also deflected by the Coriolis force in the same direction as the projectiles. This is manifested in the general picture of the global atmospheric circulation (see Fig. 3.10), in the map of currents, and in the direction of rotation of winds in cyclones and anticyclones.

Let me remind you that cyclones have low-pressure areas in the centre, whereas anticyclones have high-pressure areas. The difference in pressure in the centre and at the periphery causes a radial movement of air, to the centre in cyclones and away from it in anticyclones. This wind is deflected and twisted by the Coriolis force. In the Northern Hemisphere, the rotation of air masses occurs counterclockwise in cyclones and clockwise in anticyclones, whereas in the South Hemisphere, it occurs clockwise in cyclones and counterclockwise in anticyclones.

There is another option in which the effect is small, but lasts long enough. The Coriolis force rotates the plane of the Foucault pendulum, demonstrating that the Earth is indeed rotating. This force is also responsible for the manifestation of Baer's (aka Baer–Babinet) law, which states that in the Northern Hemisphere, the right banks of the rivers are high, and the left ones are low, and in the Southern Hemisphere everything is the other way around.

How exactly it manifests itself in all these cases can be read in more detail in the book by Varlamov, Vilen, and Rigamonti, a link to which is provided in the section "What else to read?" at the end of the book.

Are there other types of inertial forces that I have not mentioned? If the angular velocity of rotation changes over time, this leads to the appearance of the so-called Euler force. But the rotation of the Earth occurs at a constant speed and therefore no traces of this force are observed in the nature around us.

Physicists use the most unexpected reference frames in their calculations. For example, the movement of a fluid in a whirlpool or through pipes is considered in a system associated with this fluid, so that the fluid is stationary in it, and the system behaves like a drunken snake with increased activity.

1.7.7. *Newton's laws and non-inertial frames of reference*

If we add inertial forces to the real forces, then Newton's second law can be used in any frame of reference. Consider the example of a train station accelerating relative to a train. The reference system of an accelerating train that has started moving away is clearly non-inertial; therefore, inertial forces act on it, which accelerate the station with all passengers, ticket offices, stalls, a buffet, and a boss in an official hat.

In any system, the speed of the body is constant, on which no forces act, or these forces completely compensate each other so that the total resultant force is zero. And what happens if they don't compensate? In this case, the force manifests itself in acceleration. The main thing is not to forget about the forces of inertia. So, the fulfilment of Newton's first law can be viewed as a successful check that we have taken into account all the forces without overlooking the inertial forces.

In fact, the difference between the systems is not only in this formal detail. To some extent, an inertial reference frame can be considered an ideal, albeit almost unattainable, such as Plato's Atlantis, the Vikings' Valhalla, or a country with rivers of milk and jelly banks in Russia.

Many physicists believe that Newton's first law implicitly states that such an ideal exists. And we can build a system that will be inertial, albeit approximately, but with any predetermined limit of non-inertialship. So, far, the ICRS system has been best suited for this role among the existing systems. Over time, it will be replaced by another, even more accurate one. After all, physics and other natural sciences do not work with an ideal, but with the level of accuracy that allows us to explore the world around us.

In physics, when solving specific problems, those systems are popular that are similar to inertial ones, in which the inertial forces are significantly weaker than other forces acting on the body — for example, a system connected to the Earth. Naturally, I used it when giving examples from sports.

1.7.8. *Fields are sources of forces*

Now let's get back to the definition from Wikipedia. In articles about the definition of force, the fields are mentioned in some languages. Indeed, a force can act not only through contact. For example, you can drag a treasure chest from a sunken pirate ship using mechanical forces, such as ropes. But you can also put an iron shell around the chest and drag it with a strong electromagnet.

There is a force in this case, but it was caused by a magnetic field. A similar magnetic force turns the needle in our compass or rotates the shaft of an electric motor. Any child knows that a magnet is able to lift small pieces of iron such as paper clips from the floor.

Another type of field will work too. If you rub a plastic comb on something woollen, such as your own hair, then it begins to attract all sorts of small specks. The gravitational field also changes the speed of objects. Everyone who has been hit on the head by a fallen chestnut knows about it. However, I have already mentioned gravity when discussing force in sports.

How many words it took me to explain the meaning of a simple Wikipedia definition of force. In addition it is still not quite complete. I'll tell you more about the incompleteness a little further, but for now let's look for the most important of the forces.

You have seen that there are many forces in nature that act on bodies. It is good that there is a part of physics called mechanics. This is its oldest section. Mechanics explains how a body moves under the influence of various forces, and how its position, speed, and orientation in space change.

It doesn't matter what exactly gave rise to these forces. To calculate the velocity of a projectile emitted from a weapon barrel and its trajectory after a shot, it does not matter what accelerated this projectile: the pressure force of the powder gases formed during the explosion, the electromagnetic forces in the Gauss cannon, or steam in a steam catapult.

1.7.9. *Fundamental interactions*

For physics as a whole, the nature of forces is very interesting. It is clear that among them there must be forces that are not reducible to others, the most important, called fundamental. And all other forces could be reduced to a combination of fundamental ones, sometimes in a rather complicated way.

It's time to name a standard set of four types of fundamental forces in nature. These are gravity, electromagnetic interaction, nuclear forces that hold particles in the nuclei of atoms, also called strong interaction, and the so-called weak interaction associated with intranuclear processes. All of them are listed in Fig. 1.3 by the American populariser of physics Randall Munroe, taken from xkcd.com.

The meaning of this picture is that everyone knows very well about the first two types of forces, but even some physicists have difficulty when it comes to weak interaction (here the signature "Your Captain Obvious" suggests itself). They confuse it with the strong one, or simply remember that something like that exists. Among the strong interaction properties, only what is written in

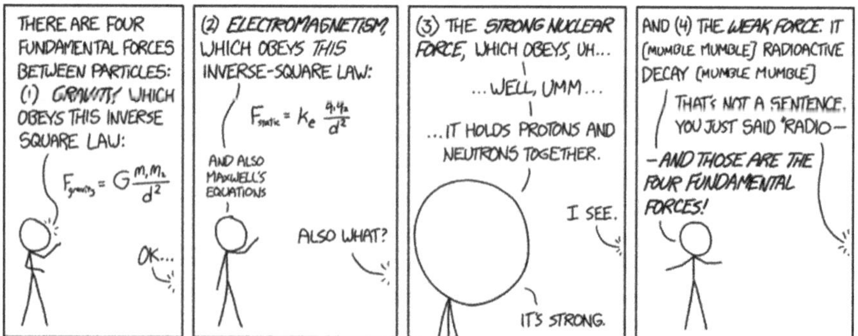

Figure 1.3. Four main interactions according to Randall Munroe's comic. Source: https://xkcd.com/1489. Creative Commons Attribution-NonCommercial 2.5 License.

Fig. 1.3 is sometimes well remembered. I will try to help you understand some of the details of these interactions in Section 6.2.

When we talk about fundamental forces, we select different types of forces that act in various ways. All other forces can be obtained by combining these fundamental ones, as in chemistry, where any substance can be described by a combination of about a hundred different atoms of chemical elements included in the periodic table.

Fundamental forces are also called fundamental interactions. For me, as a physicist, these are synonymous words. At the same time, we do not refer to other, non-fundamental forces, such as the lift of an airplane wing, as interaction. By the way, I use this example only to show how this force relates to our four (actually two of them) fundamental interactions.

We do not need gravity and weak forces, unless we want to explain what exactly compensates for the lift of the aircraft and how the aluminium atoms, from which the wing is made, appeared in nature. But electromagnetism, strong interaction, and quantum mechanics explain why matter consists of atoms, how they interact with each other to form solid (wing) and gaseous (air) substances, and why the wing is solid and retains its shape by changing the movement of air molecules around it.

Next, mechanics, or more specifically its aerodynamics section, tell you how this air movement provides lift. As you can see, there is a chain, although not very short. And all of its links belong to different branches of physics.

A little over a hundred years ago, we knew only about the first two fundamental interactions, but some of the facts known then could not be explained within the framework of this set. For example, everyone knew that the Sun is shining, but had no idea from where it gets the energy it radiates. All the options available then were considered, such as the fuel burning out permanently inside our star or compressing because of the attraction of its parts.

But in all cases, these sources of energy would not last long. Now, we know that solar energy is taken from thermonuclear reactions of the fusion of heavier elements from hydrogen, but at that time, there was not even a hint of the existence of this process.

Some new concepts had to be introduced into physics in order for all known and subsequently discovered facts to fit well into the picture of the world or to explain the results of numerous observations and experiments. Specifically, this is a strong interaction that holds together the particles in the atomic nuclei, and then a weak interaction that is responsible for the decay of some subatomic particles. Now, we do not see an urgent need to introduce new interactions, but this does not mean that it will not appear in the future.

1.8. Gravity

Humanity has known about gravity for a long time, although no one will name its discoverer. Each of us gets acquainted with the properties of gravity, mastering the difficult science of walking.

1.8.1. *The law of universal gravitation*

A simplified law describing gravitational forces, known as the law of universal gravitation, was proposed by Isaac Newton back in 1666. He claimed that the gravitational force of two bodies is proportional to their masses and inversely proportional to the square of the distance between the bodies if the latter is much larger than the bodies' sizes.

We now consider this law as an approximation that is valid for a weak gravitational field. Nowadays, we are talking about the gravitational field, and not just about the mutual attraction of bodies. This approach is preferable, especially when we study waves, including gravitational ones. However, physicists talk about the fields associated with each of the fundamental interactions.

There is a special chapter in this book (chapter 8), in which I have collected everything related to the history of physics in general and its individual branches. Among the stories told, there is no mention of the discovery of the law of universal gravitation due to a falling apple. It is so brief that I quote it right now.

This is how Newton's biographer William Stukeley described the discovery in the book *Memoirs of Sir Isaac Newton's Life* published in 1752:

> After dinner, the weather being warm, we went into the garden and drank thea, under the shade of some apple trees...he told me, he was just in the same situation, as when formerly, the notion of gravitation came into his mind. It was occasion'd by the fall of an apple, as he sat in contemplative mood: "Why should it not go sideways, or upwards? But constantly to the Earth's centre? Assuredly, the reason is, that the Earth draws it. There must be a drawing power in matter...If matter thus draws matter; it must be in proportion of its quantity. Therefore the apple draws the Earth, as well as the Earth draws the apple." That there is a power like that we here call gravity which extends its self thro' the Universe.

This idea was formulated in the form of a law, more precisely, in the form of a formula. With its help, Newton calculated the force of gravitational attraction between two balls, such as planets and the Sun. To do this, he had to create, together with Gottfried Wilhelm von Leibniz and other scientists, the foundations of what is now called integral and differential calculus.

The law of gravitation was first tested in the experiments of the British Henry Cavendish in 1798. They were carried out almost simultaneously with Coulomb's experiments, which I will discuss later, and on very similar equipment.

The forces of mutual gravitational attraction of two massive balls were a huge number of times less than the forces of electrical interaction in Coulomb's experiments, and their measurement required great skill from the experimenter. Cavendish determined the value of the gravitational constant — the most important quantity that is included in the law of universal gravitation and considered one of the few world constants.

Newton and other physicists managed to explain many of the phenomena and laws discovered earlier. For example, Kepler's laws, which describe the motions of planets, their satellites, and every cosmic little thing such as asteroids, have turned from an empirical regularity into a consequence of the laws of universal gravitation and conservation of angular momentum.

1.8.2. *Gravity and the Universe*

Consider two electrons. They attract according to the law of universal gravitation and repel because of the Coulomb force. The force of electrical repulsion is $4.16 \cdot 10^{42}$ times stronger than the gravitational force, and this truly gigantic ratio does not depend on the distance between the electrons, because both forces depend on it equally. If we take two protons, then due to the greater mass for the same charge, this ratio will decrease, but it will still remain huge.

Why do we take into account the mutual gravitational attraction of matter but neglect its electrical attraction and repulsion when studying the birth and evolution of the Universe? Electric charges are positive and negative, and two charges can attract and repel, depending on whether they have the same charge signs or different ones.

In addition, the Universe appears to be generally electrically neutral. Therefore, the total charges of galaxies, stars and other astronomical objects do not differ much from zero. So, on an astronomical scale, the situation is changing, because the positive electric charge of a huge number of protons of a star is compensated or almost compensated by a negative charge of no less than a huge number of its electrons, weakening the force of the electrical interaction of the star and, say, its planet by an even more huge number of times than $4.16 \cdot 10^{42}$.

The mass, which is always positive, replaces the charge in the law of universal gravitation.[h] The forces of mutual gravitational attraction add up and their sum could be many times over the force of electrical interaction. This explains why, on astronomical and cosmological scales, gravity becomes stronger than electrical and magnetic forces.

---

[h]There is an essential general mutual repulsion or antigravity on a cosmological scale associated with the concept of dark energy or the cosmological constant, which we will talk about in Section 1.9. But on the scale of the Galaxy or the Solar System, it is not significant, not to mention the smaller ones.

1.8.3. *Tidal forces*

The gravitational field of a massive body is always inhomogeneous in space. It falls with distance from the source. Consider the Earth, which revolves around the Sun, and three small pebbles that we need as an object to measure acceleration and force. Such objects are usually called test bodies in physics.

Place one pebble in the centre of the Earth. Ours is a thought experiment (physicists traditionally refer to them using the German word *gedanken*), so nothing prevents us from temporarily considering the planet permeable. So, let's create a small cavity around the stone, say a metre in diameter, just so that it could calmly fall on the Sun, without experiencing any forces other than gravitational ones.

We place two more pebbles on the surface of the Earth on the opposite sides relative to the centre. More precisely, they are located at points that are maximally and minimally distant from our star. At the point as close as possible to the Sun, astronomical noon will just come and the Sun will stand at its zenith. We will call this near or midday point. Midnight will come at the far point. We'll call it midnight point.

The acceleration of free fall on the Sun will be maximum for the noon pebble and minimum for the distant one. The difference between them will be small because the radius of our planet is much less than the distance to the Sun, but it is important for us that it is nonzero. The acceleration of the pebble in the centre will be close to the average for these near and midnight pebbles.

The left panel of Fig. 1.4 depicts the Earth and the forces acting on these stones. The Sun, Moon, or other sources of tidal forces are located below the figure. Therefore, the midday pebble is at the bottom, and the midnight one is at the top on the Earth's surface. The difference in the lengths of the arrows indicating the magnitude and direction of gravitational attraction to this massive body is greatly exaggerated.

If the Earth lost its mass (naturally, in our *gedanken* experiment only) and did not attract pebbles, then the distances between the near and central and far and central pebbles would increase with acceleration, which is called tidal acceleration. Its value is

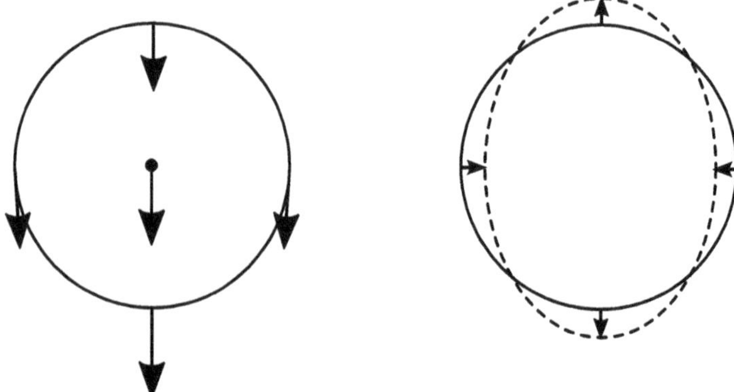

Figure 1.4.   Forces and accelerations acting on stones in different parts of an extended body are different. This results in what are called tidal forces. On the left, they are shown in the reporting system of the external massive body — the source of tidal forces (in the figure it is located below the sphere), and on the right, in the system moving with the test body in the centre of the spherical planet. Tidal forces from a massive body try to deform a spherical surface.

proportional to the radius of the Earth and inversely proportional to the cube of the distance to the Sun. This means that in the frame of reference of the central pebble, the other two would move away from it with this acceleration.

The force behind this movement is called tidal force. Under its influence, pebbles would fly away from the surface of the Earth somewhere into space.

But in real life, the Earth has mass and the planet's gravity holds all of its parts securely together. At the same time, tidal forces do not disappear anywhere; just the Earth's gravity is much greater than them. As I will discuss in Section 3.6, they are the cause of the ebb and flow. This fact gave the name to the forces, the story of which I continue.

What if we place a stone on the surface of the Earth at a point that is the same distance from the Sun as the centre of the Earth? I'm talking about a point that is the intersection of the trajectory of the centre of the Earth when it orbits the Sun with the Earth's surface. More precisely, I mean one of two points, one of which lags slightly from the centre along the orbit, and the second is ahead of it. It is easy to understand that the direction of the forces of attraction to the Sun at these points deviates from the direction of the force acting on the central pebble, albeit at a very small angle. In the left panel of Fig. 1.4, they are shown by the left and right arrows. If there were no gravity, then the tidal forces would tend to bring this stone closer to the centre of the planet.

Let's go to the reference frame of the stone in the centre of the Earth and look at the accelerations of the other stones. The forces that cause these accelerations are called tidal forces. They are shown in the right panel of Fig. 1.4. Tidal forces caused by the attraction to a massive body try to stretch the object on which they act along the direction of the external gravitational field and compress it in all orthogonal directions. Let me emphasize that this is not a special kind of force, but a consequence of universal attraction.

When one celestial body approaches another, massive enough, then the tidal forces caused by the latter can tear it to pieces. This is exactly what happened with the Shoemaker–Levy comet. In July 1994, it fell on Jupiter, but before that, tidal forces in Jupiter's gravitational field tore the comet at least 21 pieces, which fell into this gas planet and burned up there. It is not difficult to find photos of this fall, taken with the Hubble Space Telescope, on the Internet.

1.8.4. *General theory of relativity*

Albert Einstein's general theory of relativity (GR) is the modern version of the theory of gravitation. For a weak gravitational field, for example, such as in the Solar System, the predictions of general relativity hardly differ from the conclusions made on the basis of the law of universal gravitation. However, there are some minor differences. The reason for the transition from Newtonian theory to general relativity is simple — the first does not describe some effects observed in nature, and the second perfectly explains them, both qualitatively and quantitatively.

In addition, general relativity is based on the special theory of relativity, which has been verified and rechecked experimentally. If it were incorrect, then not a single particle accelerator would work, because they are designed on the basis of special relativity formulas. These are complex theories, so I will come back to them later, discussing the effects of general relativity and special relativity in Section 7.2.

Over the century of its existence, general relativity has withstood all of the numerous tests in laboratories, space, and astronomical observations. A lot of small deviations from predictions based on the law of universal gravitation are observed, which are associated with the effects of general relativity.

The triumphs of general relativity were its application to the entire Universe, which resulted in the creation of a new science of cosmology, the prediction and then the discovery of exotic black holes (BH), and the discovery in 2015 of gravitational waves predicted and described in the framework of general relativity.

In this book, I will only touch on the very basics of general relativity. Those interested in learning more, and on a popular scientific level, can read my book *How the Universe Works: An Introduction to Modern Cosmology*. In addition to the GR ideas, it tells in more detail about the evolution of the Universe, the curvature of space-time, BH, mysterious naked singularities, wormholes, and many other related phenomena and objects. A detailed reference can be found in the "What else to read?" section at the end of the book.

The ideas of general relativity have slightly changed the very approach to gravity. In it, the chosen motion of a test particle is not inertial, in which no forces act on it, but free fall, in which no forces other than gravitational ones act on the particle.

It defines the frame of reference of a freely falling observer, in which the laws of physics could be written in their simplest form. A satellite orbiting the Earth rests in such a system.

From Newton's point of view, the picture of the forces acting on the satellite or its part is as follows: in the inertial frame of reference, the forces of gravity provide a centripetal force that bends the trajectory of its motion. And in a non-inertial system associated with a satellite, the inertial forces, namely the centrifugal force when orbiting the Earth, compensate for the gravitational ones.

From the point of view of Einstein, the satellite is located in the special system of the free-falling observer and therefore no forces other than gravity or inertia act on it. Therefore, gravitational attraction becomes part of the inertial forces in this approach, which is discussed in many popular science books, but very often there is little explanation of how this statement should be understood.

Let's take another example. When a person stands on the floor, then in Newtonian mechanics it is believed that the force of his or her weight and the normal force act on this person, which are mutually compensated. Therefore, the frame of reference attached to this person is close to the selected one, i.e. inertial.

In general relativity, everything is different. A person, unless he broke through the floor or jumped, is not in the special system of a freely falling observer. So, the person would be at rest or moving at a constant speed in this frame of reference, but there is an external force that makes him or her move at an accelerated rate relative to it. This is the reaction of the floor, the only force acting on a person in the system of a freely falling observer.

As a by-product, it became clear why, in general relativity, the mass of a body included in the law of universal gravitation (called gravitational mass) must coincide with the so-called inertial mass included in Newton's second law. In Newtonian mechanics, this equality would not be necessary. More precisely, it can be used in calculations, but it is impossible to explain from the very first principles, i.e. basic concepts of physics. Indeed, in classical mechanics, gravitational mass characterizes the gravitational properties of bodies, i.e. forces of their mutual attraction, and the inertial one characterizes acceleration under the influence of external forces, and they are different things in general.

In general relativity, all bodies fall freely with the same acceleration. Naturally, this is true if the fall is really free, not slowed down by the force of air resistance or otherwise. The free fall acceleration of a body varies from point to point, depending, for example, on the distance to massive bodies, but not on the mass of the falling body. This leads to the conclusion that the gravitational and inert masses are equal. It has been verified experimentally with great accuracy for bodies, fields, elementary particles, and even quantum fluctuations, briefly speaking, for everything.

The most accurate results were obtained for the fall of ordinary macroscopic bodies. At the end of 2017, some data were provided by the experiment on board the *Microscope* spacecraft. The difference in the accelerations of gravity of titanium and platinum was less than 10^{-14} of the magnitude of these accelerations. The processing of the data of laser ranging of the Moon indicates that the relative difference between the accelerations of the Earth and the Moon in the gravitational field of the Sun does not exceed $3 \cdot 10^{-14}$. The experiment with a torsion balance gives the same constraints for a pair of beryllium and titanium.

There are also experiments at the quantum level, in which the accelerations of gravity of individual atoms of different elements or isotopes of the same element are compared. In 2020, it was experimentally proved that the relative difference in accelerations for rubidium and potassium atoms, or rather their isotopes ^{87}Rb and ^{39}K, does not exceed $2 \cdot 10^{-7}$. In all experiments, the obtained differences in acceleration are less than experimental errors. So, the experiments confirm the GR conclusion about the same acceleration of gravity for different bodies.

But let us return to our example with the GR reference frame of a freely falling observer associated with a satellite. No external non-gravitational forces act on the satellite, and an astronaut on board the satellite observes what is called weightlessness in the literature, but professionals call it microgravity. Weightlessness can be felt in any other free-falling system. Parachutists and bungee jumpers are aware of this circumstance.

But can we say that there is no gravitational field at all in this system? No, we cannot. No forces act on a point particle in its freely falling system. But there are small tidal forces that will act on parts of a non-point body, even small in size. The diameter of a spacecraft orbiting the Earth is minor compared to the distance to its centre, so these forces are quite small, but they do exist and can be measured. In general relativity, the emphasis is on the tidal forces, which are described by the equations of this theory.

1.9. Dark Side of the World

Over the past few decades, physicists have realized that our world is not only composed of ordinary matter, made of atoms, and radiation. It is also dominated by two mysterious entities, about which scientists know little more than nothing. These are dark matter (DM) and dark energy (DE). More than 95% of the mass and energy of the Universe is provided by them. These two main mysteries of modern science are conventionally combined into the general concept of "the dark sector of the Universe".

But I think that this term is not pompous enough. I also use the term "the dark side of the world". Indeed, jokes about the dark side are very common among physicists, who are also fans of the *Star Wars* films. The quote "You underestimate the power of the dark side" sounded in many lectures on DM and DE, including mine.

1.9.1. *Dark matter*

Let me tell you briefly what we know about these mysterious substances. I'll start with DM. For more than 80 years, astronomers have been constantly convinced that the masses of galaxies and their clusters, determined by their gravitational attraction, significantly exceed the masses of the luminous objects that are inside them, such as stars.

Galaxies rotate around their axis faster than if the mass was determined by their stars, gas, and dust, and the analysis of this rotation showed that some invisible mass is present far beyond their visible boundaries. Galaxies in clusters move faster relative to its centre than would follow from the mass of ordinary matter entering them. Something massive but invisible, which is responsible for this, is called dark matter.

The first to come to this conclusion was the Swiss astronomer Fritz Zwicky, in 1933. He concluded that galaxies contain many not so large non-luminous objects made of ordinary matter, the very matter that makes up the Earth, everything that surrounds us, and ourselves. Nowadays we call it baryonic matter, but at that time, the term *baryon* (when we get to the scale of elementary particles,

I will tell you that this is the name for particles consisting of three quarks) did not yet exist.

And it could not: the neutron was discovered by the British physicist James Chadwick only a year before, in 1932 (for which he received the Nobel Prize in Physics in 1935). It was too early to realize that this particle could be combined with a proton to form the concept of "nucleon", and then to add the elementary particles which had not yet been discovered at that time, and refer to this group as baryons.

Astronomers are sure of the existence of some amount of baryonic dark matter, similar to that predicted by Zwicky, but it makes up an extremely small fraction of the observed DM. Most of it is still an enigmatic non-baryonic DM, about the nature of which there are many hypotheses, but there is no certainty in any of them.

What can it consist of? Most likely, these are some particles that have not yet been discovered and that are not formed from quarks. They do not participate in electromagnetic and strong interactions, but gravitationally attract the surrounding bodies. And since they are not involved in electromagnetic interactions, they are not able to emit or absorb light. This circumstance explains why they were named as "dark matter".

It is unknown if they are able to participate in weak interactions. If they are, then theoretically DM particles could be detected in the laboratory by fixation that something unknown suddenly interacts with the detector.

The signal from this interaction — the emission of a particle, a flash, shaking of the crystal lattice, and a lot of other possible violations of the usual state — could be recorded by devices. At the same time, registering the signal is not even half the battle. It is more important to prove that it is not caused by any of the known elementary particles or something formed from them, such as an alpha particle.

1.9.2. *Hunting for dark matter particles*

Cosmic rays provide continuous bombardment with various previously discovered particles, as described in Section 2.9. Therefore, the laboratory equipment used to search for DM particles must somehow be protected from cosmic rays. This is done by placing them in underground laboratories, such as those in abandoned mines. Moreover, it is a good idea to make sure that there is almost no natural radioactivity in the surrounding rocks and soils.

The steel for the parts of the device used to detect DM is also special. Steel parts are made from steel smelted before 1945, when the first tests of atomic bombs were conducted. That steel contains no radioactive impurities, which can now be found in any modern steel due to air contamination.

I must reassure some anxiety-prone readers immediately by emphasising that I am talking about such negligible concentrations of impurities that even a thousand-fold increase in their level cannot affect your health. But sensitive detectors in modern laboratories can easily detect the signals caused by such weak radioactive additives.

Moreover, these signals will completely overwhelm any possible signals resulting from interaction with DM particles, for the sake of the search for which the whole experiment was actually started. Many problems arise during the design of the device, such as choosing its location and ensuring the absence of impurities in all of its parts. I'll talk about the search results later, in Section 6.3.

But it is possible that DM particles are also incapable of weak interactions. Then they only interact gravitationally with ordinary baryonic matter. This possibility is sometimes referred to as "mirror matter". We can only notice its presence by its gravitational attraction.

Naturally, there are enthusiasts trying to apply the hypothesis of mirror matter to everything in the world. Some of them say that inside the Sun there is a planet of mirror matter which causes tides and, in some mysterious way, is responsible for solar activity fluctuations. This idea should be attributed to fairy tales. It can be easily understood by comparing the period of revolution of a hypothetical planet with the duration of the 11-year solar cycle, or 22-year magnetic cycle of the Sun, or by realizing that the movement of this mirror planet inside the Sun will cause characteristic oscillations of its surface, which would be easily detected from observations.

Here's almost everything we know about DM. It can be added that the speed of its particles is significantly less than the speed of light. This variety is called cold. The variant with hot dark matter, the particles of which move at light or near-light speed, has been excluded based on the results of astronomical observations. The warm DM option is sometimes seriously discussed by astronomers.

There are also hypotheses in which DM does not consist of particles. Let's say, many small black holes act as these particles. Or there is no DM at all, but the laws of mechanics need to be clarified for very small forces.

The latter hypothesis is called MOND, an acronym for modified Newtonian dynamics. It can only explain some of the observational manifestations of DM, but not all of them. Instead, it requires a complete revision of the foundations of physics. This is the reason why I am not one of its supporters. Naturally, this does not exclude the possibility of the MOND theory triumphing and shaming me and other sceptics in the future.

1.9.3. *Scientific breakthroughs and controversy about new concepts*

As you can see, we have quite a lot of evidence that DM exists, but very little is known about its nature. Therefore, some are sceptical about the very idea of DM, claiming that this is not yet a theory, but its embryo, a temporary crutch that supports the building of science where it began to sag. To some extent, they are right. But if the result of the analogy is only scepticism about DM, then this is wrong. We see a pattern that is typical at the beginning of every major breakthrough in science.

I will talk in more detail about science and its rules at the end of the book, in Section 8.1, but for now, let us remember how difficult it was to incorporate into science the concepts that now underlie the scientific picture of the world. Let's take the concept of an atom. I mean the scientific concept only, not what some ancient philosophers liked to talk about.

It was created in the late 18th to early 19th centuries by physicists and chemists, among whom it is worth mentioning the British John Dalton, as well as the French Antoine Laurent de Lavoisier and Joseph Louis Proust.

Back in 1860, at their congress in Karlsruhe, Germany, chemists discussed whether atoms really exist or whether these are just some kind of crutches invented to prop up the building of science. And this happened after the discovery of the periodic law by Dmitri Mendeleev. But all the same, the disputes were serious. Some insisted that atoms do exist, and the hypothesis that bodies are composed of atoms and molecules is a serious scientific theory. Others doubted this. The supporters of the first point of view gained the upper hand in the course of heated discussions. The chemists' congress adopted the first widely accepted definition of an atom and a molecule.

Since then, scientists have learned that the atom not only consists of parts, which also consist of smaller parts — elementary particles, some of which, in turn, consist of quarks. It turned out that the atoms of one chemical element can be converted into atoms of other elements in the process of radioactive decay and synthesis. There were other details that Dalton and his colleagues did not even know about.

But all this did not abolish the ideas of the 19th-century atomists, but developed and refined our concept of atoms. It seems that with DM we are witnessing the very beginning of a similar development. However, over two centuries, the speed of gaining scientific knowledge has increased so much that it is possible that a convincing answer to the question of the nature of DM will be obtained very soon.

1.9.4. *Dark energy*

According to modern estimates, DM accounts for $25.89 \pm 0.57\%$ of the content of the Universe, baryonic matter for $4.86 \pm 0.10\%$, and the fraction of radiation is less than 0.1%. But the lion's share ($69.11 \pm 0.62\%$) is associated with dark energy. What is it?

Despite the same adjective, this mysterious entity is completely different from DM. Perhaps this is the result of the action of the so-called cosmological constant.

A century ago, in 1917, Albert Einstein applied the equations of general relativity that he had just created to the description of the entire Universe and wrote the article with which the science of cosmology began. He, like all physicists at that time, believed that the Universe is static, that is, it does not change in time. Such a hypothesis requires compensation for all forces acting on each galaxy.

In order to compensate for the force of gravitational attraction, Einstein modified his equations by introducing into them a certain force of universal gravitational repulsion, which counterbalanced the attraction and ensured the static nature of the considered model. To do this, he artificially added to them an additional term containing the so-called cosmological constant.

It soon turned out that this model was unstable, and Einstein lost interest in the cosmological constant, calling its introduction an error. Other scientists have developed dynamic models in which the Universe has expanded. The first one was proposed by Alexander Friedmann from Russia.

In his book *My World Line: An Informal Autobiography*, George Gamow wrote: "Much later, when I discussed the problem with Einstein, he remarked that the introduction of the cosmological term was the biggest blunder he ever made in his life. But this 'blunder,' rejected by Einstein, is still sometimes used by cosmologists even today, and the cosmological constant denoted by the Greek letter Λ rears its ugly head again and again and again."

Nevertheless, other cosmologists were in no hurry to abandon the cosmological constant. Despite the fact that most of them were sceptical about the existence of the Λ-term (lambda term, another name for the cosmological constant), they considered models both with and without the cosmological constant. For a long time, the latter option described all astronomical data well, but then the situation changed.

The first convincing evidence of the existence of the cosmological constant was obtained at the very end of the 20th century. This happened as a result of processing the observations of the strongest supernova explosions, more precisely, by comparing their redshifts and the light fluxes. The authors of this discovery, Saul Perlmutter, Brian Paul Schmidt, and Adam Riess, received the Nobel Prize in Physics in 2011 "for discovering the accelerated expansion of the Universe through observation of distant supernovae".

1.9.5. *Supernova explosions*

Let's clarify a little what is meant by the word *distant*. How far are they? Few have any idea what distances are in question. They are much more than the size of our Galaxy. This is because these are supernova explosions in other galaxies. In our Galaxy, similar outbreaks also occur, but rarely, on average about 10 per millennium.

But light from the last supernova in the Galaxy available for observation by astronomers, the so-called Kepler's Supernova, reached Earth in 1604 after travelling 20 thousand light-years. It was several years before the first telescope was constructed. No astronomer had seen a supernova explosion in the Galaxy in a telescope. I will tell you in more detail about the recent supernova explosions in the Galaxy and their remnants, but a bit later.

Only in February 1987 was a supernova explosion observed in the Large Magellanic Cloud, a dwarf galaxy, a satellite of the Milky Way.[i] After Kepler's Supernova, this is the closest supernova, occurring at a distance of 168 thousand light-years from the Sun.

But neither it nor the supernovae in our Galaxy are suitable for hunting DE. The supernova explosions that were used to detect the cosmological constant occurred much earlier and further away than all the ones I mentioned. Since the very first of them, the Universe has managed to expand twice. So, we are talking about events that

---

[i]A few hours earlier, neutrino laboratories recorded a couple of dozen neutrinos emitted during the collapse, even before the explosion. Astronomers are still studying the remnants of this supernova.

occurred billions of years ago at a distance of billions of light-years. Among them, astronomers are interested in certain types of supernova explosions with almost constant luminosity.

It's no secret that scientists love to classify everything. Astronomers divide all supernova explosions into five classes, and those into types. Not every flash is worthy of the hunt for dark energy. Only the worthiest, or rather the brightest, type Ia. This is a special kind of supernova. Recently, much has been said about hypernova explosions, which are especially powerful supernovae, but they are not used to search for the accelerated expansion of the Universe.

Typical explosions occur when massive stars (heavier than about 10 times the mass of the Sun) undergo core collapse due to nuclear fusion and become unable to sustain the core against their own gravity and explode, producing gold and other heavy elements during nucleosynthesis as a by-process. I mentioned this in Section 1.4.

Supernova explosions of type Ia occur in a completely different way, in a binary system consisting of two gravitationally bound stars orbiting their joint centre of mass. One of the components is a white dwarf, a small and extremely hot star. The story about them awaits you in Section 7.4.

I don't know what the reason is, whether it's an unhappy childhood or a bad upbringing, but this dwarf does not know how to behave in a decent society at all. It is a glutton and a petty thief. The dwarf steals stellar matter from its companion, devouring it greedily and increasing its mass due to gas flowing from one star to another.

Everyone knows that overeating can make people sick and cause them to burst. This is exactly what happens to our poor white dwarf. When the pressure and temperature of its core exceeds the threshold, above which carbon nuclei begin to merge, new processes of thermonuclear fusion are launched, which release even more heat.

The result is a type Ia supernova explosion that completely destroys the core, releasing a tremendous amount of energy. It is important that the flare luminosity is approximately constant, about 5 billion times brighter than the Sun.

1.9.6. *Supernova explosions and their remnants in the Galaxy*

I think that supernova explosions in our neighbourhood, in our Galaxy, in themselves deserve a little attention. In addition, I confess that I once again simplified the picture a little and wrote that the last of them, called Kepler's Supernova, was observed by astronomers back in 1604.

How often, when talking about science, you have to clarify and add something again and again! I am forced to do it here and there. The last supernova explosion for human astronomers in our Galaxy occurred at a distance of about 25,000 light-years from us in the *Sagittarius* constellation.

The supernova left a remnant in the form of a shell, expanding from the centre at a speed of about 15,000 km/s, i.e. 5% of the speed of light. It is known to astronomers as G1.9 + 0.3. Its nature was determined by a radio telescope observation in 1984. The expansion rate of the shell at 56 million kilometres per hour makes it a potential dream object for surfers and transport companies.

This value was obtained by comparing the size of the envelope, determined from the observations of the *Chandra* X-ray Observatory from NASA in 2007 and 2008. This allowed not only to determine the speed of expansion, but also to estimate when it began. And again it is necessary to clarify that the expansion began at the time of the supernova explosion, and the comparison made it possible to determine the year in which the radiation from this explosion reached the Earth. The light from it came to Earth around 1900, more precisely, in the period from 1890 to 1908.

But no telescope on Earth has detected this supernova. Why? It took place behind the centre of the Galaxy. And there is a lot of gas and dust, which weakened the light from the explosion by about a trillion times due to absorption and scattering. Naturally, astronomers did not notice it.

But how did they manage to observe the remnant of G1.9 + 0.3? After all, neither gas nor dust has gone anywhere. Really, they make any observation of this object impossible, but in the visible wave range only. But radio emission, X-rays, and gamma quanta are attenuated much less and observations are carried out with their help.

Let's go back to the explosions observed by astronomers. Shortly before Kepler's Supernova, an even brighter supernova, later named after the Danish astronomer Tyge Ottesen Brahe, was observed in the *Cassiopeia* constellation in November 1572. Brahe became famous for the greatest accuracy of astronomical observations and determination of the positions of objects in the sky in the era before the invention of the telescope.

He wrote and published in 1573 a book about his observations of the explosion titled *De nova et nullius aevi memoria prius visa stella*, i.e. "About a new star, not previously seen in the course of anyone's life or memory". Kepler, a student of Brahe, having published in 1605 the book *De Stella Nova in Pede Serpentarii*, i.e. "About a new star at the foot of Ophiuchus", used the same phrase "new star".

Hence the term *nova* arose. This is the name given to a star whose luminosity increases dramatically many times over, from a thousand to a million times. Accordingly, a supernova is a star whose luminosity increases even more, from 10^4 to 10^8 times. These intervals overlap, so in some cases astronomers need to look at the characteristics of their radiation in order to understand whether a star is nova or supernova.

Note that according to Aristotle's views, no changes are possible in the sky. I will talk about this in Section 8.2. This statement did not go well with the fact that from time to time one could see comets in the sky, which were clearly not permanent inhabitants of these places. This worried Aristotle himself, who solved the question simply. Comets were declared as some atmospheric phenomena not associated with the heavens. Therefore, the very possibility of the appearance of new stars was denied by many astronomers and astrologers. Both Brahe and Kepler were accused of confusing a star and a comet.

Because of the unconditional trust in the opinion of Aristotle, ancient and European astronomers did not mention previous supernova explosions, often brighter than the supernovae of Brago and Kepler.

Only the supernova of 1006, one of the brightest observed since the beginning of our era, is mentioned in the records of Swiss Benedictine monks from the monastery of St. Gall. But Chinese, Korean, and Arab astronomers described in detail new objects in the sky, including the supernova 1006.

The Egyptian astronomer Ali ibn Ridwan wrote that it was three times brighter than Venus or as a quarter of the Moon, and his Chinese colleagues argued that the supernova light was visible during the day and so bright that objects cast shadows. I will allow myself to doubt the correctness of this information. Have you seen shadows of moonlight during the day? I'm sure you have not. And the supernova was even fainter.

A little later, on July 4, 1054, light came to Earth from a supernova explosion in the *Taurus* constellation. Now at this place you can see the gas expanding after the explosion, forming the Crab Nebula. The light from the explosion travelled for about 6500 years, but the supernova explosion was so strong that even at such a great distance, the luminous object in the sky could be observed during the day with the naked eye for 23 days.

Supernova remnants are always very active and easy to spot. For example, the Crab Nebula is the strongest source of X-rays and gamma rays. It is located at a distance of 6500 light-years from us; its shell with a diameter of 11 light-years is expanding at a speed of 1500 km/s. There is an ionized gas inside the shell. There is a pulsar, a rapidly rotating neutron star PSR B0531+21 with a diameter of 28–30 km in its centre. It makes 30.2 revolutions per second and actively emits electromagnetic waves from the radio to gamma range.

This nebula was discovered in 1731 by its emission in visible light. It got its name from a drawing by the amateur astronomer William Parsons, 3rd Earl of Rosse, or Lord Rosse for short. The 36-inch telescope with a mirror diameter of 6 feet or 1.83 m nicknamed

"Leviathan of Parsonstown" was built by Lord Rosse and installed at Birr Castle in the Irish County of Offaly. It remained the world's largest telescope until the early 20th century.

In 1844, Lord Rosse made a sketch of the Crab Nebula visible through a telescope with a mirror of 3 feet diameter, and in 1848 he used Leviathan of Parsonstown, unique for its time, for observations. The sketch of 1844 looked like a horseshoe crab. But if the Crab Nebula was farther from us and no Leviathan could help us see it, the remnants would still not go unnoticed by astronomers.

Let's see how many times over the last millennium the light from supernova explosions in our Galaxy has reached the Earth. I have already explained that if the flash occurred in the wrong place, then it was greatly weakened due to extinction by dust and gas, and a new object in the sky could go unnoticed. But all the same, sooner or later, astronomers detect the radiation of the remnants formed after the explosion. Radio waves and X-rays from them will convey information about a cosmic cataclysm that was not noticed in time and was not caught hot on the trail. So, the list of nearby has to be complete or nearly complete.

In addition to the explosions of Ia type in 1006, 1572, 1604, and 1900 and a weaker type II flare in 1054, which formed the Crab Nebula, I can mention several more explosions known only from their remnants, including G1.9+0.3 in the constellation *Scorpio* with a neutron star inside, formed during an outburst in the 11th century; G266.2–1.2, the remnant of the outburst in about 1250 in the *Vela* constellation with the neutron star SGR 1806–20; and Cassiopeia A, the strongest galactic radio source with the neutron star CXOU J232327.8+584842 inside, formed during an explosion around 1667.

Note that the remainder in the *Vela* constellation is also called *Vela Junior*. *Vela* is the Latin name for sails, and astronomers use Latin names for the constellations. *Junior* because this remnant in the sky is projected onto a much older supernova remnant, simply called *Vela*, and formed in a type II supernova explosion from 11,000 to 12,300 years ago. There is a hypothesis that the supernova explosion that formed *Vela Junior* was due to the fact that the expanding shell from the *Vela* explosion reached a star ready to explode. So, it is possible that we are observing a kind of detonation on a cosmic scale.

We counted about a dozen explosions and remnants in the Galaxy, whose light has reached the Earth in the last millennium. And ironically, none of them were seen by astronomers through a telescope. But now they annually observe supernova explosions in distant galaxies. And they use this information to study DE.

1.9.7. *Trying to scare readers*

In the meantime, let's digress a little to a more practical question: shouldn't we expect the death of all life on Earth after a supernova explosion somewhere in the vicinity of the Sun? Naturally, not from visible light, but from the flux of gamma rays and charged particles emitted at the same time.

I'll start the answer from afar. As you know, many people like to be afraid. The success of many horror films clearly proves this. But to expect, say, an attack by hordes of zombies in everyday life is somehow strange, so you can be afraid of something that came from the depths of the sky. Alien attacks, a meteorite falling, Earth passing through a comet's tail with an immediate pestilence of all living things, etc.

You can find some of these reasons for getting scared in the book *Death from the Skies! The Science behind the End of the World* by Philip Plaits. Everything is there, from meteorites to black holes; real threats are mixed with purely theoretical ones. In reality, there is a constant threat of a strong solar flare and a magnetic storm on Earth caused by it (I will talk about this in detail in Section 3.5). What about a supernova explosion?

To begin with, let me remind you that supernovae explode permanently, but life on Earth did not stop. The analysis of the influence of various potentially dangerous factors of the outbreak was carried out by scientists repeatedly. Conclusion: the most dangerous are gamma and X-rays. If the flash is of type Ia, then it is not dangerous at a distance of more than a couple of hundred light-years (1 ly $= 9.46 \times 10^{15}$ m $= 5.9 \times 10^{12}$ mi).

Astronomers have compiled a list of stars in our Galaxy that may soon become supernovae. The closest to us are the white dwarf IK *Pegasus* B, located at a distance of only 150 ly, and the red giant Betelgeuse, located at a distance of 640 ly.

The dwarf, also known as HR 8210, will explode as a type Ia supernova, whereas Betelgeuse will explode as a weaker, type IIn supernova. Despite the fact that the minimum safe distance from a supernova is about 200 ly, there is no reason to fear. Firstly, "in the near future" on an astronomical scale may mean hundreds of millions of years, and secondly, IK *Pegasus* B by this time will move away from the Earth at a considerable distance. The explosion of Betelgeuse does not pose a danger to our existence.

Since I didn't manage to scare you with supernova explosions, I'll come from the other side. The accelerated expansion of the Universe caused by DE means that it will never stop. Different gravitationally bound groups of galaxies like the Local Group will move further and further away from each other. But there is also a hypothetical scenario in which the rate of removal begins to grow rapidly, and the scale factor will become infinitely large in a finite time.

This marks the end of our Universe, which has been given the tender nickname Big Rip. Everything, from galaxies to atoms, will be torn apart and carried away in all directions. But you can relax, because calculations show that if this scenario is realized, it will occur not earlier than in 55 billion years.

1.9.8. *Accelerated expansion of the Universe and antigravity*

By measuring the flow of energy coming to Earth, astronomers can find out the distance to a type Ia supernova explosion. Its redshift can be determined from the emission spectrum. Cosmologists had to take into account a nonzero cosmological constant in order to explain the relationship between these two quantities. It leads to the accelerated expansion of the Universe.

It can be provided not only by the cosmological constant, but also by something that works in a similar way. Physicists call this analogue of the cosmological constant by the general term *dark energy*. This is something that causes the accelerated expansion of the Universe. It can be either a Λ-term or other factors similar in action.

In any case, to explain the accelerated expansion of the Universe, it is necessary to introduce a new entity. Its reason is not clear yet, but scientists hope to bring some certainty to this question. In the meantime, we can only guess about the nature of DE. Whatever it turns out to be in the end, DE provides a general repulsion, antigravity, acting on everything in our world. It is antigravity that provides the accelerated expansion of the Universe observed by astronomers.

Physicists often use the term *negative pressure* when talking about it. What is it? If we pump a bicycle tyre or air mattress, then the air pressure hinders us. And by moving the piston, we do some work against the force of pressure. But the cosmological constant acts as a negative pressure, sucking in the piston inside. As a result, the Universe tends to expand. In addition, when the piston moves, we do not perform work, but, on the contrary, receive energy. A physicist could say that we are doing negative work.

So the Universe gains additional energy as it expands, which, according to the formula $E = mc^2$, is equivalent to additional mass. In this case, it is not only the volume of a certain part of the Universe that increases, but also the mass contained in it. In the case of a cosmological constant, these two processes balance each other so that its energy density remains constant during expansion. In the more general case of DE, it can change slightly, increasing or decreasing.

I can clarify some details after explaining the relation between antigravity and DE. The cosmological constant provides general antigravity, but of a very specific form. The antigravity caused by it does not change over time, providing a constant force that ensures the expansion of the Universe. This explains the instability of the stationary model introduced by Einstein. If the stationary universe shrinks a little, then the density of the matter filling it will increase, and with it the force of attraction, which will exceed the force of universal repulsion caused by the cosmological constant. Therefore, the universe will shrink even more. If the size of a stationary universe increases slightly, then the density of matter and the force of attraction will decrease, so the universe in this model will continue to increase in size.

But it is possible to imagine antigravity, which changes in time, for example, due to the expansion of the Universe, its cooling, or under the influence of other reasons, including those unknown to us so far. In this case, its cause is more general and we call it DE. We now believe that DE fills the Universe uniformly, but there are some hypotheses that its density may vary in different places in space.

Modern cosmology is based on general relativity or its generalizations, and on the assumptions that DE exists and the Universe is filled with baryonic matter and radiation, about which we know a lot, and DM, whose nature is not entirely clear. In addition, it contains other known particles such as neutrinos. There are many hypotheses, including that it also contains some fields that are still unknown to us. Their influence can explain, for example, the inflationary epoch in the early Universe, which is still out of the general history of development and requires separate consideration.

Further development of cosmology can explain and significantly clarify some details, but is unlikely to change the qualitative conclusion about the existence of DE and DM. It seems that in the future we will also talk about antigravity.

It is the presence of DE that determines the rate of expansion of the Universe right now and how it changes over time. Having become familiar with this concept, it's time to return to the story of evolution of the Universe.

1.10. Dark Ages and the First Stars

What happened after recombination? We can tell quite a lot about this based on theoretical calculations, but we cannot see it with our telescopes. The reason is simple — there was nothing to shine. The hydrogen and helium became too cold after cooling down, and the stars we were used to were not formed yet.

There were neither stars nor galaxies. Therefore, there was no light emission simply because there were no sources. This period in the life of the Universe is called the Dark Ages.

It ended with the appearance of the first stars. For their formation, primary hydrogen and helium had to be gathered in denser formations by gravity. After further contraction, the temperature and pressure inside the protostars reached a level at which a thermonuclear reaction became possible. It turned the clump of gravitationally bound hydrogen and helium into something completely different.

Stars lit up in the sky. It happened when the age of the Universe was from 150 million to a billion years. However, the estimate most likely should be clarified.

From the point of view of physics, the clustering of gas into protostars is a continuation of the process of increasing density fluctuations, which I spoke about. I mean those of them in which the density was increased. At the moment of recombination, the inhomogeneities were still small, but grew rapidly. The stars twinkling in the sky are a visible illustration of the existence and growth of density fluctuations in the early Universe.

The same process on even larger scales led to the emergence of galaxies and their clusters, as well as voids, that is, regions of space in which there are no or almost no stars and galaxies. Voids arose from fluctuations that reduced the density of matter in the early Universe.

Which came first, stars or galaxies? The question is interesting, but not as much as it seems at first glance. Imagine that galaxies came before. And once upon a time there was the first galaxy, still without stars. How do we know about its existence? It did not emit light and we cannot see it.

Note that the density fluctuations were of different magnitudes; therefore, the development of galaxies from them took significantly different times. The Dark Ages have ended due to sprinter galaxies that evolved from stronger fluctuations. And the galaxies that were jogging finished much later.

The appearance of the first stars led to the re-emergence of plasma in the Universe, but this time it was not evenly distributed, but concentrated in stars and other compact objects. In astronomy, this is called reionization.

1.10.1. *The chemistry of the Dark Ages*

As soon as the stars appeared, the process of stellar nucleosynthesis began, as a result of which nuclei of chemical elements heavier than lithium were formed. Before that, the Universe was filled with isotopes of hydrogen and helium, except for a very small fraction of lithium nuclei.

Helium is a noble gas; it does not enter into chemical reactions with other elements and its atoms do not form molecules consisting of several atoms. But hydrogen forms diatomic H_2 molecules. Let us recall the isotopes of hydrogen. Radioactive tritium, formed during the primary nucleosynthesis, decayed long ago and two isotopes remained, namely protium and deuterium. So, a hydrogen molecule could consist of two protium nuclei, or two protons, a proton and a deuteron, or two deuterium nuclei with two electrons around them.

The first type of molecules was found most often, while the last was rather rare. The emission spectra of these molecules are different, so astronomers can observe the emission of each type of molecular hydrogen separately. But their chemical properties are the same.

It would seem that we have exhausted all possible molecules that existed during the period of the Dark Ages, but this is not the case. Hydrogen and helium atoms cannot form a molecule, but they can form a positively charged helium hydride ion, HeH^+.

In it, two electrons form a cloud around two nuclei, a proton, and a helium nucleus. If a third electron is added to this ion, then the system decays into two atoms. However, the ion itself is stable and capable of emitting electromagnetic waves with its own spectrum. Ions with a large number of helium atoms are also possible: He_2H^+, He_3H^+, He_4H^+, He_5H^+, and He_6H^+.

Helium hydride is the strongest known acid; it reacts chemically with any substance. Therefore, it cannot be stored in any container, and any experiments in the laboratory are possible only with freshly prepared hydride. Nevertheless, these ions were investigated as early as 1925. They were obtained during the radioactive decay of tritium, i.e. not by chemical reactions at all.

Helium hydride can exist in space without a vessel. The search for radiation from such ions lasted a long time and ended in success only in April 2019.

So, I have listed all the chemical compounds that existed before the appearance of the first stars. I agree that at that time chemistry was a very simple science.

1.10.2. *Stars around us*

With the appearance of stars, the Dark Ages came to an end. But the growth of fluctuations and the creation of a large-scale structure continued after their completion. Now, the Universe is not at all uniform on the scale of typical distances between galaxies.

Our Galaxy, for example, is part of the so-called Local Group, which includes two large galaxies (the *Andromeda* galaxy, aka the *Andromeda* nebula, and Galaxy) and several dozen small galaxies, from dwarf galaxies to the more impressive Triangle galaxy, however smaller than the Milky Way.

The star formation process, i.e. the formation of new stars from gas and dust clouds, did not end with the formation of the first stars. It continues to this day. In our Galaxy, stars are formed with a total mass approximately equal to the mass of the Sun per year on average. And in galaxies with active star formation, this value can be hundreds of times higher. Naturally, stars of different masses are formed, from dwarfs to especially massive ones.

What can astronomers know from starlight? By the apparent brightness of the star and the distance to it, one can find out its absolute luminosity, i.e. energy carried away by radiation per second. Photometric observations are useful for this.

From spectroscopic observations, one can determine its spectrum and assign a star to one of the so-called spectral classes. These classes are related to the temperature of the star. Why? Surface temperature affects the emission and absorption of spectral lines of individual elements, so it can be determined from the spectrum of stars.

But the mass of stars cannot be determined by radiation. Now we know some stars close enough to us, around which exoplanets revolve and we can find out their mass from Kepler's laws. But a century ago, when the idea of the spectral classes of stars appeared, nothing was known about the planets outside the Solar System.

It was then that the so-called Hertzsprung–Russell diagram was constructed, linking the spectrum and luminosity of stars. It got its name from the names of two astronomers, the Dane Ejnar Hertzsprung and the American Henry Norris Russell, who made an important contribution to its creation. Since the distance to most of the stars was not then known, the scientists used a trick common in astronomy: they investigated the properties of a large number of stars that form a cluster. Although the distances to each star in the cluster were unknown, it was clear that they were approximately the same.

It turned out that hotter stars have a higher luminosity. However, before the creation of a complete theory of the structure of stars, the relationship of these quantities with the mass of stars was unknown. But then it appeared. What follows from this theory? Stars with a larger mass have a higher temperature and a higher luminosity, and the dependence is very strong.

1.10.3. *Stellar classification*

I give examples of stars of different spectral types. The most common M-stars, called red dwarfs, have a mass of about a third of the solar mass and a luminosity of about 4% of the solar luminosity. Their surface temperature is 2000–3500 K. Brown dwarfs are even less massive and cold, but I am interested in more hot stars.

The surface of orange dwarfs or K-class stars with a mass less than the solar mass and a luminosity less than half that of the Sun, are heated to 3500–5000 K. The Sun belongs to yellow dwarfs, which enter spectral class G. The temperature on their surface is 5000–6000 K.

More massive stars are not called dwarfs. Emission from yellow-white F-class stars with a surface temperature of 6000–7500 K is shifted towards the blue side of the spectrum compared to the Sun. Typical mass and luminosity are 1.7 times that of solar mass and 6 times that of solar luminosity.

Class A stars are called white; they have a temperature of 7500–10000 K. Mass and luminosity exceed those of the Sun by about 3 times and 80 times. Even hotter are blue-white stars of spectral class B with a temperature of 10–30,000 degrees.

And class O blue stars are claiming the title of record holder. They are very rare, one star in millions of stars of other spectral types. But they are very massive, with more than 50 times the mass of the Sun, and hot, with the surface temperature of 30–60,000 degrees.

Their radius exceeds that of the Sun by 15 times. The high temperature and surface area provides strong radiation. So, the luminosity of these stars is adequate, approximately one and a half million times more than that of the Sun.

The life of these stars is bright, but brief. The blue star has a lot of hydrogen reserves, but it spends it at the speed of a drunken merchant on a spree. So, it doesn't last long, only several million years, which is a short time on an astronomical scale. And then there's bankruptcy, the explosion of a new or supernova, i.e. the end is no less bright than the entire wild life.

On the other hand, the M-class curmudgeons manage to stretch their initially small reserves for a long time. By astronomical standards, these red dwarfs have not spent much of their time since their formation, with everything remaining in the Motherland's bins. They are not up to revelry with a spectacular suicide in the finale.

But let's return to the birth of a new one, to star formation. If there is a burst of star formation in some region and many new stars appear, then there are stars of all spectral types among them. The most massive and hottest of them emit UV radiation. From it, astronomers can easily detect these areas. However, because UV emission does not reach the Earth's surface, it must be observed from a satellite.

Over time, the most massive stars will complete their life cycles, and the stellar population will be left without its brightest representatives and begin to live a measured, routine life. Like our Galaxy. The starburst is over.

The story of the past of the Universe is often illustrated with popular pictures that can be found on the Internet, such as on the NASA website. All evolution of the Universe is depicted as a part of a glass: a birth from quantum fluctuations, the inflationary stage, relic radiation, the Dark Ages, the first stars, and structure development. The bending of the glass walls represents the accelerated expansion of the Universe caused by dark energy.

2 THE SUN AND SURROUNDINGS. SCALE: SOLAR SYSTEM

Our Galaxy is not a record holder, but it is not some kind of dwarf galaxy either. It is spiral, the Solar System is located in one of the arms at a distance of about 30,000 ly from the centre. And there, as expected in a decent society, there is the supermassive black hole *Sagittarius A**, so we are not ashamed in front of the inhabitants of other galaxies. I will talk about this hole in Section 7.3. For obvious reasons, we cannot see and photograph the Galaxy from the outside. In Fig. 2.1 instead of it you can see a photo from the "fashion magazine for galaxies".

The Hubble Space Telescope has photographed the spiral galaxy NGC 2985 in the constellation *Ursa Major*, which should be recognized as a very beautiful, almost perfectly symmetrical representative of this class of galaxies. Our Galaxy is clearly not so symmetrical, because about 10 billion years ago it absorbed a smaller galaxy from its surroundings, damaging not only its reputation but also its symmetry.

However, let's move on to places closer to us, to our star. There are many diverse stars in the Universe, but the Sun is of particular interest to us. And it is disappointing. It is a yellow dwarf with a mass only twice that of a typical star in our Galaxy. On the other hand, it's a star that's not exotic, but cosy and quite safe. And the neighbours are decent. They do not explode like supernovae, do not

Figure 2.1.   Spiral galaxy NGC 2985. Source: ESA/Hubble and NASA.

irradiate us with gamma rays,[a] do not climb into our Solar System, but keep at a distance. Not every planet system is so lucky. Considering that humanity, despite all the enthusiasm associated with space travel, is doomed for a very long time, if not always, to be tied to the Solar System, it is worth looking around to know what it is.

---

[a]More precisely, they irradiated, but rarely. Strong gamma-ray bursts from supernova explosions occur 2 or more times in a billion years. Some scientists consider them to be the cause of the Ordovician-Silurian extinction, during which more than 60% of marine invertebrates died about 450 million years ago.

2.1. The Solar System: Planets and Beyond

Let's start with an inventory of the property. We have one yellow dwarf star, scientifically G2V star, four smaller planets close to it with a hard surface, namely Mercury, Venus, Earth, and Mars, and four large gas planets far from the star — Jupiter, Saturn, Uranus, and Neptune. There are eight major planets in total.

Until 2006, Pluto was also considered a planet, but the International Astronomical Union demoted it to a dwarf planet. Since 2009, it is again considered a planet, but only in the states of Illinois, where its discoverer Clyde Tombaugh comes from, and New Mexico, where he worked. Illinois authorities even designated March 13, 2009 "Pluto Day".

The two groups of planets are separated by an asteroid belt. It contains the dwarf planet Ceres. Another asteroid belt, called the Kuiper belt, is located beyond Neptune. This is the region of the Solar System from the orbit of Neptune (30 AU from the Sun) to a distance of about 55 AU from the Sun. Let me remind you that the astronomical unit (AU) is the average distance between the Earth and the Sun; 1 AU $\approx$ 149,597,870.7 km $\approx$ 92,955,807.3 mi. The hypothetical Oort cloud — the source of long-period comets — is located even farther from the Sun at a distance of up to 50–100,000 AU.

In the Kuiper belt, several large objects can be distinguished that do not reach the large planet rank, the so-called plutoids. They are headed, of course, by Pluto himself, followed by Erida (it is not in the Kuiper belt, but in a larger, so-called scattered disc with its ice objects), Makemake, and Haumea.

Pluto has five satellites, Haumea has two, and the rest have one each. This was once considered the basis for considering Pluto to be a real planet, but after it was discovered that even some asteroids have satellites, the weight of the argument dropped dramatically. For example, the American spacecraft *Galileo* in 1993 discovered that the asteroid Ida, with a diameter of only 32 km, has a satellite with a diameter of 1.4 km, called Dactyl.

Several more bodies in the Solar System are in line to receive the status of a dwarf planet. It is Sedna from the Oort cloud, which was named after the Eskimo goddess of sea animals, and a dozen smaller bodies. Triton, the largest moon of Neptune, is supposedly just a plutoid captured by Neptune.

In our small inventory, we have many asteroids, comets, as well as planetary satellites. Of the satellites, two can be distinguished, in which the ratio of their mass to the mass of their home planet is maximum. They are the Moon and Pluto's satellite, Charon. Each of them and their planet revolve around a common centre of mass. Therefore, the Earth with the Moon is sometimes called a double planet, as well as Pluto with Charon.

The gas giants also have rings, especially pronounced around Saturn, which consist of small particles ranging in size from a few millimetres to 10 kilometres, and mostly made from dust and ice. The photographs of the rings of Saturn obtained by the Cassini space mission are simply mesmerizing (see Fig. 2.2).

Is there something of value lying around in the far corner? Could we have missed something? Naturally, the Oort cloud, Kuiper belt, and scattered disc probably have many undiscovered objects, including possible plutoids, not to mention every little detail.

Could a star be hiding there, for example, the notorious Nemesis, which the press periodically frightens us with? This hypothetical dwarf star allegedly orbits the Sun at a distance of 50,000–100,000 AU and periodically arranges mass extinctions on the Earth. It is very unlikely.

Even the faintest star, such as a brown dwarf, has a mass substantially greater than that of Jupiter. A brown dwarf is a cross between ordinary stars and giant planets. Such a substellar object has a mass in the range from 0.012 to 0.008 solar masses, i.e. from 12.5 to 80 masses of Jupiter. If it is not very far from the Sun, then the effect of its attraction on the orbits of the planets would be noticeable.

But let the star hide in the far outskirts. All the same, it should emit, albeit not in the visible range, but infrared rays. Astronomers have long taken images of the entire sky in several ranges of infrared wavelengths and have not found any mysterious source in the Solar System.

Figure 2.2.   Rings of Saturn and its icy moon, Mimas. Photo: Space Science Institute/JPL-Caltech/NASA.

Maybe they lacked sensitivity? To answer this question, it is enough to remember that astronomers find planets in other star systems, the so-called exoplanets. I repeat: planets, not stars, and far away, not right under our noses. And they have found red, white, brown, and other dwarfs for centuries. How can astronomers miss such a close celestial object of their usual type if they find them in the distance?

In fact, it is possible. Let me remind you that we can only detect a part of the exoplanets. Some of them are more convenient for detection by the methods used for this. For example, they are very massive or they periodically block part of the light from the star when orbiting it.

So a not very massive planet may remain undetected for a long time if we do not know exactly where to direct the telescope to search for it. However, a very small celestial body simply will not be considered a planet; perhaps it will be called a plutoid.

Could it be somewhere on the outskirts of the Solar System that there is still an undiscovered sufficiently massive planet, much heavier than Pluto and other plutoids? There are astronomers who claim that such a planet exists, its mass is 5–10 terrestrial ones, the average distance to the Sun is 400–800 astronomical units, and the angles of inclination and elongation of the orbit are small. But until astronomers see it through telescopes, they will not believe in its existence.

We know all the parameters of the Sun and major planets, and with great accuracy. They show that almost all the mass of the Solar System is concentrated in the Sun (more precisely, 99.86% of the total mass), but its angular momentum is inferior to the angular momentum of the planets. According to this indicator, the Sun loses to Jupiter by about a hundred times.

The Sun is more massive, but the distance to Jupiter is significantly greater than the radius of the Sun. The orbits of the planets lie in planes close to each other (deviations do not exceed 6°) and they all turn in the same direction. When viewed from the Sun's North Pole, the planets rotate counterclockwise. Therefore, the rotational moments of the planets are summed up, forming a clearly nonzero angular momentum of the Solar System, preserved from the moment of its formation.

2.1.1. *Dimensions of the orbits of the planets: The Titius–Bode rule*

Are there laws without theory? Yes, there are some. They are called empirical (that is, based only on experience) patterns. Some of them, over time, received a theoretical explanation and passed into the category of ordinary laws. These include Mendeleev's periodic table of chemical elements or the formula for the frequencies of the spectral lines of the hydrogen atom, explained in the framework of quantum mechanics.

But there are still unexplained patterns. An example is the Titius–Bode rule, which describes the distance from the Sun to various planets in the Solar System in astronomical units (AU). It can be written as follows: the average distance of the planet to the Sun is 0.4 AU $+ N \times 0.3$ AU. Here N is an integer equal to 0 for Mercury and 2^n, where n is an integer, for other planets. As a result, we get a set of distances of $0.4, 0.7, 1, 1.6, 2.8, 5.2, 10.0, 19.6$, and 38.8 AU, which correspond to Mercury (real average distance 0.39 AU), Venus (0.72 AU), Earth (1 AU), Mars (1.52 AU), the asteroid belt or the hypothetical ruptured planet Phaethon (2.2–3.6 AU), Jupiter (5.20 AU), Saturn (9.54 AU), and Uranus (19.22 AU). Neptune with an average distance of 30.06 AU falls out, but Pluto with its 39.5 AU, which is not considered a major planet, roughly corresponds to it. The reason for this pattern is still unknown.

Since we are talking about the planets, I immediately warn readers that there will be no detailed excursion around them. After all, space missions flew to each of them, providing us with a plethora of pictures, videos, and information; Mars rovers moved across Mars, and even Pluto was honoured with its mission, like the comet Churyumov-Gerasimenko. And there are also satellites, including large ones, asteroids, and comets, including those that pose a hazard to the Earth.

So planetary visits could easily double the size of the book. I include the most interesting things about them in Chapter 3, dedicated to the most important planet for us, i.e. the Earth. There is also some room for a separate Section 3.6 about the Moon. So you will learn about diamond rains on the gas giants, but a little later.

2.2. How Do Celestial Bodies Move?

Having counted the planets and making sure that there is no shortage, it's time to discuss exactly how the celestial bodies move. The simplest laws of celestial mechanics were discovered by the German astronomer and physicist Johannes Kepler.

2.2.1. *The movement of a light body in the gravitational field of a heavy one*

I'll start with the simplest case. There is a massive body. It is point-like or spherically symmetric, like a ball. If it is a star or a gas planet, then we will assume that it does not rotate around its axis and its symmetry will not be broken by centrifugal forces wishing to slightly distort the shape of the ball, turning it into an ellipsoid flattened along the axis of rotation. The second body is also a point-like or a very small ball, but it is much lighter, so we will neglect its mass in comparison with the mass of the first. There is nothing else in the Universe. The second body moves in the constant gravitational field of the first body.

A frame of reference attached to a massive body and not rotating will be inertial in this case. We use it to describe the trajectory of the second body.

Even Newton proved that the trajectory of a lighter body in this case must have the form of an ellipse, parabola or hyperbola. These three curves are called conical sections and are mathematically the closest relatives.

Each of them has specific points, called the focal points of this curve. It is in the focus of the trajectory of the light body that the immobile massive body is located.

If the trajectory has the shape of an ellipse, then the light body will orbit around the heavy one along the same path all the time, making each turn in the same period of time, called the period of revolution. It will never leave the system.

This is roughly how planets move around single stars or satellites move around planets, such as the Earth around the Sun and the Moon around the Earth. Maybe because in addition to the two bodies in the system, there are also other objects that slightly distort this ideally periodic motion. Moreover, there is also the effect of general relativity, which will be discussed later.

2.2.2. *Ellipse and its properties*

An ellipse is a circle flattened in some direction (or elongated in a direction perpendicular to it). The degree of oblateness is characterized by a dimensionless quantity called the eccentricity of the ellipse. If the circle is not flattened at all, then it remains a circle. This is a special case of an ellipse with zero eccentricity.

And if you flatten it infinitely strongly, then the circle will turn into a straight line segment. This is another limiting case of an ellipse with an eccentricity equal to 1. The eccentricity of any ellipse lies in the range from 0 (circle) to 1 (line segment).

In addition, the ellipse has one remarkable property. It has two points called foci, and at any point on the ellipse, the sum of the distances to the foci is constant. This property is used to draw a beautiful regular ellipse on a piece of paper. We take two pins or needles, stick them into the points where the foci of the ellipse are located and from the thread we make a closed loop, the length of which exceeds the distance between the pins doubled. We put the loop on the pins and the point of a pencil or pen. We take the pencil to the side so that the thread stretches, forming a triangle.

Then we simply move the pencil so that the thread remains taut all the time, drawing a line with it on paper. It turns out to be an ellipse because the sum of the distances to the pins equals the length of the thread minus the distance between the pins, and these values do not change when drawing. If both foci coincide, then you get a circle. If you spread them to the maximum distance that the length of the thread allows, then the ellipse will turn into a segment.

The distance from the focus to the point of the ellipse varies from point to point, unless the ellipse is degenerate into a circle. The point at which it is minimal is called the periapsis or pericentre, and the point at which it is maximal is called the apoapsis or apocentre. Anyone who has drawn an ellipse at least once knows that both of these points lie on the major axis of the ellipse passing through its foci.[b]

However, the words *periapsis* and *apoapsis* are very rare. For the orbits of celestial bodies, names are used associated with what exactly the light body revolves around. If around the Earth, like the Moon and artificial satellites, then this is perigee and apogee,

---

[b]In astronomy, it is referred to by the tricky term "line of apses".

because in Greek the Earth is called *Gaia* (Γαία). If we are speaking about the planets and small bodies around the Sun (*Helios*), then this is perihelion and aphelion. If an exoplanet revolves around a star or two stars orbit around a common centre of gravity, then astronomers speak of periastron and apoastron, because *astrum* is a star in Latin.

2.2.3. *Orbital speed*

Back to moving along an elliptical trajectory. Various bodies, from a satellite to a stone thrown in a vacuum, can move at such an orbit. Naturally, sometimes we see only part of it. The trajectory of a stone thrown into the distance is a small section of an ellipse in the absence of air and its resistance. Over time, it would reach the surface of the Earth, into which the thrown object usually falls. But if you throw it horizontally at a speed greater than a certain minimum, called the orbital speed or the first cosmic velocity, then it will revolve around our planet, turning into an artificial satellite. For the Earth, the orbital speed is approximately equal to 7.9 km/s or 28,500 km/h.

I am sure that the story will only benefit if it is accompanied by examples. On the Internet, it is easy to find Newton's drawing, with which he illustrated the possibility of moving in a circular orbit around the Earth. There is a cannon on a high mountain that shoots horizontally. As the initial velocity increases, the projectile flies farther, and after reaching the first cosmic velocity, it flies around the Earth. Out of respect for the great physicist, let's take this drawing as a basis.

In Fig. 2.3, at point A, the Isaac Newton cannon is stationary. The gravitational field of the Earth outside the planet coincides with the field of a material point in its centre. It is marked with an asterisk in the figure. Apart from a massive centre, a cannon, and a projectile, there is nothing in the entire Universe. The cannon shoots a projectile horizontally.

There is no atmosphere, no air resistance, so the projectile flies according to the laws of celestial mechanics, unless it hits the

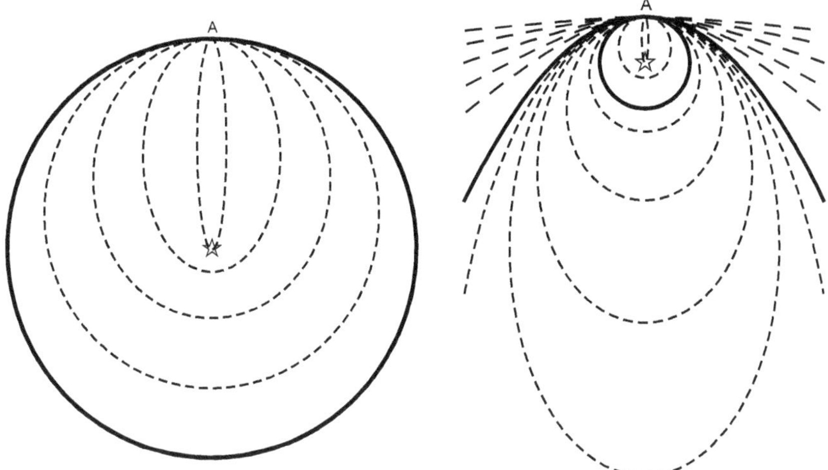

Figure 2.3.   Trajectories of bodies moving in the gravitational field of a massive point body, the position of which is indicated by an asterisk. At point A, their speed is directed horizontally. As the initial velocity increases, the orbits will be ellipses with decreasing eccentricity. Upon reaching the first cosmic speed, the orbit will become circular. The trajectories are elliptical for the initial velocities between the first and second cosmic speeds, while the massive body is in nearer focus. When the escape velocity is reached, the orbit becomes a part of a parabola along which a body can fly away from a stationary one. A further increase in the initial velocity corresponds to hyperbolic trajectories.

ground. But for now, let's mentally squeeze the Earth into a point and temporarily forget about colliding with the surface.

If the projectile is simply released without initial velocity, then it will fly to the centre and hit it. Its trajectory is a straight line segment with an eccentricity of 1. Let me clarify that the initial velocity can be associated with rotation. When we release a stationary stone from our hands, standing at the top of the Leaning Tower of Pisa, we usually say that it falls in a straight line. Indeed, it would have been the same segment of a straight line, if not for the rotation of the tower together with the Earth. Therefore, in a frame of reference that does not rotate with the Earth, the trajectory of a falling stone is part of an ellipse, and in a frame of reference tied to the foot of the tower, the falling stone will slightly deviate to the east.

Now let the cannon shoot and give the projectile a speed lower than the orbital speed. The projectile will fly in an elliptical trajectory. The left panel of Fig. 2.3 shows several dashed ellipses, all of them corresponding to the orbits of the projectile at different initial velocities. The higher this speed, the larger the size of the ellipse and the lower its eccentricity. If we count from the centre, then the figure shows orbits with eccentricities equal to 0.99, 0.8, 0.5, and 0.2.

Each of these ellipses has two focuses. In this case, the centre of the Earth is in the farthest focus from the gun. This means that the gun is located at the apogee of the orbit. The perigee is closer to the centre than the cannon.

Let us now release a projectile with an initial velocity equal to the first cosmic velocity. In this case, the eccentricity of the orbit drops to zero, the two foci of the ellipse merge together, and it turns into a circle. It is drawn with a solid circle in Fig. 2.3. The speed of motion of the projectile along it is constant, so the first cosmic speed is equal not only to the initial velocity of the projectile, but also to its speed at any point.

If I added the surface of the Earth in the form of a circle located slightly below the circular orbit and cut off all dashed orbits after they reached the surface, it would get a Newtonian drawing. Indeed, when the projectile reaches the surface of the Earth, it stops moving along the trajectory, regardless of whether it hits it, gurgles into the ocean, or drowns in a swamp.

However, I did not end these curves and did it on purpose to discuss one nuance. Imagine that a group of mad scientists in secret laboratories made a projectile from a substance unseen in nature, which quietly penetrates any of the Earth's rocks without experiencing any resistance. Such a formulation of the problem is quite suitable for a thought (*gedanken*) experiment.

Well, if you want the question to have at least some physical meaning, then we could discuss the motion of a clot of dark matter flying through the location of the gun and through the gun itself with a certain horizontally directed speed. Dark matter passes through ordinary matter without resistance, but is attracted to it — in other words, it is affected by its gravitational field. I talked about DM properties in Section 1.9.

So, either our secret projectile or a clot of dark matter quietly passes through the surface of the Earth and goes deep into its bowels. I am sure that some of the readers decided that it continues to move along the dashed elliptical trajectory shown in Fig. 2.3. But no! The fact is that the projectile's acceleration in the gravitational field, i.e. the acceleration of free fall, in this case, will be provided not by the entire mass of the spherically symmetric planet, but only by those parts of it that are closer to the centre of the Earth than the projectile.

Therefore, the closer to the centre, the smaller it is, and in the very centre the acceleration of gravity vanishes. And the trajectories in Fig. 2.3 are calculated as if the acceleration were provided by the entire mass of the Earth. That is why they incorrectly describe the motion inside the planet.

If the density of the Earth's interior were constant (which is not the case, as you will learn in Section 3.3), then the magnitude of the acceleration due to gravity inside the Earth would be proportional to the distance from its centre. Indeed, in this case, the mass of a part of the Earth inside a sphere, concentric to the globe, would be proportional to the cube of its radius. According to the law of universal gravitation, this mass should be divided by the square of the radius, which gives the mentioned dependence.

Inside such a homogeneous Earth, a clot of dark matter[c] would also move along an elliptical trajectory, but not along the dashed ellipse, but along a slightly different path. In this ellipse, which describes the underground part of the trajectory inside the homogeneous Earth, the centre of the Earth is not in focus, but in the centre.

Perhaps some of you are familiar with the problem of the motion of a clot of dark matter released without initial velocity. No? So this is the very often considered problem in which a stone is thrown into a well drilled through the Earth along its diameter. In

---

[c]I am aware of a certain schizophrenicity of such a task, but in physics it is customary to think about all kinds of thought experiments, and not just realizable ones.

this case, either the Earth does not rotate or the well is dug along the axis of rotation.

The questions in the problem may be different, but in any case, the stone reaches the second edge of this well on the opposite side of the Earth. And the ellipse in Fig. 2.3 at zero initial velocity would be reduced to a straight line segment to the centre of the Earth.

2.2.4. *Escape speed*

Who said that the cannon on the mountain can't fire a projectile at a speed exceeding the first space speed? It can, because our imagination has no limit. But since this case forces us to go beyond Newton's drawing, we will have to draw it separately, in the right panel of Fig. 2.3. The point is that we need to change the scale, because the trajectories of the projectile will now move away from the Earth at significantly greater distances. So, for some trajectories, only a part of them will fall into the drawing.

If a stone is thrown at a speed greater than the so-called escape speed or second cosmic velocity, then it can fly away from a massive body infinitely far. The second cosmic velocity exceeds the first one $\sqrt{2} \approx 1{,}414$ times, so for the Earth it is 11.2 km/s. In this case, the speed of movement decreases with distance from the attracting body.

Naturally, motion along an infinitely long path requires infinite time, so it will certainly not be periodic, like a motion along an ellipse. In this case, the trajectory will be a piece of a curve called a hyperbola.

This curve also has two foci, the difference in distances to them is constant at any point on the hyperbola. Far from the foci, the hyperbola becomes close to a straight line. Bodies that have flown into the Solar System from interstellar space move along the hyperbola.

The Sun is at one of the focal points of this curve. So, an interstellar object far from our star flies almost in a straight line, and in the gravitational field of the Sun turns and flies out in a different direction. The deviation angle is the greater, the closer to the Sun the body flies and the lower its initial velocity. The shape of the hyperbola is determined by its eccentricity. It is always greater than 1 for a hyperbola.

A special parabolic trajectory lies between the elliptic and hyperbolic trajectories. In order for the body to fly along it, it must be given the second cosmic speed exactly. Slightly less speed and the orbit is already elliptical; slightly more — hyperbolic. In this case, the rate of removal of the body does not simply fall over large distances, but tends to zero. The eccentricity of the parabola is 1.

I left in the picture a solid circular orbit and a couple of dashed ellipses from the left panel. As the initial velocity of the projectile increases, it will move along the ellipses also drawn by the dashed line, but outside the circular orbit. How are these ellipses different from previous ones? The centre of the Earth is in the nearer focus for them and the cannon is now located at perigee.

The apogee, like the entire trajectory, lies outside the Earth. As the initial speed increases, the size and eccentricity of these ellipses increases. In the right panel of Fig. 2.3, you see orbits with eccentricities of 0.2, 0.5, 0.7, 0.8, and part of the orbit with an eccentricity of 0.9.

When the initial velocity reaches the second cosmic one, the ellipse degenerates into a parabola with its unit eccentricity. This path is drawn with a solid line. The projectile moves away from the Earth at any distance moving along it.

Naturally, it will also fly away when its initial velocity is greater than the second cosmic one, but it will fly away along a hyperbolic trajectory. The dashed lines with longer strokes drawn in the right panel of Fig. 2.3 are the parts of hyperbolas that correspond to the eccentricities of 1.5, 2, 3, 5, and 10 (in the figure, they are arranged in the order from bottom to top). When flying past the massive body at the same distance from the centre of gravity, the higher the initial velocity, the greater the eccentricity of the orbit and the smaller the deflection angle of the light body.

One can understand that a cannon shot can theoretically send a spacecraft on a flight to the Moon or to distant planets, but it is impossible to launch an artificial Earth satellite in such a way. The point is not even in its deceleration in the atmosphere, but in the fact that the elliptical trajectory is closed and the projectile from the cannon, fired at a speed lower than the second cosmic one, will sooner or later return and hit the very gun which sent it flying.

Here is the basic information about the possible trajectories of a light body in the gravitational field of a massive one. Kepler's laws tell us not only about orbits, but also about the change in the speed of bodies during their movement, in particular about the periods of revolution of planets and satellites.

2.2.5. *Motion of several massive bodies*

We know practically everything about the motion of two bodies in the case when their masses are comparable. In this case, the centre of mass of two objects is stationary in a certain inertial reference frame. Therefore, objects move synchronously and in opposite directions, if you look from the centre of mass. The ratio of distances to bodies is always constant and inversely proportional to the ratio of their masses. In the frame of reference of the centre of mass, the velocities of bodies are always directed oppositely, and their ratio is also inversely to the ratio of masses of the two bodies.

The distance to the heavier body and its speed are less than to the lighter one. In this case, both bodies can move either along ellipses (including a circle and a straight line segment), hyperbolas, or parabolas. Naturally, both bodies move along similar trajectories.

But if there are three or more bodies, then it doesn't look good. Scientists over the centuries have not been able to find a solution to the three-body problem, and it looks like it will never be found. It is clear that with the help of a computer we can calculate the motion in any particular case, but the general formula of this motion would delight physicists much more.

Since there is no correct solution, false ones are proposed instead. When many bodies move, their centre of mass is motionless, so many people are mistakenly convinced that the Sun and the planets are orbiting in ellipses around the centre of mass of the Solar System, which is called the barycentre in astronomy. Alas, this is a delusion.

The planets attract each other. These additional forces distort those ideal elliptical trajectories, which would be the case if the Solar System consisted only of the Sun and one planet. These distortions, displacements, and rotations are very weak, but from them you can estimate where the planet is located, and which gravitational field distorts the ideal trajectories of its neighbours.

So astronomers predicted the position of the planet Neptune by its gravitational influence on Uranus. This was independently done by the Frenchman Urbain Jean Joseph Le Verrier and the Englishman John Couch Adams. The German astronomer Johann Gottfried Galle directed the telescope to the area of the sky where an undiscovered planet should be, according to Le Verrier's calculations, and discovered it. It took him less than an hour.

As a result of the constant action of such disturbances, the orbits of the planets, including the Earth, very slowly change their parameters, from eccentricity to size and orientation in space. The characteristic time of such changes is tens of thousands of years, and they affect the climate due to fluctuations in the amount of solar radiation reaching the Earth.

The result is a kind of cycle, astronomical and climatic, named after the Serbian astrophysicist Milutin Milanković. The current global warming is happening much faster and cannot be explained by Milanković cycles.

If a light body approaches a massive one, for example, if an asteroid flies near Jupiter, its orbit can deviate greatly. Space missions launched from the Earth use the passage near the Moon and planets for the so-called gravitational manoeuvre to change the speed of the spaceship.

For example, the ship which is heading towards the Sun must considerably decrease its initial velocity relative to the Sun, which is equal to the Earth's orbital speed, in order to get closer to the star. In some space missions, it flew first to Jupiter, made a manoeuvre in its gravitational field, decreasing its orbital speed, and only then flew to the Sun. Naturally, without extremely accurate calculations, the whole undertaking would be doomed to failure.

Our Sun is a single star; it has no companion. This is good, because the stability of the orbits is bad in the system of binary stars (a star system consisting of two stars orbiting around their common barycentre) or multiple star system; a planet can either approach the star or move away from it, which is dangerous for life on it. Note that, according to astronomical terminology, double or triple stars means that two or three stars appear to be close together in the sky as seen from the Earth. They can be at very different distances and do not orbit around a common centre of mass as binary stars do.

2.2.6. *Perihelion shift*

In the Solar System, the orbits of the planets are very close to ellipses and would coincide with them, if not for the attraction of other celestial bodies, as well as the impact of the effects of general relativity. The shift of a perihelion is one of these effects. This is a slow rotation of the orbit as a whole in its plane around the Sun, known as the apsidal precession in astronomy. It was noted for Mercury by Urbain Le Verrier even before the creation of general relativity.

Inspired by the success of the search for Neptune, Le Verrier theorized the existence of another planet, Vulcan, between Mercury and the Sun. Some astronomers have even reported seeing this planet near the solar disc. But as time went on, the telescopes improved and it became clear that these messages did not correspond to reality. No Vulcan was discovered then or ever since.

On the other hand, general relativity predicted the apsidal precession or perihelion advance of any body orbiting a more massive one, including Mercury. Calculations have shown that the relativistic precession value of 0.43 arc seconds per year is added to the 5.32 arc seconds per year precession due to perturbations from the other planets. The contribution of the Sun's nonsphericity is negligible. The overall effect of about 570 arc seconds per century is in excellent agreement with the observations.

A smaller precession was found for Venus and some satellites, all of which also confirm Einstein's theory.

2.2.7. *Exoplanets and exocomets*

The time has passed when people wondered whether other stars have planets or the Solar System is unique in the Universe. After the discovery of exoplanets, this issue was resolved. As of June 2021, the existence of 4758 exoplanets in 3517 planetary systems has been reliably confirmed, of which 783 have more than one planet. So the Solar System is definitely not the only system that has planets, including those capable of becoming habitable and carrying living or even intelligent beings.

In 2019, half of the Nobel Prize in Physics was shared by two Swiss astronomers, Michel Mayor and Didier Queloz, for the discovery of one of the types of exoplanets orbiting stars similar to the Sun. It is worth noting that astronomers have discovered not only exoplanets, but also exocomets, i.e. comets flying in other stellar systems. Naturally, fewer of them are known, simply because comets are smaller than planets.

How are exoplanets found? Two methods are more commonly used. Sometimes planets pass through the visible disc of a star and slightly reduce its apparent luminosity. Or they reduce the radio emission from the far pulsar. It is important that this happens at constant time intervals corresponding to the periods of exoplanet orbital.

The second method involves the fact that the planet and the star revolve around their centre of gravity. Therefore, the star moves along a small ellipse and the change in its speed relative to the Earth can be measured using the Doppler effect. As a result of processing data on changes in this speed, it is possible to determine the number of large planets in the star system and their parameters: mass and distance from the star.

2.3. Electricity, Magnetism, and Their Relationship

On the scale of the Solar System, one cannot ignore the second of the fundamental interactions, the electromagnetic one. Therefore, I have to tell you something about its foundations. This is necessary, in particular, to understand the magnetic fields of the Sun and

Earth and their influence on the movement of electrically charged particles.

In general, on astronomical scales, the magnetic interaction is often stronger than the electric one, because strong currents caused by the movements of positively and negatively charged particles can flow through electrically neutral bodies. So, magnetic fields play a very important role in stars and more exotic astronomical objects like quasars or the pulsars, described in Section 7.1.

2.3.1. *Long road to electromagnetism*

Mankind knew about individual examples of electrical interactions even in antiquity. Everyone has seen lightning. For a long time, the main method of generating electric charge was friction or the discharging of electric fish. So, look at Fig. 2.4, which reminds us of electrification during friction, and consider how this phenomenon ultimately gave rise to the world of devices that use the laws of electromagnetism.

Figure 2.4.   The pieces of styrofoam stuck to the cat's hair due to static electricity. Photo by Sean McGrath distributed under a CC-BY 2.0 license.

The first discoveries in the field of electricity and magnetism are described in Section 8.2. Be that as it may, the idea of electric attraction and repulsion of magnets and the Earth's magnetic field was fully formed by the 18th century. Let's add to it the concept of a positive and a negative electric charge and its conservation, and we get all the prerequisites for the study of electricity to move to a higher, quantitative level. In other words, scientists moved from concepts to numbers.

The first quantitative law associated with electrical interaction was Coulomb's law, named after the French engineer and physicist Charles-Augustin de Coulomb, who discovered it. He proved that the force of interaction of two electrically charged balls, the distance between which is much greater than their size (it is very close to the force of interaction of two point charges), is directed along the line connecting their centres.

Charges of the same sign repel, whereas charges of different signs attract. The magnitude of this force is proportional to the product of two balls' charges and is inversely proportional to the square of the distance between them.

This law made it possible to give the first definition of electric charge and to introduce the units of measurement of this quantity. But scientists had to go a long way before getting to Coulomb's experiments. I will tell the story of the discovery of the basic laws of electrical and magnetic interactions separately in a chapter dedicated to these issues: Section 8.2. Unlike gravity, I will not limit myself to a couple of names.

The story will be replete with the names of many people who have made their own significant contributions to the knowledge of these entities, which were initially mysterious for scientists. Little is written in textbooks about the early stages of the history of knowledge of electricity and magnetism, but their importance should not be underestimated.

A breakthrough in understanding the nature of electromagnetism came after scientists unexpectedly discovered that an electric current creates a magnetic field, and a change in a magnetic field generates an electric field. Before that, the very idea of the similarity of these forces seemed strange, because there were so many differences between their manifestations.

Many bodies can be electrified by friction, but none of them can be magnetized in this way. Iron is sometimes magnetized when hit hard, such as with a sledgehammer,[d] but electrification on impact was not observed.

The French physicist and astronomer Dominique François Jean Arago had some vague ideas that magnetism is somehow related to electricity. While participating in the work of the expert commission to find out the causes of shipwrecks, he noticed that after a strong storm at sea, iron objects on board were strongly magnetized. Even compass needles were remagnetized sometimes. After the storm, they sometimes pointed in opposite directions. So the storm clearly had a powerful magnetic effect on the ship and its iron parts. Arago suspected that the reason could be due to lightning, that is, electricity. But he did not share such an unusual guess with anyone until September 4, 1820.

On that day, Arago, the scientific secretary of the French Academy, introduced the academicians to the latest discovery of the Dane Hans Christian Ørsted (or Oersted) at a meeting of the Academy. Ørsted found out that an electric current creates a magnetic field around the wire through which it flows.

This not only demonstrated for the first time the connection between electrical and magnetic phenomena, but also made it possible to measure the strength of the electric current flowing through the wires. Physicists began to measure the angle of deflection of the compass needle located near the wire, and ended up with a special device — an ammeter, which is still used today.

Then the Englishman Michael Faraday showed that a change in the magnetic field causes the appearance of an electromotive force that generates an electric current. Thus, electricity and magnetism, considered very different things for centuries, suddenly became

---

[d]The mechanism of magnetization is explained in Section 4.1. It is related to the influence of the Earth's magnetic field.

close relatives. And if we talk about practical applications, then Faraday's discovery led to the emergence of electric generators, motors, and transformers.

2.3.2. *The emergence of the concept of a field as a foundation of modern physics*

Faraday put forward the idea of a field. It is so familiar to modern physics and was quite revolutionary for his time. Instead of saying that two charges of the same sign repel each other in accordance with Coulomb's law, physicists say that one charge creates an electric field around it. This field acts on the second charge, causing a repulsive force.

It would seem that there is not much difference in these explanations, but soon physicists were faced with situations in which there are fields but no obvious sources of them. For example, if we look at the distant Sirius at night, our retina reacts to the electric field of light from this distant star.

Where are the electric charges that caused this field? If they exist, then they are very far away, on Sirius. More precisely, they existed 8.6 years ago, when the light that recently fell into our eyes was emitted by the alpha star from the constellation *Canis Major*.

The compass needle shows us the direction of another field, the Earth's magnetic field. It creates a torque that turns the needle. The compass is in our hands, and the source of the field is located deep in the bowels of the Earth.

Suppose science is faced with a fundamentally new interaction that physicists want to study. It can manifest itself qualitatively, for example, as a radioactive decay. Or it can cause a mechanical force that can be measured. Scientists familiar with the concept of a field in the latter case try to determine from experiments the laws that describe two aspects of the interaction process.

They ask themselves two questions. First: what, and according to what laws, causes the field that characterizes this new interaction? Second: how is the force resulting from the action of the field related to the magnitude of the field and the parameters of the object on which this field acts?

Take electrical interaction as an example. An electric field is caused by the action of an electric charge. After Coulomb's experiments we know the formula for its strength. And we know exactly how the electric field acts on the charges in it: the force is equal to the product of the field by the value of the charge.

In Section 8.2, you will see that this is enough to introduce a new unit of physical quantity called electrical charge. When two bodies interact electrically, the charge of the first one determines the strength of the electric field created by it, and the charge of the second one determines the force with which this field acts on it.

Let's move on to the magnetic field. Imagine a person who knows nothing about magnetism but has access to physics reference books. Let's open some of them and find out that a magnetic field is what arises around electric currents and permanent magnets. If you are not too lazy to look up what a permanent magnet is, you will find out that it is what is capable of creating a magnetic field around itself.

From these definitions, without having any other information, it is impossible to understand what a permanent magnet is and what exactly it causes around itself. If the word "sepulka" says something to you, then this is a classic example of the situation described by Stanisław Lem.[e] If you do not try to come up with a definition, but ask yourself the two questions mentioned above, then the picture becomes simplified.

The strength of the magnetic field, especially its direction, can be found with a small compass. Section 8.2 tells how William Gilbert, using a compass and a model of the Earth he had made,

---

[e]In the story "Voyage Fourteenth" from *The Star Diaries* by the Polish science fiction writer Stanisław Lem, it is written:

"The Encyclopaedia Cosmica gives the following definitions on the subject Sepulka — pl: sepulki, a prominent element of the civilization of Ardrites from the planet of Enteropia; see "Sepulkaria."

Sepulkaria — sing. sepulkarium, establishments used for sepuling; see "Sepuling"

Sepuling — an activity of Ardrites from the planet of Enteropia; see "Sepulka" "

called the *terella*, had proved four centuries ago that the Earth is a huge magnet.

The magnitude of the magnetic field strength can be found by measuring the moment of force acting on the compass needle. But it is better to measure the force acting on a wire with an electric current. It is named after the famous French physicist André-Marie Ampère and is proportional to both the current and the strength of the magnetic field.

This allows us to answer the second question, about the relationship between force and field parameters. The answer to the first question of how a current or the movement of an electric charge causes this field has long been received. For the current, this is the law discovered by the French Jean-Baptiste Biot and Félix Savart, to which list in a number of countries is added another famous French scientist, the Marquis Pierre-Simon de Laplace.

These are a lot of advantages of introducing the concept of a field. Much to the surprise of physicists, it turned out that the field has many properties of bodies. For example, it has its own energy, momentum, and angular momentum. Without knowledge of their existence, it is easy to get into trouble when analysing the simplest situations.

In Fig. 2.5 you can see two electrically charged particles and their directions of movement. The strength of their electrical interaction is easily found from Coulomb's law. It obeys Newton's third law: the force of action is equal to the force of reaction.

If the charges of the particles have the same sign, they repel; if they have opposite signs, they attract. The force of electric attraction

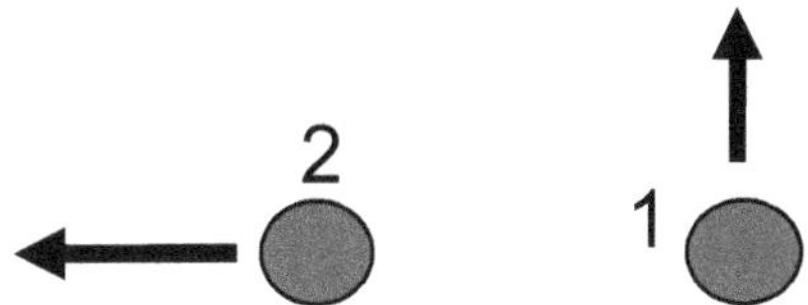

Figure 2.5.   In the magnetic interaction of two charged particles, the action force is not equal to the reaction force.

or repulsion acting on particle 1 is opposite to the force acting on particle 2.

What about magnetic interaction? The formula for the magnitude and direction of the magnetic field of a moving charged particle is easy to find in the reference book. I simply indicate that at the point where particle 2 is located, the particle 1 creates a field perpendicular to the plane of the drawing. It deflects the motion of particle 2 upward or downward depending on the signs of the charges of the particles.

Particle 2 does not create any magnetic field at the point where particle 1 is located, and, accordingly, creates no magnetic force acting on it. As you can see, in this example the force of action is not at all equal to the force of reaction.

Was Newton wrong? No, otherwise his third law would have long been excluded from textbooks. So what's up? This is not an easy task and it can puzzle many professional physicists. I begin the answer by the formulation of Newton's third law: in a closed system, the sum of all forces acting on each object included in this system is equal to zero. If the system consists of only two objects, this will give us the standard formulation that the force of action is equal to the force of reaction.

But who said that the system depicted in Fig. 2.5 consists of two objects? Naturally, everyone can see two particles on it and there is no third one nearby.

However, there is a third, hidden participant in the system. This is a magnetic field. It has energy and momentum and is also acted upon by force. The sum of the forces acting on two particles and the magnetic field is zero in full accordance with Newton's laws. But how do you understand this example without knowing about the field's existence?

I hope that I was able to convince readers that the idea of a field, which so unexpectedly burst into physics, has not only firmly established itself there, but has long become an important part of this science. So I continue about a specific field that describes the phenomena that are now called electromagnetic.

2.3.3. *Maxwell's equations*

The experiments of Oersted and Faraday showed that electric and magnetic fields are two faces of a single electromagnetic field, which resembles the two-faced Janus. They are not only sources of each other, but also transform one into another when the frame of reference is changed.

The very concept of an electromagnetic field was finally formulated by the Scottish scientist James Clerk Maxwell. He also derived the equations describing this field that subsequently received the name of their discoverer. These equations were written in their modern form by Oliver Heaviside and other scientists. The Maxwell system of equations not only firmly entered the university physics course, but also penetrated a little into popular culture. You can buy a T-shirt printed with these equations. Some people even get themselves tattooed with it.

Maxwell's equations include quantities that characterize the electric charges of bodies and electric currents arising from the motion of these charges. But they have no magnetic charges and magnetic currents, because no such charges exist. Naturally, scientists carefully checked whether magnetic charges, also called magnetic monopoles, could be found anywhere, but all attempts failed.

For example, if such a monopole flew to us from outer space, then in the Earth's magnetic field it would rush to the magnetic pole, either north or south. There is an ocean near the North Pole, but Antarctica is located near the South Pole. So, scientists looked for traces of monopoles in samples of Antarctic ice. No monopoles were found in particle accelerators, cosmic rays, and many other processes.

Let me note an amusing detail: the famous theoretical physicist Paul Dirac not only proved that if there were magnetic monopoles, then their magnetic charge should be proportional to a certain minimum value, but also determined the magnitude of this quantum. So, physics can sometimes draw conclusions even about what is not in nature.

2.3.4. *Charge movement in a magnetic field*

In our time, when the word *electromagnetic* is very common, some people, on the contrary, need to explain the differences between electricity and magnetism. One of the differences, namely the absence of magnetic monopoles, I have already noted. I'll tell you about the second most important, and all the others can be considered to some extent a consequence of these two.

If an electric charge is placed in an electric field, then force acting on it is directed along the field. In accordance with Newton's second law, it causes the acceleration of a charged body. In a parabola, an electric charge moves in a uniform electric field in the same way as a thrown stone moves in a uniform gravitational field.

Now let's remove the electric field and place the charge in a magnetic field. If it is motionless, then no additional magnetic force acts on it. But how does it move if its velocity is nonzero? In this case it is acted upon by a force named after the famous Dutch physicist Hendrik Antoon Lorentz, winner of the Nobel Prize in Physics for 1902.

This force is proportional to the particle speed and perpendicular to both it and the orientation of the magnetic field. The Lorentz force causes the particle to move with acceleration, but due to its direction it cannot change the energy of the charged body. Therefore, the modulus of the charge speed vector does not change with time, only its direction.

In a uniform magnetic field, the charge generally begins to rotate along a helical line with an axis directed along the field. It is often called a spiral, but for mathematicians these are completely different curves. The threads of screws and drills, springs, spirals fastening the pages of a notebook, DNA, and RNA have the form of a helical line. It is a combination of uniform movement along the field and rotation with constant angular velocity around a circle in a plane perpendicular to it. It can degenerate into a circle perpendicular to the direction of the field and into a straight line along the field.

If the field is not very uniform, then the charged particles rotate around the field lines of force, moving along them. In Section 3.4, you will learn that this is exactly how the particles of the solar wind behave in the Earth's magnetosphere, occasionally indulging us with the spectacle of the aurora. And in Section 6.2, I will tell you how they try to keep plasma in peculiar traps with the help of a magnetic field in order to get a controlled thermonuclear reaction.

Many people know that powerful magnets are used in synchrophasotrons and other charged particle accelerators. And some people are sure that the particles in accelerators are accelerated by the magnetic field. This is not so; the magnetic field alone cannot accelerate anything. It is needed in order to deflect charged particles, turning their path into a circle. The particles are accelerated due to the force produced by an electric field.

An accelerated charge emits electromagnetic waves. So it emits waves when it is moving either in an electric field, or in a magnetic field, or it is simply slowing down inside some kind of medium that decreases its speed. The radiation generated by the rotation of charges in a magnetic field is called synchrotron radiation. Its generation is a very common process in space.

Synchrotron radiation occurs when high-energy particles move in accelerators, in which they fly in a circle in a magnetic field, for example in synchrotrons, which gave their name to this kind of radiation. It leads to unnecessary energy losses. However, scientists began to create the so-called photon factories for the generation of synchrotron radiation, mainly X-rays. It is used in medicine, physics, biology, chemistry, materials science, and other fields.

I described the basic properties of electricity and magnetism. There are no magnetic charges and electric charges can be positive and negative. If the body is isolated, then its full electric charge is conserved. But nothing prevents us from getting two bodies with positive and negative charges from an uncharged body if the sum of these charges is equal to zero. It is enough to rub a plastic comb on wool or hair to charge both the comb and the hair with charges of different signs. And here again we look at Fig. 2.4.

2.4. Waves

I have already told you a little about classical electromagnetic waves from radio to gamma rays. But much can be added, so

I continue my story. I would like to point out that waves are a much more general concept. There are waves of a completely different nature, like waves on water, in an elastic medium (sound, seismic waves, shock waves), string vibrations, and much more. Scientists even talk about heat waves. Physicists divide waves into two fundamentally different types, transverse and longitudinal.

2.4.1. *Longitudinal and transverse waves*

To begin with, I name a typical longitudinal and transverse wave. Sound in the air is a longitudinal wave. When it propagates, air molecules move back and forth along the direction of sound propagation. And the vibration of a long, stretched string is a transverse wave. Displacements and speeds of individual sections are perpendicular to the string along which these vibrations propagate.

What is the difference between these two cases? The direction along the propagation of the wave is uniquely determined, but we cannot say this about a perpendicular direction. A lot of perpendiculars can be drawn. If the wave is transverse, then it can be displaced in any direction orthogonal to the direction of propagation. If the string is stretched vertically, then it can be pulled towards the north, or it can be pulled towards the west. It turns out that any permissible vibration of the string can be obtained as follows: hit the string in the north, wait a fraction of the vibration period, and hit it in the west. Both blows must be made with a certain force, more precisely with a given ratio of the two displacements of the string.

In oscillation theory, this will be called the two amplitudes of the basic oscillations and the phase shift between them, but I will not delve into this jungle. The main thing is understanding that transverse vibrations can have the same frequency but differ in the direction of displacement. This is called polarization.

Sound in a gas or liquid is clearly longitudinal. But vibrations in solids are more complicated. You can hit one end of a long rod lengthwise so that a compression wave travels along it from one end to the other. It is easy to guess that it is longitudinal.

But you can hit this end from the side. In this case, a wave of bending and displacement goes along the rod and obviously it is transverse. This example shows that sometimes different types of waves can propagate in the same medium, usually at different speeds.

For example, longitudinal seismic waves in the bowels of the Earth propagate faster than shear waves. But if the waves reach the liquid core, then the transverse vibrations can no longer pass further and only longitudinal vibrations emerge from the core. Actually, therefore, seismologists learned about the existence of the core and were able to determine its size.

2.4.2. *Electromagnetic waves in vacuum: The relationship between a frequency and a wavelength*

An alternating electric field creates an alternating magnetic field next to it and vice versa. As a result, a diverging spherical electromagnetic wave, in which a field of one type always generates another, propagates in all directions from the source.

Suppose we are looking at a very distant source, say, the star Sirius. We see the light that was once emitted from its surface. Since Sirius is far away, the part of the wave reaching the Earth is practically indistinguishable from a simple wave called a plane one. Roughly speaking, this is a wave in which electric and magnetic fields and the energy associated with them move in a certain direction, called the direction of wave propagation and these directions are parallel for different parts of the wave. In our case, they are directed from Sirius to Earth. Displacements or fields, which they carry, are the same at any point in the plane perpendicular to this direction. This is why the wave is called plane one.

Consider a monochromatic electromagnetic wave with a fixed frequency. This can be a signal from a radio transmitter or light emitted at a certain transition of an electron inside the atom shell of a particular chemical element. At any fixed point through which the wave passes, both electric and magnetic fields change sinusoidally with this frequency.

The whole pattern of the fields moves in the direction of wave propagation at a constant speed, called the phase velocity of the wave. It depends on the medium in which the wave propagates and, to a lesser extent, on its frequency. In vacuum or emptiness, all types of electromagnetic waves propagate at the same speed. This is one of the fundamental physical constants denoted by the Latin letter c and is equal to 299792458 m/s. It does not depend on the frequency of the emitted light and on the speed of its source.

This is the maximum speed at which anything can move in our world. All attempts to detect the existence of hypothetical superluminal particles named tachyons have failed. I'll return to this question in Section 7.2.

Now let's cast a magic spell and our wave stops and freezes. Naturally, this is nothing more than a thought experiment, but it helps us figure out the wavelength. Let's measure the electric and magnetic fields at an arbitrary point and start moving from this point in the direction of the wave propagation, frozen after a successful spell. Soon we will find another point in which the fields will be exactly the same as in the initial one. Then second, third, etc.

Why am I sure of this? Simply because they were exactly the same in the initial point before one, two, three, or more periods of field change. During this time, they shifted towards the direction of propagation by one, two, etc., wavelengths. From this, it is easy to conclude that the speed of the wave is equal to the ratio of its wavelength to the period of oscillation. So, the product of frequency and wavelength is equal to the phase velocity

2.4.3. *Electromagnetic waves in vacuum: Polarization*

Let's cast an anti-freeze spell and start the wave again. Let us choose an arbitrary point that we have already become fond of and place there a device that registers in real time the change in the magnitude and direction of the vectors of the strength of the electric and magnetic fields.

A study of its readings shows that these directions are always perpendicular to each other and to the direction of wave propagation. Not surprisingly, this is precisely what follows from the corresponding solution of Maxwell's equations.

More precisely, the strengths of the electric field, the magnetic field, and the direction of propagation form what mathematicians call the right-handed triple of vectors. These are three mutually perpendicular vectors taken in a certain order, the same as the x, y, and z axes in the Cartesian coordinate system. A set of vectors along the x, z, and y axes, taken in this order, form a left-handed triple. These two sets are obtained from each other by mirroring or simultaneously changing the direction of all the three axes. The last operation in mathematics is called inversion. An example of it is shown in Fig. 8.8.

Since both fields are directed perpendicular to the direction of propagation, then a plane monochromatic electromagnetic wave belongs to transverse waves. So it must be polarized. If in our thought experiment the device fixes exactly how the electric field strength vector changes in time at any fixed point, then the end of the electric field vector forms an ellipse that I talked about in Section 2.2. But first, I want to clarify how exactly it is obtained.

So, I choose any suitable scale, with which I convert the electric field strength (in the SI system, it is measured in volts per metre) into segments of a certain length (in the SI system, it is measured in metres) and keep the direction of the vectors. Say, an electric field of 1000 V/m corresponds to a segment of 1 metre. That is, having measured at some point the field of 100 V/m, directed towards the star *Algol*, I draw on paper a 10 cm long arrow directed towards *Algol*.

In the next measurement, I get the field 73 V/m in the direction of *Aldebaran* and draw an arrow 7.3 cm long, appropriately oriented. All arrows start at the same point.

I draw segments on paper that correspond to measurements, not forgetting to note the time of their taking (in a thought experiment, everything is permissible, so my measurements take place instantly). The reader may doubt that all the arrows will be drawn exactly on the sheet and will not need to go beyond it. No, I will get a drawing, not a sculpture. I won't need the third spatial coordinate. Indeed, in a transverse wave, the vectors of the strengths of both fields lie in a plane orthogonal to the direction of wave propagation. I just placed a sheet of paper parallel to this plane.

In an ideal monochromatic wave, the end of the arrow demonstrating the current electric field strength draws an ellipse. In this case the angular velocity of rotation of the arrow is constant. It can be obtained by multiplying the frequency of the wave by 360° or 2π radians. Rotation can be in any direction, clockwise or counterclockwise. Physicists say that such a wave is elliptically polarized.

And how does the magnetic field strength behave? Maxwell's equations answer this question. It also draws an ellipse in a plane perpendicular to the direction of wave propagation. It is like an ellipse for an electric field, i.e. it has the same eccentricity and the same ratio of the lengths of the major and minor axes. But it is rotated at an angle of 90°, so that the vectors of the strengths of the electric and magnetic fields and the direction of propagation form a right-handed triplet.

So, the ellipse described by the magnetic field is uniquely obtained from the ellipse of the electric field and vice versa. To describe any of them, you must indicate the size of its major and minor axes and their directions.

However, any ellipse can degenerate. In one limiting case, it turns into a straight line segment. This polarization is called linear. In a linearly polarized plane electromagnetic wave, the electric field is always directed in a certain direction or in the opposite one, and the magnetic field is always directed perpendicular to both directions of the electric field and the direction of propagation. Linearly polarized light is obtained by passing any light through a device called a polarizer.

In the second limiting case, the ellipse goes into a circle. This light is called circularly polarized. Circular polarization can be left or right. In a wave with left-handed polarization, the vectors of the electric and magnetic fields rotate counterclockwise when looking towards the wave. Right-handed polarization means that the rotation occurs clockwise if the observer is located on the side towards which the wave is directed. Circularly polarized waves can be obtained by passing the linearly polarized wave through a special device.

How do the fields change in a linearly polarized plane monochromatic electromagnetic wave (every word is important here)? On the Internet, it is easy to find animated pictures for this case; they clearly show how the sinusoids drawn in the figure move forward.

And in a wave with circular polarization, the end of the vector corresponding to the electric field will describe a helical line, similar to the screw. The vector of the magnetic field strength will also describe a helix rotated by $90°$ relative to their common axis. The whole picture will move along the direction of this axis.

In the general case of elliptical polarization, the picture of the electric field will be similar to the circular one, but with a slight change. It is flattened in some direction. The magnetic field is obtained by turning this flattened helix $90°$. By combining two polarized waves of the same frequency, propagating in the same direction, you can get a wave with an arbitrary polarization.

Perhaps the reader has some questions related to the polarization of an ideal plane monochromatic electromagnetic wave. A simple experiment can help answer them.

Take a thread, and at one end, tie a small weight, such as a nut or paper clip. Attach the second end to something so that this home-brew pendulum can swing freely in any direction. Then, deflect it or push it aside.

What do you see? The weight begins to describe an ellipse, the centre of which is below the suspension point. The dimensions of its axes are determined by the initial deflection and the speed of the pendulum, so that the eccentricity of the ellipse can be different. It can move in a circle with zero eccentricity. In this case the thread, when moving, forms the surface of the cone. Or it can oscillate in some vertical plane, like a pendulum, so the ellipse degenerates into a straight line segment, acquiring a unit eccentricity.

An arrow directed from the centre of the ellipse to the weight constantly changes its length (if the eccentricity is not zero) and direction. Let this arrow describe the vector of the electric field in the vertically propagated wave. What about magnetic? It is described by another horizontal arrow, always perpendicular to the electric one.

It is easy to understand that the weight can move both clockwise and counterclockwise and the direction of the long axis of the ellipse can be an arbitrary one on the horizontal plane. It is easy to see the cases corresponding to the left and right circular polarizations, linear polarizations and the most general form of elliptical polarization.

What are the disadvantages of this simple demo? Over time, due to friction, the trajectory of the weight will approach the centre, which is not the case for an electromagnetic wave propagating in a vacuum or a transparent medium. In addition, we are looking at the change in fields at one point, so the wave propagation is not particularly visible.

2.4.4. *Natural light*

All this is great, but only for an ideal world. Real light emitted by different atoms has slightly different frequencies. There are many reasons for this, but one of them is the most fundamental. This is the so-called uncertainty principle, a cornerstone of quantum mechanics, which I discuss in Section 5.3.

As a result, light from different atoms cannot interfere with each other. And by combining two polarized waves from these atoms, you cannot get a polarized wave. A mixture of many waves from different atoms gives us what is called natural unpolarized light.

We can get a natural light from the Sun, Sirius, or an incandescent lamp. If we mix a polarized and an unpolarized light, we get a partially polarized light. Natural light can be partially polarized after reflection, refraction, or scattering. If the light hits the water surface at an angle of $53°$ to its normal, then the reflected light is linearly polarized. When the angle of incidence is different from $53°$, then the reflected light is partially polarized, as well as the refracted light.

Light from the sky, or rather light scattered in the atmosphere by fluctuations and dust particles, is also partially polarized. The light from the clouds is unpolarized. If you get a polarizer in your hands, look through it at the sky and twirl this simple device. See for yourself that I have not deceived you.

Bees and some other animals can orient themselves by the position of the Sun in the sky, even if it is hidden behind clouds. In doing this, they are helped by their vision, which is able to determine the polarization of light. By looking at the sky between the clouds, they are able to understand where the Sun is.

Scientists are sometimes no more stupid than bees and have also learned to obtain information from the polarization of light. There is a branch of optics called the polarimetry, the purpose of which is to determine the degree of polarization of the light of the observed object. The devices allow us to find out what percentage of its light is polarized and how exactly, and determine the polarization parameters of this light.

This is especially important in those cases when we can only observe an object, and we need to learn as much as possible about it. Information about the polarization of light will not be superfluous. Also important is information on its spectra in different wavelength ranges, from radio waves to gamma rays.

But not only the ranges are different, but also the types of astronomical observations. The simplest ones are just observations of the object: Did Mars or Jupiter suddenly disappear, or is everything all right there? Has a new comet appeared in the sky? How many sunspots and their groups are on the Sun today? One can determine the exact position of the object in the sky or measure its brightness. These are tasks for positional astrometry and photometry.

You can get the emission spectrum of this object, and from it determine its chemical composition, temperature, and much more. This is spectroscopy. It is possible to measure the degree and parameters of the polarization of light reflected from an object; this will give information about the properties of its surface or particles in its atmosphere. This is a polarimetry. Spectral polarimetry tracks polarization in different spectral ranges. In general, all available parameters of electromagnetic waves are used in astronomy or remote sensing.

2.4.5. *Why do we need high antennas and TV towers*

In the early period of the development of radio communications, it was believed that the antenna should ideally have size of at least a quarter of the emitted wavelength, which was clearly impossible for the long-wave range. But they tried to make the antennas large and began to use tall structures like the Eiffel Tower, which was originally built for the exhibition and had no practical use, except as a stand for a restaurant with an observation deck. Then, for the antennas of radio transmitters, they began to build special structures like the *Funkturm Berlin* (Berlin Radio Tower) or the radio tower of the Comintern.

For short and ultra-short waves, large antennas are not necessary, as well as for television. Over time, radios got rid of the long antennas installed on the roof or hung around the room. Even mobile phones have lost their short protruding antenna. They now only have a small antenna hidden inside.

But high TV towers are still needed. For what? The fact is that if the transmitter operates at a high frequency, in the range of FM radio, television, VHF, cellular communication, or GPS, then its signals can be caught only in the line of sight. You do not receive an FM signal in Tokyo from Los Angeles and even from Osaka. But the signal from the satellite is easy to catch using the appropriate antenna.

The higher the transmitting antenna is located above the ground, the longer is the line of sight. Simply because the Earth is round. As you move away from the transmitter, it disappears behind the horizon. Therefore, high TV towers perform the same role as the towers of beacons: they increase the signal's line of sight. For the same reason, cellular antennas are placed at a not very large distance from each other and, if possible, on the tops of mountains and hills or on tall structures.

2.5. Formation and Evolution of the Solar System

2.5.1. *Gas and dust cloud collapse*

According to modern views, the Solar System was formed from an interstellar molecular cloud several light-years across, consisting mainly of hydrogen and helium, but enriched in heavier elements left over from the stars of previous generations. Due to the growth of instability, the denser part of the cloud began to gather together, forming the basis for further evolution. Other stellar systems were formed from the rest of the cloud.

The central part of this cloud compressed and collapsed, forming a protostar, which began to glow when it reached a temperature of several thousand degrees. The energy for its heating was taken due to the potential energy of the gravitational interaction, in other words, due to its parts approaching each other. It happened about 4.6 billion years ago, but one must understand the low accuracy of this estimate.

The temperature and density in the centre of the protostar increased until the processes of thermonuclear fusion began. Hydrogen began to be transformed into helium and then into heavier elements because of the stellar nucleosynthesis. The energy released in this process went into the heating of the newborn star, which we now call the Sun.

The increase in plasma pressure in the centre and the radiant pressure of light stopped the collapse process, maintaining all layers of the Sun in a state of equilibrium. The star began the process of evolution, which will continue for a long time.

But not all the substance of the cloud of gas and dust has gathered in the place of formation of the protostar. This was prevented by the rotation of the cloud, the angular velocity of which increased with compression due to the conservation of the angular momentum of the system. As a result, the cloud took the form of a protoplanetary disc with the Sun in the centre.

The outer regions of the disc remained relatively cool. After the Sun began to shine, it warmed up these areas, but rather moderately. The farther away from the Sun the part of the disc was, the lower its temperature was.

2.5.2. *Formation of planets*

But the disc itself was not stable. Its parts were attracted to each other, collided, large parts rubbed against small ones, bringing them together. Due to hydrodynamic and gravitational instabilities, separate compactions began to develop in the disc, which became local centres of planet formation from the matter of the protoplanetary disc. And even after the protoplanets were formed, they still interacted with each other while changing at the same time.

Naturally, scientists have theories describing the formation and evolution of planets, taking into account many factors. Dust particles formed larger formations — planetesimals, which collided with each other and with smaller objects. Suffice it to say that all bodies in the Solar System have experienced billions of years of bombardment by fairly large asteroids. Many of those having a hard surface but no dense atmosphere have perfectly preserved traces of ancient cosmic collisions.

Mercury is said to have lost its mantle after such a collision. It is a widely known hypothesis that the Moon was knocked out of the Earth in a collision with the hypothetical protoplanet Theia. I will talk about it in Section 3.6. It is possible that the strong inclination of the axis of rotation of Neptune to the plane of its orbit is also the result of an ancient collision.

The theory of planetary formation explains why terrestrial planets, small and with a solid surface, were formed near the Sun, while larger gas giants were formed far from it. The difference is because of ice, whether water, methane, or ammonia, which evaporated under the influence of solar radiation near the star, but did not evaporate far from it. The border of the two zones, called the inner and the outer Solar System, runs just between the orbits of Mars and Jupiter.

The theory of planetary formation takes into account not only collisions of particles, but also other processes, for example, the interaction of particles with the remnants of a protoplanetary disc and a planetary disc formed at an early stage of evolution. It was confirmed by studies of the nucleus of the comet Churyumov-Gerasimenko, on which the descent module Philae of the Rosetta spacecraft landed. It turned out that the core consists of at least two parts of different origins, which have been combined into a single complex core.

2.5.3. *Cosmic prerequisites for the origin of life*

The author of this book has made every effort not to get involved in questions related to biology, since they deserve no less than a separate book, especially when it comes to such difficult and painful issues for some people as the origin of life. Nevertheless, at this point, I am simply forced to deviate from this principle.

It is difficult to deny that there is life in our Solar System, and that this life is intelligent. This may seem like a miracle and something incredible, if not for the considerations associated with the so-called anthropic principle, described in more detail in Section 8.3. Its essence can be summarized very briefly: if there were no intelligent life, there would be no one to complain that there is no intelligent life in the Solar System (or in the Galaxy, or in the Universe).

Simply because there would be neither the author of a book complaining about this fact nor readers who can share the bitterness of the non-existence of life. And if there is an author and readers, then intelligent life definitely exists. It is clear that this philosophical principle does not tell us anything about whether there is life in the systems of other stars.

For a long time, scientists did not know if there were such systems with planets in other stars at all. But nowadays astronomers have discovered a lot of them. In some of them, conditions, such as the temperature, intensity, and radiation spectrum of a star, can be favourable for life of the terrestrial species. But this does not mean that there is life there. So what do we know about the possibility of its appearance?

There is a hypothesis of *panspermia*, suggesting that life was brought to Earth from somewhere in space. This is a very old idea, put forward by the Greek philosopher Anaxagoras in the 5th century BC, or maybe someone else earlier. It was quite popular in the 19th century. However, the proposed answer to the question of the origin of life is nothing more than a deception. After all, if life is brought to Earth, say, from some other planet, the question naturally arises as to how it appeared there. More precisely, how life first appeared in our Universe, even if later it could somehow spread in it. And since our Universe did not always exist (remember the Big Bang), then the beginning of the chain of the propagation of life must be obligatory. This brings us back to the original question of how life originated.

I do not get into the issues of the emergence of life from organic substances, evolution, and so on, especially since biologists and, perhaps, chemists understand them better than me. Let's focus on what preceded this. I mean the process called chemical or prebiotic evolution. This is the emergence of organic substances from inorganic molecules under the influence of natural factors. Let's just mention two facts that are related to this issue and are directly related to comets — typical objects of the Solar System.

The first is that two groups of scientists discovered on the surface of the Churyumov–Gerasimenko comet the precursors of the sugars necessary for the existence of life forms known to mankind. The second is associated with processes on an artificial comet created in the laboratory.

Scientists took dust particles composed of silicates (or carbon), which were surrounded by an ice shell of a mixture of water (H_2O), methanol (CH_3OH) and ammonia (NH_3), placed in a room with low pressure and a temperature of $-198°C$ and irradiated with ultraviolet light. Then the "comet dust" was heated to room temperature (after all, comets also heat up as they approach the Sun). It was possible to find several different monosaccharides, including ribose, which is part of the RNA of living organisms, in what was formed as a result of the chemical reactions.

So, it was possible not only to discover some precursors of substances that form living organisms, but also to reproduce the possible process of their formation in nature.

2.6. How is Heat Transferred?

Life on our planet is provided by the flow of solar radiation. This radiation is associated with the flow of heat from the centre of the

Sun to its surface. Then it is transferred to the Earth. To understand how this whole chain works, it is worth discussing how heat is transferred. There are three main mechanisms that are important in this process: thermal conductivity, convection, and thermal radiation. In addition, it is worth mentioning the mechanism using the heat of vaporization. Naturally, you can occasionally find other, much more exotic mechanisms.

2.6.1. *Three main heat transfer mechanisms*

Thermal conductivity is the transfer of heat from warmer parts to colder ones by virtue of the thermal movement of molecules. In other words, heat is transferred, but what transfers this heat remains immobile for us. We do not see individual molecules making thermal motion and do not consider a transferring heat substance to be mobile.

The simplest example is a house wall in winter. It is cold outside, but warm inside because the heating is on. Heat passes through the wall to the outside, heating the surrounding space, but the walls are stationary. We poured warm soup into a bowl and the bowl got hot due to the thermal conductivity of the material that it is made of.

Thermal conductivity is the main mechanism for transferring heat through a solid, in which molecules cannot move far. If you burn yourself with hot tea or a silver spoon sticking out of a cup with this tea, then feel free to blame the thermal conductivity. Naturally, this isn't your fault, right?

A cloud of aromatic hot steam rises above the soup plate. Hot air rises above the fire. The smoke from the chimney rushes into the sky. This is already the second heat transfer mechanism — convection. In this case, the transfer of heat is accompanied by mechanical movement, jets, and flows.

The air above the fire heats up, becomes lighter than cold air, and rushes upward. It is convection that provides traction in the pipes of stoves and fireplaces, like much else in nature — from the transfer of energy inside stars to the formation of winds and clouds in the atmosphere. Convection occurs not only in air or gas, but also in liquid. Boiling or about to boil water is ready to confirm this.

If, when brewing a magic potion, we stir it in the right direction, say, clockwise, then this is also convection, even if not natural, but forced. Blowing on the very tea that not only burned us, but also persuaded a silver spoon to enter into a criminal conspiracy with him, we use convection as a weapon of retaliation, diverting energy from the cup of tea into the surrounding space.

The last of the three main heat transfer mechanisms is heat radiation. If we kindle a huge fire, then we will feel this mechanism with the whole body. And in summer, on the beach or in the desert, it's hard not to notice the Sun's heat. In addition, each of us, at least once in his life, set something on fire with the help of the Sun's rays collected by a large magnifying glass. It greatly increases the efficiency of solar heating of the area on which the rays are focused.

The method of transferring heat by radiation is important primarily when the body is heated to a high temperature. If you turn on an incandescent light bulb, then its spiral heats up so much that we immediately notice it, or rather just see it. After all, it is for this that incandescent lamps were invented.

In this case, an incandescent lamp emits most of its energy not in the form of visible light, but in the form of infrared radiation. If the temperature of the heated body is increased further, then ultraviolet radiation will be added to the infrared and visible radiation. Sunburn reminds us of UV rays. After all, the Sun differs from the filament of an incandescent lamp not only by a greater intensity of radiation, but also by a higher temperature. But the X-ray radiation of the Sun is non-thermal and occurs due to other mechanisms.

2.6.2. *Heat transfer in vacuum*

It remains to recall when thermal radiation is the only heat transfer mechanism. It happens in a vacuum where the other two don't work. There are simply no molecules there that could move somewhere or make thermal vibrations. The Sun and other stars emit energy through thermal and non-thermal radiation. Any body cools down in space only due to this mechanism. This process is rather weak for moderately heated bodies.

Therefore, the removal of heat from spacecraft and even astronauts in spacesuits working in outer space is not an easy task. The writers have frightened readers for so long with the terrible cold of space emptiness that many do not even realize that an astronaut would simply boil in his spacesuit without a cooling system.

On Earth, we also use vacuum to suppress two heat transfer mechanisms. It is this which is located between the two walls of the thermos, designed to maintain the desired temperature of drinks and dishes.

There are many time-honoured ways of keeping something cool or warm. For example, wrap it up in a blanket or other thermal insulation material. We regularly use this trick in the winter to keep our bodies from getting cold in the cold. But inside of us there is a built-in heater that runs on the energy of eaten steaks, pizzas, and salads.

A heated pot of porridge will cool down, despite any wrapping, but this will happen in a much longer time than without wrapping. However, no one would call a blanket a thermos.

The thermos must necessarily have a glass or stainless steel flask with double walls, between which air is pumped out. It is because of this that the original temperature of its contents remains for a long time inside the thermos.

The fact is that in vacuum, heat losses associated with thermal conductivity and convection are minimal but nonzero because of the neck of the thermos and the fasteners that connect the flask to the body. Radiation losses remain, which are quite small due to the low temperature of hot or cold drinks. Especially when compared to the temperature of the surface of the Sun or the spiral of an incandescent lamp turned on.

The thermos was patented in 1903 and was a slightly improved Dewar flask, invented in 1892 by the Scottish physicist James Dewar for storing liquefied gases at low temperatures. Dewar himself did not patent his invention, since he did not think about its use in everyday life. However, this situation is not uncommon in scientific research. Many familiar things and devices appeared as by-products or technical applications of scientific developments.

2.6.3. *Heat transfer by evaporation and vapour condensation*

An additional method of heat transfer is associated with the heat of vaporization. In that very cloud of hot steam above the bowl of soup, the heat spent on converting part of the water in the soup from liquid to gaseous state is hidden. The process of evaporation and subsequent condensation of steam in the form of precipitation plays an essential role in the mechanism of climate and weather formation on our planet.

I will describe one curious utilitarian application of this method of heat transfer. Imagine baking a turkey or a large whole chicken in the oven (not in the microwave). Heat penetrates into the inside of the carcass due to thermal conductivity, i.e. pretty slowly. So there is a risk of either overdrying the surface of the turkey due to baking for too long, or getting a poorly baked middle.

The so-called heat pipe can come to the rescue. It is a piece of a thin metal tube, closed at both ends, and partially filled with a low-boiling liquid. The device is forcibly inserted into the carcass so that one end of it is close to the middle, and the other protrudes outwards. The outer end should be lower than the inner one. The turkey or chicken is baked with tubes inserted into it, which are then removed, washed, and set aside until next use.

How does it work? The liquid inside the heat pipes collects in their lower parts, which are located outside the baked piece. A liquid inside pipes has a low boiling point. Therefore, it quickly heats up in the oven and evaporates. A steam rises through the heat pipe to the inner, colder parts of the carcass. There it condenses and flows down the walls in the form of a liquid. And the heat released by condensation warms the poultry from the inside, ensuring a more even roast.

It would be strange if such a wonderful device was used only in the kitchen. Naturally, this is not the case. Heat pipes are used, in particular, in the cooling systems for computers and laptops, and less often for smartphones, but first of all in spacecraft. This heat transfer mechanism is also used in steam cooking. In this case, no technical devices are needed, because steam participates in the cooking process in the most natural way.

Having tasted the baked turkey and washed it down with tea, on which we blew correctly and ensured its optimal temperature, we can proceed to the bowels of the Sun.

2.7. The Sun: Structure, Activity, and Influence on the Earth

For us, there is no more important object in the sky than the Sun. Exploring the CMB, distant galaxies, or even Mars can be exciting, but to be honest, these are far from the day-to-day interests of humankind. Quite another matter is the Sun, which not only keeps the Earth in orbit, but also supplies it with heat and energy.

It is no coincidence that all types of power plants, except for nuclear (and possibly thermonuclear in the future), ultimately use the energy that came from the Sun. It depends on the Sun how plentiful the harvest will be and how satisfyingly we will eat.

It is clear that many astronomers are engaged in the study of the Sun. More precisely, this was started even before the appearance of astronomy itself, because it is impossible not to notice the Sun in the sky on a cloudless day. It is also impossible not to notice, for example, a solar eclipse, when the Moon covers the Sun or part of it.

2.7.1. *How the Sun is studied*

For centuries, observations of the Sun were purely visual, and then more and more powerful telescopes began to be used. With the help of spectrographs, we obtained a spectrum of solar radiation and, from the emission and absorption lines of individual elements, we learned the chemical composition of the Sun. The study of the polarization of light from the Sun has provided a lot of additional information, including the strength of the magnetic field of individual parts of the Sun.

It hardly needs explaining why the telescopes that are used to observe the Sun differ from those that observe stars, galaxies, or planets. Not only because they make observations during the day, and the others at night. The light fluxes from the Sun and other stars are too different.[f]

It is clear that progress does not stand still. In 2019, the first observations with a four-metre solar telescope in Hawaii began, and soon they plan to commission a telescope of a similar size in the Canary Islands. Prior to that, the largest solar telescopes had mirrors up to 1.6 m in diameter. And some of their nocturnal counterparts boast mirrors over 10 m in diameter.

From observing only visible light, astronomers moved on to observing the Sun in the entire range of electromagnetic radiation, from radio waves to gamma rays. The detectors went beyond the atmosphere to spacecraft. A space solar observatory can observe our star around the clock, without fear of clouds and fog. A ground-based one cannot observe at night, when the Sun and the telescope are on opposite sides of the Earth. You don't have to wait for a solar eclipse to photograph the Sun's corona — its outer regions. It is enough to use a coronagraph device, in which the radiation of the central, brightest parts of the Sun is obscured by a special screen.

And if the space solar observatories are put into orbit around the Sun, they make it possible to view it from different angles. You can simultaneously observe the surface of the Sun from two spacecraft orbiting in the same orbit, but very far from each other, getting something like a stereo image of what is happening on this surface. No wonder a pair of such satellites was named *STEREO*.

Another advantage of space missions is the ability to look at the Sun from an unusual angle. From the Earth, we look at it in one plane, close to the equatorial plane of the Sun. If a flare or prominence is in the centre of the visible solar disc, then we see it in the direction radial to the Sun.

---

[f]The bearded joke states that you can only look at the Sun through a telescope twice: with your left and right eyes. But it has nothing to do with real solar telescopes.

The directions to the Earth and to the centre of the Sun will be opposite. If the flare is on the limb, that is, at the edge of the visible disc, we will look at it from the tangential direction. The directions to the Earth and to the centre of the Sun will be perpendicular. This makes it possible to understand the spatial structure, although observations in the two directions are not carried out simultaneously (as on the *STEREO* satellites).

But everything that happens near the poles of the Sun (although much less interesting happens there than at its equator), we see only in the tangential direction. Therefore, the opportunity to look at the Sun along the axis of its rotation is very tempting.

For all the time, only the *Ulysses* space mission could have done this. In addition, it measured the properties of the interplanetary medium outside the plane of the Earth's orbit — the ecliptic. The mission was challenging. To approach the Sun, the device flew in the opposite direction from it and used Jupiter's gravitational field to reduce its angular velocity relative to the Sun and to exit the plane of the Earth's orbit.

Other satellites are constantly measuring the parameters of particles that the Sun emits outward in different directions, the so-called solar wind. This data is also used to predict space weather, which will be discussed later. They also measure the flow of energy from the Sun, which is characterized by the so-called solar constant, which is now equal to $1367\,\text{W}/\text{m}^2$.

What is it? This is the power falling upon the unit area of a surface perpendicular to the direction of the Sun and placed outside the Earth's atmosphere, but at the same distance of 1 AU from the Sun, as the Earth. No solar battery can provide more energy, even during the day and with the Sun at its zenith. Due to the absorption of light by the atmosphere and the far from 100% efficiency, it cannot give out even that much, but we are still talking not about green solar energy, but about its source.

The total radiation power of the Sun, $3.83 \cdot 10^{26}\,\text{W}$, can be obtained by multiplying the solar constant by the surface area of a sphere with a radius of 1 AU. Naturally, if we are sure that our star emits energy equally in all directions, both in the plane of the ecliptic and perpendicular to it. However, if there are differences, they are hardly significant.

In recent decades, one branch of solar physics called helioseismology has provided a wealth of data. As everyone who looks in its direction knows, the Sun is opaque. It consists of dense plasma that absorbs and re-emits light, just as the Universe behaved before recombination.

A photon born in the interior of the Sun can make its way to the surface for hundreds of thousands of years (and 10 million years from the core), being constantly absorbed and re-emitted, and then reach the Earth in just (150 million km)/(300 thousand km/s) = 500 seconds, that is, over 8 minutes. Therefore, we cannot look into the bowels of the Sun, but we can perfectly see its surface.

We also cannot look into the bowels of the Earth, but the science of seismology has provided us with information about what is inside, based on the characteristics and time of propagation of seismic waves (elastic waves, sound waves in the ground) in different directions and places. Similarly, helioseismology obtains information about the inner regions of the Sun from the oscillations visible on its surface. Since the Sun is not solid, in addition to acoustic waves, there are also surface gravitational waves,[g] similar to waves on the surface of terrestrial reservoirs.

2.7.2. *Internal structure of the Sun*

Let's see what we know about the Sun as a result of centuries of observations. The fact that the source of the Sun's energy is a fusion reaction became known less than a century ago. We will discuss it in Section 6.2. The time has passed, when in order to estimate the time of existence of our star, physicists figured out how many tons of anthracite[h] have to be burned per second to provide the observed

---

[g]This term is often misunderstood. Waves of this type are so named because the vibrations occur due to the action of gravity. They have nothing to do with the recently recorded gravitational waves described by Einstein's general theory of relativity.

[h]Burning the finest coal was chosen because it was the most efficient energy production process known at the time.

solar energy flux. Now we know the corresponding fusion reactions (there are more than one) in all details.

These reactions take place not within the entire volume of the Sun, but only in its core, the radius of which is 4–5 times less than the radius of our star. The core serves as a kind of furnace that heats the entire Sun. Its volume is not very large. A sphere with a radius 0.2 of the solar radius has a volume of $0.2^3 = 0.8\%$ of the solar one.

But it contains from a third to half the solar mass. This is not surprising, because the density of matter in the centre of the Sun reaches 150–$160\,\text{g/cm}^3$, which is not only 150 times higher than the density of water, but significantly exceeds the density of all heavy metals known to us.[i] There is also a very high pressure (about 25 petapascals) and a temperature of 15 million degrees Kelvin. At the edge of the core, at a distance of a quarter of the Sun's radius from the centre, the density drops to $2\,\text{g/cm}^3$.

As the hydrogen in the centre turns into helium, the equilibrium that maintains the size of the core shifts slightly, and every 100 million years the brightness of the Sun rises by about 1%. In just 4.5 billion years of its existence, it has added about 30% to the original luminosity.

The Sun is rotating. Its core is so dense that it spins like a top or a car wheel. All parts of the core have the same angular velocity. Such a rotation is called solid, because this is the only way that solid bodies can rotate. Helioseismology provided us with data on the core of the Sun and its rotation.

The stove, that is, the core, is surrounded by the so-called radiant transfer zone, which transfers energy from the core to the outer regions of the Sun. It extends to 0.7 solar radius, while the temperature drops to 2 million Kelvin, and the density at its outer border is $0.2\,\text{g/cm}^3$, five times less than that of water. The zone of radiant transfer also rotates rigidly with the same angular velocity as the core.

The temperature inside the zone changes with distance from the centre, but not fast enough for convection processes to be included

---

[i]But this is a fluff compared to the density of white dwarfs (about $10^7\,\text{g/cm}^3$) and neutron stars (about $5 \cdot 10^{14}\,\text{g/cm}^3$).

in the heat exchange. Energy is transmitted by emitting and absorbing photons. It takes a photon, on average, 170,000 years to travel across the radiant transfer zone. Then it enters the convective zone, where energy is transferred mainly due to convection, and this process is significantly accelerated.

The regions external to the radiative transfer zone do not rotate like a rigid body, but with different angular velocities. This rotation is called differential one. It is easy to observe it by stirring tea in a cup. Naturally, this is possible because the tea is liquid. More precisely, this is possible because it is not solid, because the gas can also rotate at different angular velocities. The outer parts of the Sun rotate differentially, too.

Areas near the solar equator complete a revolution around the Sun's axis in 25.38 days, and near the poles in 34.4 days. In addition, the angular velocity of rotation varies not only with latitude, but also with distance from the centre. It is clear that it is impossible to put together the parts with solid rotation and the parts with differential rotation without some layers sliding relative to others. The part of the Sun lying on their border with a width of about 4% of the radius is called a tachocline. It plays an important role in the generation of the solar magnetic field. It is inside the tachocline that the transition from solid to differential rotation occurs.

2.7.3. *The surface of the Sun and processes on it*

Outside the tachocline, there is a convective zone. It stretches almost to the solar surface. But what is this surface? Everyone knows that the Sun does not have a solid surface. And everyone who has seen the Sun knows that this surface is clearly visible.

Astronomers talk about the Sun's photosphere. This conditional border of the Sun is placed where an observer sees it. The Sun becomes almost transparent to visible light rays outside it. But the Sun can be considered opaque inside the photosphere. The photosphere temperature is about 5800 K, and the density is $0.2 \, g/m^3$, which is significantly less than that of the Earth's atmosphere. Only 3% of the hydrogen in the photosphere is ionized, and most of it is present in the form of atoms.

As the name suggests, there is intense convection in the convective zone, accompanied by the motion of the substance and its mixing. It occurs in the form of columns, forming a peculiar structure on the surface, called granulation. Hot plasma rises up the centre of the column, spreads, cools, and drops down along the edges of the granule — the visible tops of individual columns. The typical granula diameter is about 1000 km.

Granules are dynamic formations, constantly appearing, changing, and disappearing. Their lifetime is from 8 to 20 minutes. In addition to the characteristic cellular structure of the photosphere, granulation manifests itself in horizontal flows with velocities from 300 to 500 m/s for supergranules. Supersonic aircraft have such speeds on Earth. The speed of motion of a substance in granules is determined by the Doppler effect. It is maximal in the centre and falls towards the edges of the granule. The process is somewhat reminiscent of boiling soup over low heat. It's better to look at it not even in a photograph like Fig. 2.6, but in a video that is easily found on the Internet.

Flares and quieter processes, such as sunspots or ejections of prominences, regularly occur on the Sun's surface (see Fig. 2.7). All of them are studied with great interest by astronomers. The huge energy of flares has been stored before them as the magnetic field energy. It is released as a result of the restructuring of the system of currents and magnetic fields.

2.7.4. *Solar cycles and magnetic field*

In general, the magnetic field plays a crucial role in all processes in the outer regions of the Sun. The 11-year cycle of solar activity is well known, during which the number of sunspots changes. In fact, this is the period between two reversals of the solar magnetic field, half of the 22-year period of its direction change.

The source of this field is the electric currents in the plasma. They change over time and with them, the magnetic field changes. The magnetic poles on the Sun switch places approximately 11 years after the previous exchange. This period of time is considered to be the period of solar activity change. The characteristic

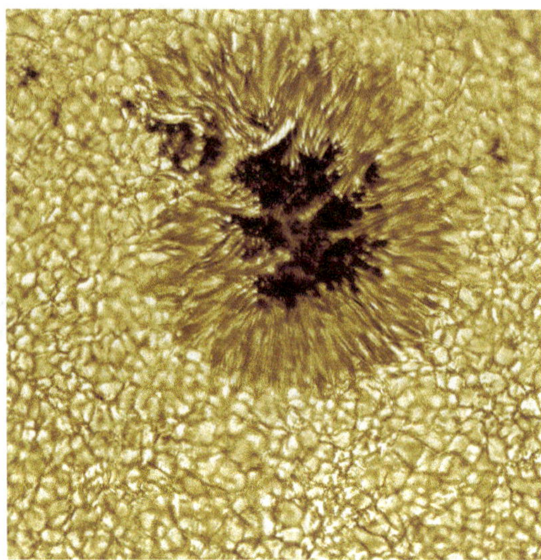

Figure 2.6. Sunspot surrounded by granulation. Photo from Vacuum Tower telescope (Wikipedia). This image is in the public domain. Source: nasaimages.org.

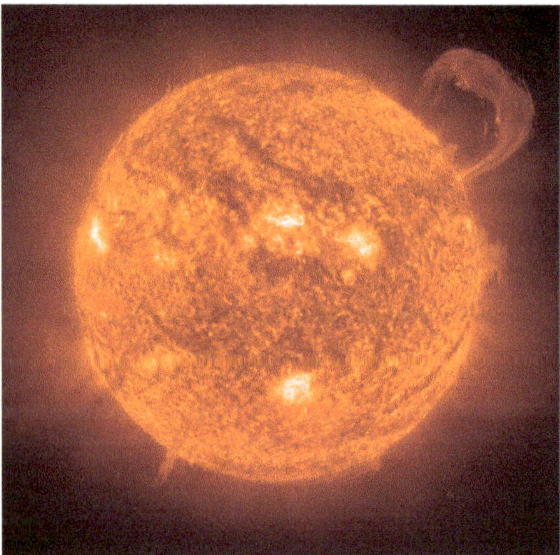

Figure 2.7. The Sun is photographed all the time in different wavelengths. The photo in the far-ultraviolet range in the helium line with a wavelength of 304 Å was taken from the *SoHO* (Solar and Heliospheric Observatory) spacecraft, launched in 1995 by NASA and ESA. Prominences or plasma ejections and the structure of the photosphere are visible. White spots correspond to areas with a particularly strong magnetic field.

period of the solar magnetic field oscillations is about 22 years or two solar activity cycles.

There are theories describing the cycles of solar activity and the change of the solar magnetic field, but they are still insufficiently developed and do not provide quantitative estimates and predictions. Nevertheless, I will tell you the basic ideas of the so-called solar dynamo theory. It explains the generation of the magnetic field in cosmic bodies by the currents in their core caused by their rotation. Dynamo processes take place inside the Earth, giant planets, stars, etc. There is no dynamo on the Moon, Venus, and Mars, so their magnetic fields are very weak.

What about the Sun's magnetic field strength? At the level of the photosphere, it is not very strong; it is, on average, of the order of several gauss, i.e. several times more than on the surface of the Earth. But in some areas, it is much higher. As a result, the energy stored by this field is enormous and can be released during flares and other processes, accompanied by a restructuring of the system of currents.

Strictly speaking, the solar magnetic field is not so easy to describe. So, it makes sense to consider it as a combination of two fields: toroidal and poloidal ones. Behind these tricky terms, there are fields similar to the field of a bar-shaped permanent magnet or a coil with current and a straight wire with current. Due to processes in the depths of the Sun, one field generates another and vice versa, so they periodically change their magnitude.

This is possibly due to the differential rotation of the Sun's surface. At the moment of activity minimum, the magnetic field of the Sun is similar to the field of a straight bar magnet or to the magnetic field of the Earth. The magnetic lines of force are extended along the meridians. Physicists call it a dipole field, and heliophysicists call it a poloidal one. The lines of force rotate with the surface of the Sun. But due to the fact that the rotation is faster at the equator, they gradually bend, winding around the Sun with each revolution. And there are a lot of turns. In 11 years, the number of revolutions with a monthly period exceeds a hundred.

Consider the lines of force of the magnetic field under the surface of the Sun in the period of maximum solar activity. Their segments can go out to the surface, forming arches above it. They are connecting sunspots — darker areas on the Sun's surface. The magnetic field in the spots is much stronger than the average one, and its strength can reach several thousand gauss.

The activity of the Sun is traditionally characterized by the Wolf numbers introduced by the Swiss astronomer Johann Rudolf Wolf. This is the sum of the number of sunspots visible on the Sun and tenfold the number of sunspot groups. If there are no spots on the Sun, then the Wolf number is zero; if there is one spot, then 11 (one spot and one more group consisting of this spot are taken into account). The Wolf number rarely exceeds 250; it is small quite often. It is clear that this is a rather subjective assessment and lately some astronomers prefer to use more objective indicators.

Each solar cycle is unique. There were periods when activity almost stopped. During the so-called Maunder minimum (from 1645 to 1715), sunspots almost disappeared on the Sun. This was accompanied by abnormal cold weather in Europe and North America; therefore, this time interval is sometimes called the Little Ice Age. During the Dalton minimum (1790–1830), there were significantly fewer sunspots than in the middle of the 20th century.

At the very beginning of the 2020s, solar activity was also at a minimum. The reason for these variations is still unknown. Cycle durations also vary, so predicting the properties of a new cycle is a favourite pastime for heliophysicists.

Note that they do this not only out of curiosity. Solar activity affects economic indicators such as yields and prices. A possible reason for this is a change in the level of solar cosmic rays, affecting the state of the Earth's atmosphere at high altitudes, its circulation, climate, and further along the chain.

Some enthusiasts are sure that solar activity determines the entire life of mankind, from periods of war and peace to small everyday details, such as the level of consumption of mustard. However, an elementary check without data manipulation smashes this reasoning to smithereens.

But the Sun does not end with the photosphere. Around it stretches a layer 2000 km thick, called the chromosphere, and then begins the corona or the outer solar atmosphere. The more distant regions are called the heliosphere, and it extends beyond Pluto's orbit.

2.8. Beyond the Sun

2.8.1. *Solar wind and corona*

The uppermost, rarefied, and hottest layer of the Sun's atmosphere, consisting of plasma, is called the *corona* (Latin for "crown", in turn derived from Ancient Greek $\kappa o \rho \acute{\omega} \nu \eta$, which means garland or wreath). A stream of ionized particles (mainly helium–hydrogen plasma) flowing from it into the surrounding space at a speed of 300–1200 km/s is called the solar wind.

In solar flares, streams of photons, ranging from gamma or X-ray ranges to radio waves, as well as streams of charged particles, are thrown in different directions. In addition, giant clumps of plasma, called coronal mass ejections (CMEs) and shown in Fig. 2.8, can fly out of the solar *corona*.

The word *giant* is used for a reason; CMEs are much larger than the Earth, although smaller than the radius of its orbit. However, they are not much smaller, but only approximately three times less. They are similar in shape to a croissant; as they move away from the Sun, the curvature decreases and the shape approaches that of a sausage. In the period of maximum solar activity, up to three or four CMEs are emitted per day, and sometimes up to seven, in the period of the minimum — one in five days or even less often.

The lines of force of the CME magnetic field are directed along its axis; electrons and ions rotate around them along a helical line. I talked about this in Section 2.4. The field prevents them from leaving the CME, which in this case is similar to magnetic traps for confining plasma during thermonuclear fusion on the Earth, which I will talk about later in Section 6.2.

The time it takes for the CME to reach the Earth depends on its speed and ranges from 13 to 86 hours. These are the boundaries of

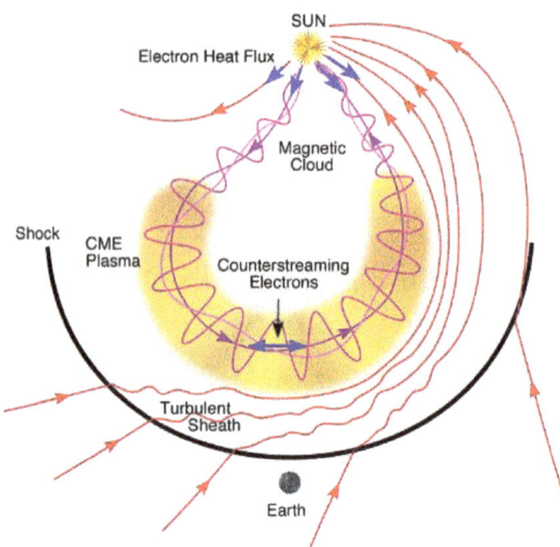

Figure 2.8. Coronal mass ejection (CME) has the shape of a croissant. In the figure, the CME is moving towards the Earth. The lines of force of the magnetic field begin or end at the Sun, which, like everything else in nature, has no magnetic charge. They twist due to the rotation of the Sun. When a magnetic cloud is ejected, it forms plasma CME escaping from the star. In front of the Sun is a spherical shock wave. If the CME hits the Earth, then a magnetic storm occurs on it, the strength of which depends on many factors, including the direction of the CME magnetic field. Extracted from: https://eclipse2017.nasa.gov/coronal-mass-ejections.

the interval, that is, the minimum and maximum recorded values of the CMEs' flight time to the Earth. Usually this time is 2–3 days.

But not all CMEs threaten the Earth equally. The most dangerous for the inhabitants of our planet are those that occur near the centre of the solar disc; then the emission is directed mainly at us. You can clarify this statement by taking into account two factors. To begin with, during a flight of 3 days, that is, $1/120$ of a year, the Earth will shift in orbit by an angle of $360°/120 = 3°$, so the CME must shoot ahead of time to "hit" it. In addition, the CME does not fly quite in a straight line. It is influenced by the interplanetary magnetic field, and the lines of force of this field twist like a spiral, forming an angle to the direction to the Sun.

So, for the Earth, the CME is most dangerous if formed during a flare in the equatorial region of the Sun, shifted to the right from the centre by about 1/5 of the visible disc. Flares at the edge of the disc (limb) are dangerous due to their electromagnetic radiation, more specifically X-ray and far-UV radiation. However, any flare that we see on the Sun provides a flux of such radiation. How dangerous are arriving CMEs? This is highly dependent on the orientation of their magnetic field, which interacts with the Earth's magnetic field, and this orientation is difficult to determine in advance.

But the Sun threatens us with not only CMEs, but also other types of emissions or transients, as astronomers call different streams of particles from the Sun. Among them, there are fast currents, magnetic clouds, and other dirty tricks, ready to arrange a magnetic storm on the Earth. We will talk about what all these streams and CMEs can do on the Earth. But a stream of particles constantly flies from the Sun, forming the solar wind, which I have already mentioned. The solar wind plasma moves away from the Sun at supersonic speed.

2.8.2. *New space missions to explore the Sun*

At the end of 2019, NASA released the first data from the *Parker Solar Probe* space mission, launched in 2018 to explore the Sun at close range, up to ten solar radii, more precisely 6.9 million kilometres. It was especially difficult to protect scientific equipment from extreme temperatures and strong radiation. The project cost NASA $1.5 billion and was named after the renowned astrophysicist and solar explorer Eugene Newman Parker.

What did the *Solar Probe* find when it approached our star a little closer than the orbit of Mercury? Curious behaviour of the magnetic field, associated with the fact that so-called reconnections of the field regularly occur in it, and the rearrangement of its structure, in which a lot of energy stored in the magnetic field is released. Reconnecting can be compared to short-circuiting an electrical circuit, but only very roughly.

It found that the solar wind near the Sun "blows in gusts", which have a complex structure. But on the way to the Earth, traces of the structure disappear due to the fact that waves of different lengths propagate in interplanetary space at slightly different speeds. Flares are detected in the *corona*, which are not visible on the Earth. Some more details are important for the physics of the Sun, where in the near future we should expect a flurry of new works devoted to the analysis of the data from the *Parker Solar Probe*. The *Parker Solar Probe* approached the Sun at the distance 10.5 million kilometres (6.5 million miles) at perihelion on 29 April 2021, improving more than fourfold the previous record of approaching our star.

A major milestone and new results from NASA's *Solar Probe* were announced on December 14 in a press conference at the 2021 American Geophysical Union Fall Meeting in New Orleans. For the first time in history, a spacecraft "has touched the Sun". The *Parker Solar Probe* has flown through the Sun's *corona*.

In early 2020, the *Solar Orbiter* from the European Space Agency (ESA) was launched in addition to the *Parker Solar Probe*. It should get closer to the Sun than Mercury, but further than the *Parker Solar Probe*, only to 60 solar radii. Its highlights are different. The spacecraft was launched into an orbit strongly inclined to the plane of the ecliptic. How much? At 24°, but an extended mission is planned, in which the inclination will be increased to 33°. This will allow us to observe areas near the pole of our star.

Joint observations with the *Parker Solar Probe* are also planned. However, the first results have already been published. And they relate to the heating mechanism of the solar *corona*. Its temperature reaches about a million degrees, while the temperature of the photosphere is about 6000 K.

The *Solar Orbiter* approached the Sun at a distance of 77 million kilometres, twice as close as the Earth. The very first pictures taken by a telescope operating in the far-ultraviolet range showed an image of our star at a record magnification. It turned out that its entire surface is covered with small active regions, which seem to provide heating to the corona. They are too small to be observed from a satellite orbiting the Earth.

The Earth and other planets are orbiting the Sun inside its *corona*'s extension. How can they not slow down and fall on our star? And why haven't we boiled yet? With distance from the Sun, the temperature of the interplanetary environment first rises, reaching 2 million degrees, and then falls. It reaches only 100,000 degrees near the Earth's orbit. But even this, it would seem, is clearly enough for cooking barbecue from the Earth and its inhabitants. Not really, the temperature of the environment is high, but the density of particles is extremely low, so the planets heat up weakly, and the resistance force is negligible.

2.8.3. *The Solar System boundary*

The heliosphere is the outermost layer of the Sun. It is the cavity formed by the Sun in the surrounding interstellar medium filled with fast-moving solar wind plasma. Can it be considered part of the Sun? On the one hand, the Earth and other planets are orbiting inside it and the heliosphere does not greatly affect their movement; on the other, the solar wind and interplanetary magnetic field (IMF) are generated by the Sun. IMF is a part of the Sun's magnetic field that is carried into interplanetary space by the solar wind. So choose the answer to your own taste.

The outer edge of the heliosphere is called the heliopause. Some believe that it should be considered the edge of the Solar System. In 2012–2013, the *Voyager 1* spacecraft crossed the heliopause at a distance of 121 AU from the Sun and became the first spacecraft to go beyond the limits of our planetary system. Then *Voyager 2*, launched back in 1977, did the same. Both spacecraft continue their missions. Initially calculated for 5 years, they dragged on for 42 years.

Due to the fact that the reliability of these spacecraft turned out to be higher than the calculated one, NASA faced an unplanned problem, although not the only one. The devices cannot receive energy from solar panels because they are too far from the Sun.

The *Voyager* spacecraft are equipped with generators using the plutonium-238 radioisotope, but due to its relatively short half-life (87 years), the generators now generate 40% less heat than immediately after launch. This is no longer enough for heating and powering all scientific equipment.

Four instruments operated on board *Voyager 2* in 2020 — two plasma sensors, a magnetometer, and a low-energy particle detector. The heating system of the cosmic ray sensor was turned off at the beginning of summer 2019. By the way, the readings of this particular sensor testified to the crossing of the heliopause. By midsummer, its temperature dropped to $-60°$C, but it still worked despite the fact that it was not originally intended for use at such low temperatures.

On January 25, 2020, due to a lack of energy, the automation turned off all scientific instruments, but three days later, NASA engineers remotely, from a distance of 18.5 billion kilometres, re-launched the scientific equipment and one of the two systems that consume a lot of energy.

But it is possible that the first man-made object that crossed the heliopause and went beyond the Solar System was not the *Voyager 1* spacecraft at all. It could be the usual steel plate cap which was welded over the borehole. In 1957, the United States conducted a series of nuclear weapon tests in Nevada called Operation Plumbbob.

On August 27, an explosion was made in the borehole. Its 900-kilogram steel plate cap was thrown at a speed of at least 66 km/s or 240,000 km/h, which is six times the escape speed on the Earth. Since preliminary calculations assumed a tremendous speed, the cap was filmed with an experimental high-speed camera to measure this velocity.

What is the further fate of the steel plate? No one knows. It and its parts were not found, and no one could study experimentally the behaviour of bodies flying in the atmosphere with such a tremendous speed, either then or now. Some believe that the lid just evaporated, and others believe that it flew out in some form from the Earth's atmosphere, and then from our planetary system.

One thing is clear: it could not reach the stars. Its speed is about 4500 times less than the speed of light, and the distance to the nearest star is about 4.3 light-years. So the flight time can be estimated at $4.3 \cdot 4500 = 19350$ years.

2.8.4. *Guests from other worlds*

What about objects flying into our system from interstellar space? In recent years, astronomers have discovered the first two of such bodies. This is the *Oumuamua* object (the Hawaiian word meaning a messenger from afar) discovered in 2017 and a comet discovered in 2019 and named after its discoverer Gennady Borisov. Their high speeds and hyperbolic trajectories indicate that they are guests from other stellar systems.

Natural curiosity immediately compels us to ask where they come from exactly. This can be assumed by calculating the trajectory of a comet or asteroid within the Solar System and tracing the movement of the object and nearby stars after reversing the direction of time.

Most likely, Borisov's comet flew from the star Ross 573 and reached us after a journey of 910,000 years. The next option is the star GJ 4384. The calculations are complicated by the fact that comets and other small celestial bodies do not move quite along the trajectories predicted by mechanics for the movement of a material point. They are affected by reactive forces caused by the outflow of gases and other matter from their surface. These are very weak forces, but with a long action they can affect the trajectory; therefore, without taking them into account, it is impossible to reliably calculate where a guest came from.

2.9. Cosmic Rays

In addition to particle fluxes and radiation from the Sun, the Earth is also irradiated by galactic cosmic rays emitted by other objects in our Galaxy and even other galaxies. Penetrating into the heliosphere, they reach the Earth and are partially absorbed by the atmosphere. Therefore, the intensity of cosmic rays increases with height.

As a matter of fact, they were discovered in 1912 with the help of equipment that was raised by balloons to a great height. The Austrian (later American) physicist Victor Francis Hess was awarded the 1936 Nobel Prize in Physics for the discovery of cosmic rays. Together with others, he proved that the radiation that ionizes the atmosphere is of cosmic origin.

Here is an interesting detail. Hess shared this prize with the American Carl David Anderson, who was awarded it for the discovery of the positron in 1932. It was the first detected antiparticle and the first elementary particle discovered by Anderson in cosmic rays.

Later, he and Seth Henry Neddermeyer discovered another type of elementary particle called muons, also in the cosmic rays.

2.9.1. *Antiparticles*

Let's digress a little and talk about antiparticles and annihilation. Almost every elementary particle in existence has an antiparticle. This is its double with the same rest mass, but opposite signs of all charges, including electric.

An antiphoton completely coincides with a photon that does not have any charges; it is the same photon. Therefore, the photon is the so-called truly neutral particle. There are a couple of similar particles, but I'll talk about them later.

All other particles have corresponding antiparticles. When a pair of particle and antiparticle meet, they can annihilate, i.e. disappear, leaving behind photons or other particles. Before annihilation, they approach each other. Their velocities are equal and opposite to each other in the frame of reference of their centre of mass, but change in time due to the electric attraction of particles with a nonzero charge. If two photons are formed after annihilation, their frequencies are equal and they are emitted in opposite directions in the above-mentioned reference frame.

At the energies of particles are large, tens and hundreds of different particles can be formed. If the relative speed of approach of a particle and an antiparticle is very close to the speed of light, then some among newborn particles may have a greater rest mass than initial particles. Similar reactions occur in some accelerators. There is also a reverse process: a particle–antiparticle pair can be formed by the interaction of two gamma quanta with high energies.

The particle is called a component of the pair which is more common in the world around us. The antiparticle of an electron is called a positron; many antiparticles are named simply by adding the prefix "anti" to the name of the particle. There are antiprotons, antineutrons, etc. Antimatter is formed from antiparticles only. It is described in detail in Section 6.1.

Naturally, the lifetime of a free antiparticle is the same as that of a particle. Therefore, the positron does not decay by itself. However, there is usually some kind of electron next to it and they annihilate, while positrons are not so often found near electrons. As a result, we live in a world where there are many electrons and protons but very few positrons and antiprotons.

So, positrons and muons were discovered in cosmic rays. It was a good source of high-energy particles before the development of accelerators. As a matter of fact, it hasn't gone anywhere. In cosmic rays, there are particles with ultra-high energies that cannot be achieved at any of the existing or planned accelerators. Most likely, accelerators will not reach this energy level in the foreseeable future. The Universe is the poor man's particle accelerator, as the saying goes.

But these particles from outer space themselves usually do not reach the Earth's surface. The cosmic ray flux at sea level is about 100 times less than the primary cosmic ray flux. On the other hand, a particle with very high energy, hitting an atom of one of the atmospheric gases, causes a whole cascade of secondary particles, the number of which increases during subsequent collisions with the atmosphere. This cascade reaches the ground. By the distribution of secondary particles, specialists can determine the energy and directions of arrival of the very primary high-energy particle, from which the cascade began.

The lion's share of cosmic rays are protons, a couple of percent are alpha particles (nuclei of helium atoms), and less than 1 percent accounted for heavy nuclei, electrons, positrons, and all sorts of exotics. In galactic cosmic rays, there are also nuclei of various chemical elements, heavier than helium, such as carbon, oxygen, magnesium, silicon, and iron.

Among the secondary particles, there are nuclei of radioactive isotopes with a short half-life. These include ^{7}Be, ^{32}P, ^{38}S, and a number of others. Some of them exist in the Earth's atmosphere thanks to cosmic rays only. The concentration of the ^{7}Be isotope with a half-life of 53.3 days is an excellent indicator of the intensity level of these rays.

2.9.2. *Radiation background*

Cosmic rays, both solar and galactic, as well as the decay of radioactive elements create a natural background radiation on the Earth.

Cosmic rays provide about a sixth of the total radiation dose. Much greater is the contribution of the radioactive noble gas radon, emitted from the ground and some building materials.

The flux of galactic cosmic rays that bombard the Earth is approximately isotropic and constant in time and before entering the Earth's atmosphere amounts to about one particle per square centimetre per second. It interferes with experiments to search for many exotic nuclear reactions, especially those caused by weakly interacting particles, so such experiments are usually carried out in underground laboratories located in abandoned mines or tunnels under mountain ranges. Underwater detectors are also used, particularly in experiments in which water reacts with particles trying to detect.

Galactic or intergalactic cosmic rays accelerate due to supernova explosions and when interacting with their remnants and with shock waves. It is curious that high-energy particles of cosmic rays cannot propagate over extremely large distances because of the interaction with the relic radiation photons. The source of ultrahigh-energy cosmic rays cannot be located more than 150 million light-years away.

2.10. Solar System on the Scale of the Earth and the Universe

2.10.1. *Scales of near-space and the Solar System*

Since we live in the Solar System, it is worth considering how it looks on our scale ladder. It is clear that it is much larger than the Earth and much smaller than the Universe, but how much? Surprisingly, the answer to this simple question leads to some interesting conclusions.

Let's look at Fig. 2.9 in which some characteristic scales associated with the Earth, the Solar System, the Galaxy and the Universe are plotted. Immediately I draw your attention to the fact that the distances and sizes are given on a logarithmic scale. They are taken in light-years and those that are smaller are taken in light-hours (lh), minutes (lm), seconds (ls), and even milliseconds (lms). So the

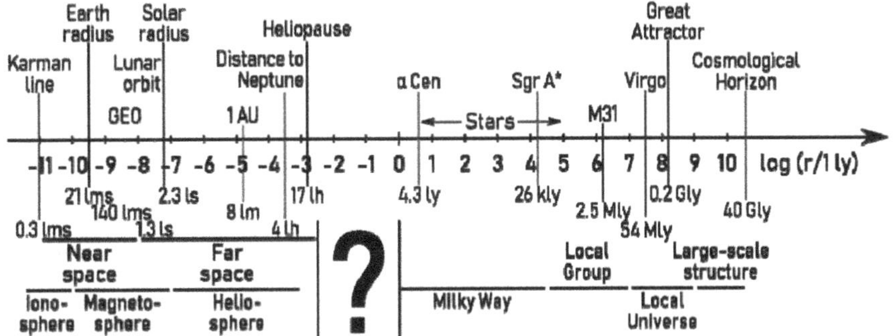

Figure 2.9.   The cosmic scales in one diagram.

number "0" on the scale corresponds to one light-year (1 ly), which is approximately equal to 9.5×10^{12} km $\approx 6 \times 10^{12}$ mi, or 9.5 trillion (I mean a million or 1000 billion, and not a British million million) kilometres in our usual units of length. Standard prefixes are used for large scales. 1 kly = 1000 ly, 1 Mly = 10^6 ly, 1 Gly = 10^9 ly.

Each shift to the right by one division increases the size 10 times, so that the numbers "1" and "2" on the scale correspond to 10 light-years and 100 light-years. A shift to the left by one division decreases the size by a factor of 10, so that the numbers "−1" and "−2" on the scale correspond to 0.1 ly and 0.01 ly.

Having dealt with the distance scale in Fig. 2.9, you can take a closer look at the scales highlighted on it. On the left, there are the relatively small scales associated with the Earth. They are opened by the Karman line — the conditional border of the cosmos. This is the height above the surface of the Earth, above which space begins. There is some debate as to where it is best to do it, but for the drawing, we used the most common height of 100 km. Then we see the radius of the Earth, i.e. the characteristic scale of the Earth as a planet. We will talk about it in Chapter 3.

The abbreviation GEO denotes an altitude of 35,786 km above sea level, at which satellites fly in geostationary orbits. Satellites make a full revolution around the Earth in exactly one day, like the Earth itself below them. And since they fly in the plane of the equator, from the surface of the Earth they seem to be motionless at one point in the sky. These satellites are important for communications and signal transmission, particularly satellite television, as well as in monitoring the planet's surface, atmosphere, and oceans, missile warning systems, and other purposes.

Due to the properties of such an orbit, a satellite dish aimed at a geostationary satellite constantly looks in its direction, which makes it possible to relay television and radio programmes throughout the Earth. The signal delay time associated with the travel of radio waves to the satellite and back is twice as long as 140 ms, which is a little bit.

Therefore, the space for a satellite in a geostationary orbit is very valuable, and sites in GEO are distributed between countries on request, and for specific satellites, and not by the "could be used" principle. When a satellite comes to the end of its life, then with the help of the remaining fuel it is taken from the geostationary orbit to a close orbit for spent satellites, officially called "cemetery", more precisely, "burial orbit".

Then we see the radius of the Moon's orbit. Let me remind you that this is the farthest distance a human being has moved away from their home planet. But even this, humanity cannot afford too often. After the lunar expeditions of the Americans under the Apollo programme in 1969–1972, no human has set foot on the surface of our natural satellite during half of a century. The distance to the Moon is 1.3 light-seconds. This leads to a 2.6 s delay of the signal from the Earth to the Moon and back. This is already a significant value, which makes the remote control of the lunar rovers quite complicated. Figure 2.9 also shows two characteristic dimensions related to the Earth, namely the radii of the ionosphere and magnetosphere, which will be discussed in Sections 3.3 and 3.4.

Let's move on to the scale of the Solar System. The radius of the Sun is about twice the radius of the Moon's orbit. Nothing strange, but not everyone knows about this detail. The distance to the Sun is 1 AU or 8 light-minutes. If the Sun suddenly explodes, people on the Earth will see the flash only after 8 minutes. Other planets of the Solar System are located at distances comparable to an astronomical unit. You were told about the Titius–Bode rule in Section 2.1.

The most distant of the planets, Neptune, is 4 light-hours away from us. Therefore, the signal delay becomes so large that it excludes the possibility of remote control of any vehicle moving at a significant speed along a route not laid out in advance. Spaceships are moving fast, but along calculated trajectories. The Martian rovers are moving as NASA wishes, but very slowly. Flight details for the Martian helicopter were planned and communicated in advance.

Our Solar System extends to the heliopause, the outer border of the heliosphere. The dimensions of the heliosphere are approximately 17 light-hours. Let me remind you that the spacecraft launched by mankind have not only visited all the planets, but also gone beyond the heliosphere.

2.10.2. *Objects within the Galaxy*

Let's go beyond the Solar System. The star closest to us is located in the constellation *Centaurus*. *Alpha Centauri* is a triple star system composed of a binary system orbited by *Proxima Centauri*, currently the nearest star to the Sun. In the picture, the *Alpha Centauri* star is designated as α Cen. It is 4.3 light-years distant.

Think how much larger this distance is than the size of the Solar System! Compare the 4.3 years that light flies towards us from the *Alpha Centauri* with the 4 hours or even 17 hours that it travels from Neptune or inside the heliosphere. The corresponding distances in Fig. 2.9 are marked with a question mark because we don't know much about interstellar space. And there is not much from the known that can really interest readers.

The scale of our Galaxy begins. There is the supermassive black hole *Sagittarius A** near the Galaxy centre, which I discuss in Section 7.3. This is why the distance from us to the centre of the Galaxy is designated as Sgr A* in Fig. 2.9.

And since the stars in the Galaxy are located both near the Sun and beyond the centre, the distances to the stars of the Milky Way galaxy are very different. In the figure, the inscription "stars" refers to the distance to stars only in our Galaxy. The more distant ones enter into some other galaxies. For example, they could belong to the M31 galaxy, also known as the *Andromeda Nebula*. The distance to it is also indicated in the figure. This is one of the large galaxies closest to us.

2.10.3. *Jet propulsion*

Whether humanity can fulfil the old dream of science fiction writers and first fly to other worlds and then conquer them? This question naturally arose after I started talking about other stars and planets. I will soon outline my view on this problem, but for now I want to explain how exactly you can travel to other planets and stars. So far, all successes in this direction have been associated with the missiles that use the principle of jet propulsion.

Many people have heard about it, but not everyone knows what it is. I will try to cover the basics of jet propulsion without going into the technical details.

Let's assume that the spacecraft is frozen motionless somewhere in the depths of space. The team is in place, the fuel is in the tank, the cook is in the galley, the engines are checked, and you can start. What force allows the ship to gain speed in the desired direction?

Let me give you a simple, albeit somewhat unusual, analogy. A gang of vile scoundrels left you in the winter on the ice of a frozen pond, in the very middle of it. The ice is perfectly horizontal and perfectly slippery, so you won't be able to push off from it.

You lie motionless, but you need to somehow get to the saving shore. What to do if you cannot count on help? Besides, there is no wind that could blow you away; you have neither a rope with a hook at the end nor other equipment.

The solution is pretty straightforward. If you cannot push off from the ice, then you need to push off from something else. Let's say, from your own boot or felt boot. You need to remove it from your feet and push it with all your strength in the direction opposite to the place where you want to get.

Let's say you can throw it with a speed of $5\,\text{m/s} = 18\,\text{km/h}$. Not so much, but let's take into account the unusual conditions. A pair of felt boots weighs from $400\,\text{g}$ (for children) to $2\,\text{kg}$. Let one felt boot weigh $1\,\text{kg}$. You and the felt boot form a closed system in which the horizontal components of the total momentum cannot change. The position of the centre of mass of the system of you and the felt boot remains motionless in the middle of the lake. If your felt boot gets the momentum of $5\,\text{m/s} \times 1\,\text{kg} = 5\,\text{kg} \cdot \text{m/s}$, then in accordance with the law of conservation of momentum (and with Newton's third law), you get the same momentum, but directed in the opposite direction.

Let your weight, together with clothes and another felt boot, be equal to $100\,\text{kg}$. Then, you will begin to move at a speed of $5\,\text{kg} \cdot \text{m/s}$: $100\,\text{kg} = 0.05\,\text{m/s}$. If the coast is at a distance of $100\,\text{m}$ from you, then in $2000\,\text{s}$ you will reach it. Naturally, if you did not throw the felt boot in the most successful way, then in addition to the forward movement, it will provide you with some spin. This slows down your rescue slightly and makes the slide less comfortable.

This is an example of using the principle of jet propulsion. Part of the system is given a speed in the opposite direction, and the rest of the system gets the speed increment in the desired direction. And it doesn't matter what part it is.

Its mass and the speed of its movement relative to the rest are important. You could have thrown not a felt boot, but a mitten, a hat, or even a coin from your wallet, but the speed of your sliding would be different. You can increase the speed by throwing one more mass, for example, a second felt boot.

Likewise, a rocket can accelerate and eject molecules of hot gases (conventional chemical rockets in which fuel burns in an oxidizer), ions accelerated by an electric field (ion engine), or air or water accelerated and ejected from a nozzle (turbojet engine).

2.10.4. *A pessimistic view of interplanetary travel and space expansion*

Let's discuss the prospects for interstellar travel. They are more than bleak. I start with the journey to the nearest stars. If we could travel at the speed of light, we could fly to the star *Alpha Centauri* in 4.3 years, plant a flag there, take a selfie, explore everything that comes close to hand, go back during the same 4.3 years, and return to Earth 9 years after the start. But the speed of available rockets is catastrophically less than the speed of light, so the journey will take more than one millennium.

This is today's estimate for the case if the rocket is capable of making this path at all. Let me remind you that you must first gain speed, and then slow down in front of the *Alpha Centauri* star. And then accelerate again, but towards the Earth, and brake again. Let's take into account that the rocket engine is very inefficient.

In order to accelerate a rocket to a certain speed, an engine must not spend a certain amount of fuel, but a certain percentage of the rocket's total mass. Suppose that 9/10 of the total mass should be spent on acceleration to the star (a very optimistic estimate for today). This means that 1/10 will remain after the initial acceleration. After braking in front of the *Alpha Centauri*, it will remain only 1/100, after accelerating back to the Solar System it will be 1/1000, and after braking at the Earth, only 1/10000 of the original mass.

And besides the fuel, you must also have a crew with supplies for the millennium of its trip, an engine, and, in fact, the spacecraft itself. It is clear that it is much easier to send a robot, but even for this case, the calculation looks very pessimistic.

The reader may not believe, hearing from me, that the cost of accelerating 9/10 of the original mass of the ship is a very optimistic estimate for today. I confess that I lied a little. I even lied very much.

This is a wonderful, incredible, fantastically optimistic assessment for today, and anyone who is familiar with the so-called Tsiolkovsky equation can easily be convinced of this. Konstantin Tsiolkovsky published it in 1903, but in 1810 the British mathematician William Moore did the same, and I'm not sure who the first discoverer of it was.

So, let our rocket start not even from the Earth, but from somewhere from the border of the heliosphere and fly to *Alpha Centauri*. The point of such a launch is not to fly 4.3 light-years, but 17 light-hours less, to minimize the influence of the Sun's gravity, which reduces the rocket speed in comparison with the Tsiolkovsky formula.

Our rocket has a rocket engine that ejects exhaust at a speed that we denote by the letter u. This speed for chemical rocket engines is now significantly less than $5\,\text{km/s}$, but let's accept this clearly overestimated value.

If the rocket spends on acceleration so much of the working fluid that its mass decreases by N times, then the rocket will reach the speed $v = u \cdot \ln(N)$. Let me remind you that "ln" denotes the natural logarithm.

In my optimistic estimate, $N = 10$ and $v = 5\ln(10)\,\text{km/s} = 11.5\,\text{km/s}$. This is not only far from the speed of light, but it is also 26,057 times slower. So such a rocket will fly to the destination point for more than $26057 \cdot 4.3$ years ≈ 112000 years. Too long, isn't it? Even if scientists strained themselves and tenfold the exhaust speed, the estimate of time would decrease only tenfold.

But we have another reserve — the increase in N. Let's try to achieve a speed of at least $1/10$ of the speed of light due to it, which gives a one-way travel time for more than 43 years. A simple calculation shows that for $u = 5\,\text{km/s}$, the required value of N will be written as a number with 2605 digits before the decimal point. I think that there is no need to continue further.

But for those who are especially persistent, I repeat that the ship must accelerate twice and decelerate twice, which only requires a decrease in the initial mass by N^4 times. And this is already a number with 10423 decimal places. In other words, out of every million tons of starting mass, no more than 10^{-10411} g will return back to the Earth 90 years later.

If the mass of the returning ship is equal to a single electron mass of 9.1×10^{-31} kg, then the question can be formulated as follows: how many galaxies like ours with a mass of 9.6×10^{41} kg will it take to form its initial mass? Could there be enough material for this from all galaxies located in the observable part of the Universe? However, the number N^4 is so large that even without calculation it is clear that the correct answer is no.

At the same time, in reality, electrons as a component of cosmic rays are flying from *Alpha Centauri* to us and from us to it. It's just that the mechanism of their acceleration has nothing to do with jet engines; they do not slow down near the target and do not return back.

The systems of the Alpha star in the constellation *Centaur* have long been reached by electromagnetic waves generated by man, streams of antineutrinos from our atomic reactors, etc. But the term "space travel" means something completely different. As we can see, this is not all. Let me remind you that so far a person has only travelled to the Moon, i.e. to a distance of 1.3 light-seconds. At the same time, huge resources were invested in the preparation of the flight. Would anyone want to spend money on an expedition to *Alpha Centauri*, which may return in a millennium or two?

Science fiction writers describe humanity scattered across different stars, or even different galaxies. They trade, fight, and sometimes have a common government. You don't have to be a wise man to understand that this is all feverish delirium. What kind of union can there be if the answer to the request from the Earth to *Alpha Centauri* does not come earlier than in 8.6 years? Even asking for urgent help will become obsolete, let alone just chatting. What kind of war can there be if the cost of a flight is many times greater than all possible profits and trophies?

Well, technical progress does not stand still, and after some time, humanity may have spacecraft that are superior to modern ones, as much as an aircraft carrier is superior to the simplest boat. Maybe this will change everything? The answer is no.

In this case, we can still talk about an expedition to the nearest stars, but not about the exploration and conquering of the Galaxy. This has been known for a long time. The Polish science fiction writer and philosopher Stanisław Herman Lem wrote about this in detail in his wonderful book *Summa Technologiae* back in 1963, shortly after Gagarin's flight.

Physics claims that matter, energy, or information cannot travel faster than light in a vacuum. If this is the case, in order to get to the *Sagittarius A** black hole in the centre of the Galaxy, it is necessary to spend at least 26,000 years and then the same amount of time to return. This means that the flight will take more than 52,000 years. Because of the relativistic effects, it will be a shorter period of time for the crew, but it will take 26,000 years for humanity on the Earth!

Who will need to send an expedition which will certainly not return for the next 50,000 years? Let me remind you that a vehicle for it has yet to be invented in the distant future. And the expedition to the *Andromeda Nebula* will return no earlier than 5 million years later, even with a magical broom flying at the speed of light.

But let's say physics figures out how to move faster than the speed of light. Science fiction writers have a couple of favourite words for this: "subspace passage", "null transportation", and "flying through the wormhole". Their meaning is simple: to remove the restriction on the speed of movement.

Surprisingly, even then, humanity will not conquer the Galaxy, as the same Lem argued in his book. To put it all quite simply, there are too many stars to just visit even in our Galaxy, not to mention the Universe as a whole. Just to visit, not to fight, to trade, or to enter into love affairs, as science fiction writers tell us about.

I do not go on with all arguments. All the same, there are romantics who still dream of domination over the Universe. However, let's think about the other side of the problem. If humanity is unlikely to visit the nearest stars and certainly will not visit distant ones, then it is just as difficult for aliens to visit us. And why did the evil reptilians from the centre of the Galaxy suddenly decide to fly to the Earth on their flying saucers and steal a couple of civilians here? They are simply unaware of the existence of a human civilization.

At this moment, they can theoretically observe light and radio waves emitted from the Earth about 30,000 years ago, during the Upper Palaeolithic period on our planet. Neither radio nor television had yet been invented at that date, and not a single construction the size of an apartment building had been built. Why would reptilians rush, tails upward, to their flying saucers and further to the Earth?

2.10.5. *Distant galaxies and the cosmological horizon*

We have not yet reached the right edge of Fig. 2.9. We turn to the scale of the Local Group, which includes dozens of galaxies, mostly dwarf ones. It includes the *Andromeda* Galaxy and the Milky Way, which I have already mentioned. All these galaxies do not fly apart from each other due to mutual attraction.

We talk about the Local Supercluster of Galaxies or the *Virgo* Supercluster on a larger scale. The name is due to the fact that the centre of this supercluster is observed in the constellation with the Latin name *Virgo* used by astronomers. In Fig. 2.9 it is designated as *Virgo*. Our Galaxy, the Sun and the Earth are falling towards the centre of this cluster due to its attraction, against the background of the general expansion of the Universe.

However, this is not the strongest centre of attraction, which is usually called an attractor. About 200 million light-years away, there is an excess of matter called the Great Attractor. It is much more massive than the local supercluster.

We fall on it too. And practically in the opposite direction, there is another attractor located in the constellations named *Perseus* and *Pisces*. Further, you can find even larger superclusters, for example, the Shapley concentration. The set of attractors can be enumerated for a long time. This is the beginning of a scale on which the large-scale structure of the Universe is manifested.

This is the result of the development of those small deviations in the density and velocity of matter after the Big Bang which I described in Section 1.2. At http://www.ifa.hawaii.edu/info/press-releases/local_void/ you can find videos and other materials about the large-scale structure on a scale of 200 million light-years, prepared by R. B. Tully.

The rightmost mark in Fig. 2.9 is called the "cosmological horizon". This is the radius of the part of the Universe that we, in principle, are able to observe.

Its existence stems from two facts that are known separately: the finiteness of the age of the Universe and the finiteness of the speed of light. The combination of these factors leads to the fact that light from especially distant objects would have to travel to us for longer than the Universe has existed, which is, of course, impossible. Light from points more distant than the cosmological horizon could not reach us. We can observe objects at distances up to 40 billion light-years. We cannot see anything that is further away, and a hypothetical astronomer from those places cannot see us either.

3 GLOBAL PROBLEMS. SCALE: EARTH

3.1. Climate, Seasons, Day and Night

Let's move on to the scale of our planet. The Earth is in the habitable zone of the Solar System, so the climate here is suitable for the existence of life. Indeed, we are lucky: in many places, it is suitable for quite a comfortable life. I want to mention three important facts that will help us understand some of what we observe around us.

First: the globe is round. Second: it rotates around its axis, which is inclined at an angle of $23.5°$ to the direction perpendicular to the plane of its orbit, that is, the ecliptic. Third: the Earth is moving in an ellipse that is fairly close to a circle. The eccentricity of the Earth's orbit is very small, only 0.0167. We constantly observe the consequences of these trivial facts in nature.

3.1.1. *Climatic zones, change of day and night*

To begin with, the Earth is a ball or almost a ball. From whatever side the Sun shines, there always is a place on it where the Sun is at its zenith. The energy flux per second per unit surface area in this place is slightly less than the solar constant due to the loss of radiation in the atmosphere for absorption and scattering. If there are clouds in the sky, then the flux is lesser, but for now, let's consider the cloudless sky over the entire planet.

All other points of the Earth's surface are at this moment, almost at the same distance from the Sun, because the Earth's radius is significantly less than AU. But illuminance, that is, the total luminous fluxes incident on a surface per unit area, are significantly different in these points. There is a night in half of the planet. On the illuminated side, the lesser the flux, the farther the source of radiation is from the zenith. More precisely, the energy flux per unit area is proportional to the cosine of the angle between the zenith and the direction to the Sun.

Now let's take into account the rotation of the Earth. It is responsible for the change of day and night. And the tilt of the Earth's axis leads to the change of seasons.

Let this axis be perpendicular to the plane of its orbit. Then the lengths of the day and night would always be equal on the entire surface of the Earth. The height of the Sun at noon would be maximum possible and it would correspond only to the geographical latitude of the point of observation.

Let me specify that this would be absolutely true only if there were no refraction of light in the atmosphere, that is, the bending of light rays due to the decreasing of the atmosphere density with height. On a planet without an atmosphere, everything would be exactly as it is written above. On the real Earth with its atmosphere, the influence of refraction exists, but it is not very noticeable. Refraction causes the day to be slightly longer than the night, with this effect intensifying as you approach the poles. At the very pole of a hypothetical Earth with an atmosphere but without an axis tilt, the Sun would be visible all the time. It would be circling low above the horizon. Further in the text, I do not make a reservation about the absence of refraction, but this is always implied.

So, the angle between the zenith and the position of the Sun at noon would be equal to the latitude of the place of observation, both north and south. And the cosine of this angle, that is, the cosine of latitude, would determine the magnitude of the solar energy flux at noon. The average flux in one day would also be determined by the latitude and would decrease with distance from the equator.

We would get climatic zones: a warm tropical or equatorial zone, two cold polar or arctic zones, and temperate and subtropical zones in between each of the two hemispheres. And only at the equator could the Sun be at its zenith; this would occur every day at noon.

But the Earth's axis is not perpendicular to its orbit. Now it is oriented towards the North Star (*Kinosura* or alpha *Ursa Minor*). Due to the influence of other celestial bodies, this direction changes, but very slowly, at a speed of about 0.5 degrees in 100 years. Five thousand years ago, the "polar" star was *Tuban* (alpha *Dragon*), in 2000 years it will become *Alrai* (gamma of *Cepheus*), and in 12000 years — *Vega* (alpha *Lyra*). But on the scale of human life, we can assume that the direction of the Earth's axis is practically unchanged.

3.1.2. *Changing seasons, polar day, and polar night*

It is the tilt of the axis that causes the change of seasons on the Earth, as shown in Fig. 3.1. To be convinced of this, consider a situation opposite to that considered before. Suppose that the axis of the hypothetical planet would lie in the plane of its orbit, similar to the axis of Uranus, which is deviated from the plane of its orbit by

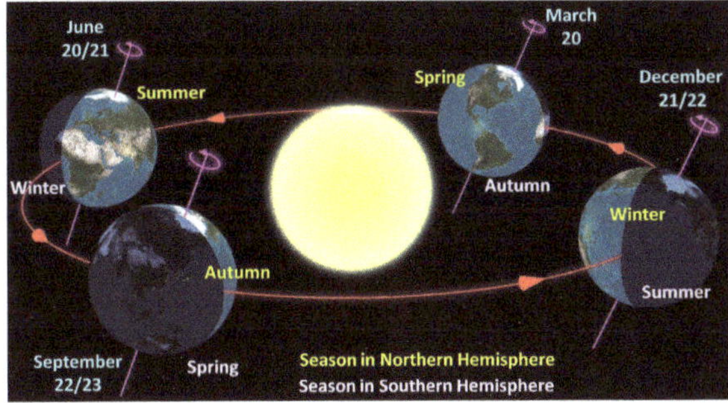

Figure 3.1. The tilt of the Earth's axis leads to changes in the seasons. At the time of the summer solstice in June there is a polar day in the Arctic, and a polar night in Antarctica, in December it is vice versa. The boundaries of the regions where the polar day and night can be observed are called the polar circles. This file is made available under the Creative Commons CC0 1.0 Universal Public Domain.

less than 8°. Such a planet is curious because at any point of it twice a year the Sun is at its zenith.

Why do the seasons follow each other? When the North Pole of this hypothetical planet with an axis in the plane of the ecliptic is pointing at the Sun, a polar day is observed throughout the Northern Hemisphere and a polar night in the entire Southern Hemisphere. It is the middle of summer in the Northern Hemisphere and the middle of winter in the Southern Hemisphere, respectively. We call this day the summer solstice.

Time passes, the planet turns in its orbit and the direction to the Sun rotates relative to the almost constant direction of the axis. At the North Pole and near it there is still a polar day, at the South Pole there is a polar night, but around the equator, day and night are already replacing each other.

When the angle between the axis and the direction to the Sun reaches 90°, the vernal equinox occurs. Day is equal to night on the entire planet. After this moment, a polar night begins at the North Pole, which lasts six months, and a polar day begins at the South Pole. When the Sun is over the South Pole, the Winter Solstice comes, and after another quarter of a year, comes the autumn equinox.

At every point on the planet's surface, two seasons are clearly distinguished: summer and winter. The inclination of the Earth's axis is not so great, the angle between it and the perpendicular to the plane of the ecliptic is 23.5°. So we have four seasons: summer, winter, and between them, spring and autumn. In summer, there are places where the day lasts a long time, but not the twenty four hours. The duration of a summer day is longer than the duration of the night, and in winter, days are short and nights are long.

If you look at the *globus*, you can see the parallels of 23.5° north and south, which are called the Northern and Southern Tropics. They are the boundaries of the tropical region in which the Sun can be observed at its zenith on a certain day. It is impossible in London or in any other place north of the Northern Tropic or south of the Southern Tropic.

The parallels $90° - 23.5° = 66.5°$ are called the Northern or Arctic and the Southern or Antarctic Polar Circles. They are tilted at an angle of $23.5°$ from the poles in the Northern and Southern Hemispheres and are the borders of the circumpolar zones, in which there could be polar days and polar nights. There is a temperate climate zone between the tropic and the Arctic Circle. If the tilt of the planet's axis exceeded $45°$, it would not exist.

The last factor — the elongation of the Earth's orbit — leads to the fact that the distance between the Earth and the Sun and, accordingly, solar radiation, change throughout the year. The Earth is located at a minimum distance of 147 million kilometres on January 2–5, and at a maximum of 152 million kilometres between July 3 and 7.

Therefore, the solar radiation flux is maximum in January and minimum in June, which softens the temperature contrast between summer and winter in the Northern Hemisphere (thus, summer is slightly cooler and winter is slightly warmer due to this factor) and enhances it in the Southern Hemisphere. In addition, because of a change in the angular velocity of the Earth's orbital motion, astronomical spring and summer (from spring to autumn equinox) are now 8 days longer than autumn and winter in the Northern Hemisphere, and vice versa in the Southern Hemisphere. However, due to the fact that the Earth's orbit slowly rotates with a period of about 21,000 years, this advantage of the Northern Hemisphere will not be eternal and will go to the Southern one in about 10,700 AD.

3.2. Phase States of the Substance

Before getting to know our home planet, Earth, I will tell you about the most common aggregate states of matter: solid, liquid, gaseous and plasma. All of them are present on the Earth, surrounding us from all sides. I have talked about the Sun with its plasma, and here, I am going to talk about the solid surface of the Earth's land, the liquid ocean, and the gaseous atmosphere above them.

In physics, for the sake of rigour, they prefer to talk about the phase state of matter; this is a more general concept, but we do not make a distinction between these two terms for now. I want to talk about phase transitions between them, metastable states, a critical point, and many other related things.

3.2.1. *Solid state, liquid state, and phase transitions between them*

The most important and most available liquid on the Earth — water — will help us learn about the state of aggregation. Its chemical formula is H_2O and, like all other chemicals, it can be in various states of aggregation. Let's start the study with two states that humanity has always known about, so they do not have a discoverer. They are the liquid and solid states.

Pour water into a glass, filling it halfway. You can immediately see several properties of the liquid. It takes the form of the vessel into which it is poured. The entire bottom half of the glass is filled with water. But there is no water in the upper half.

The liquid has a surface. If it is in a gravitational field and surface tension forces have little effect, then the surface is flat. More precisely, it is almost flat, but at the very walls, the water level rises slightly due to the fact that water wets the walls.

However, in some cases, the shape of the surface may be different. For example, in zero gravity, water droplets will take a spherical shape, and raindrops do not differ much from a ball. If a unit drop of water falls into a bathtub or basin, then its shape is not the shape of the vessel. The surface tension keeps it compact. It also significantly affects the behaviour of small water droplets. But temporarily forget about surface tension, focusing on volumes of water significantly larger than a drop.

A fluid retains its volume if its thermal expansion is neglected. When we tilt the glass a little, the water flows from one place to another, filling that part of the vessel in which there was air before.[a]

---

[a]The French scholar Marc-Antoine Fardin from the Université de Paris obtained in 2017 the Ig Nobel Prize in Physics and 10 trillion Zimbabwean dollars. His study "On the rheology of cats" states that cats can be considered both solid and liquid due to their ability to take the shape of the vessel in which they lie.

The volume of water remains the same, but its form changes. We can stick our finger in water without using too much force. And then pull it out and the surface of the water regains its former shape. For a more viscous liquid, it takes longer to recover the shape, and in some cases much longer.

Throw an ice cube into another glass. This is the same water, but in a solid state. The cube does not fill the bottom half of the glass. If the glass is tilted, the cube can move to its lowest point, but the shape of the cube doesn't change. So, a solid has a shape and retains its volume, if we neglect the thermal expansion.

It has a surface. The surface can be of any shape. When placed in a container or vessel, a solid retains its shape. Attempts to shove a finger into a piece of ice will clearly not be completed successfully; the state is called solid precisely for this reason. But ice can be broken into pieces that won't come together on their own.

We pour some water into the third glass and throw a piece of ice there. Let's wait long enough for thermal equilibrium to be established in the glass, but not long enough for all the ice to melt (or not all the water is frozen if the experiment is carried out in the winter in a forest in Lapland). The first thing we see is that some of the ice has melted (or some of the water has frozen). That is, part of the water has changed its state of aggregation.

If you dip the thermometer in a mixture of ice and water, it shows a temperature of $0°C$. Let's clarify: at standard atmospheric pressure. In this case, both water and ice have such a temperature. This is the temperature of the water–ice phase transition at a pressure of 1 atm.

So, ice heated above $0°C$ will melt, and water cooled below $0°C$ will freeze. At a temperature of $0°C$, a phase transition occurs with a change in the state of aggregation. Physicists call it a first-order phase transition. What does this mean?

3.2.2. *Long-range order in crystals and quasicrystals*

To answer this question, let us remember how a liquid is structured and how a crystalline body is structured. Ice is a crystal. Crystalline bodies (both forming one large single crystal and several polycrystals consisting of many small crystals) have a crystal lattice and the atoms forming it are arranged in a certain order (this is called long-range order).

The lattice is usually not quite perfect, but in general, the order of arrangement of individual atoms in a small part of a single crystal can be easily extended to the entire crystal. For ice, which is familiar to us, the molecules form a hexagonal lattice, shown in Fig. 3.2. Its unit cell has the shape of a prism with a base in the shape of a regular hexagon.

The time has come to clarify what is meant by long-range order. This is easiest to do by considering the analogy of a crystal on a plane, i.e. a two-dimensional crystal.

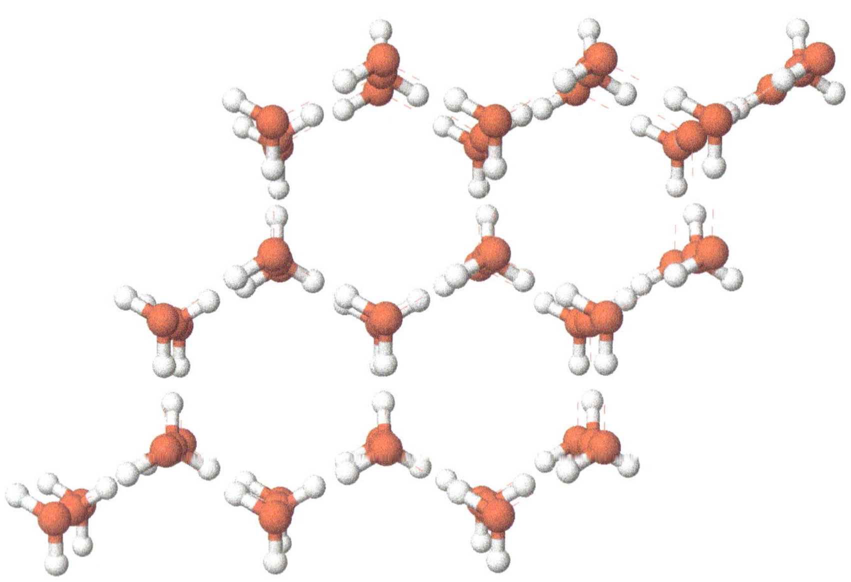

Figure 3.2. Ice molecules form a hexagonal lattice. Red balls — O atoms; white balls — H atoms. The crystal structure is similar to the structure of diamond: each H_2O molecule is surrounded by four molecules closest to it, located at equal distances of 2.76 Å from it at the vertices of a regular tetrahedron. This file is licensed under the Creative Commons Attribution-Share Alike 3.0 Unported license by Danski14.

If we want to lay out the floor with tiles of the same shape and size, then this is easy to achieve if the tile is in the form of a square, regular triangle, or hexagon. The tiles cover the entire plane without gaps or overlaps. Having chosen one tile, we can call it the unit cell of the crystal. By shifting this cell, we get the position of any other tile. But with a regular pentagon, this trick will not work.

Note that, in addition to shifts, the resulting floor pattern has some additional symmetry, in that it transforms into itself after a certain transformation. This symmetry corresponds to the symmetry of the unit cell. For a square unit cell, this is a rotation by $90°$, $180°$, and $270°$, and for a triangular tile, by $120°$ or $240°$, and we can rotate a hexagonal tile by $60°$, $120°$, $180°$, $240°$, and $300°$. Physicists in such cases speak of axial symmetry of the fourth, third, and sixth order. In addition, the resulting mosaic can be translated into itself by simply reflecting it in a correctly positioned mirror.

We can lay out the floor with rectangular tiles. It allows mirror symmetry and $180°$ rotations, i.e. it possesses axial symmetry of the second order. It also allows reflection from a mirror in the middle of the cell oriented parallel to any of the sides, i.e. it has two axes of reflective symmetry.

You can lay out the plane with parallelograms or rhombuses. In the first case, mirror symmetry disappeared; in the second, we can speak of symmetry with respect to reflections from a mirror located along any of the diagonals of the rhombus.

It is not difficult to lay out the plane with tiles of the shape of crosses. But the unit cell should be the smallest possible. In reality, the unit cell is not a cross, but a square. We could group 5 squares into a cross-like shape, but this does not make it the unit cell.

Let's return now to the regular pentagon, which is not suitable for our mosaic. I clarify that it is impossible to fill a plane with the same pentagons or heptagons. But if you combine them with other tiles, such as triangular tiles, you can get a beautifully paved mosaic floor.

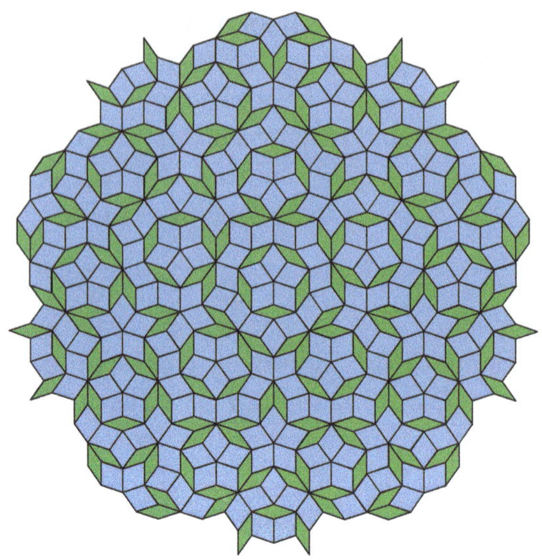

Figure 3.3. Quasicrystals have long-range order, but they have no unit cell. Extracted from Wikipedia. This work is in the public domain.

An example of such a floor is shown in Fig. 3.3. The resulting pattern does not form a real crystal, but its relative, called a quasicrystal. It does not have a unit cell, but it does have a long-range order. Such quasicrystals we have learned to obtain from alloys. In 2011, the Israeli Dan Shechtman received the Nobel Prize in Chemistry for the discovery of quasicrystals.

Many crystals are anisotropic, that is, their properties and the parameters characterizing them change depending on the direction. This applies to many different physical quantities. For those hungry for details, I'll mention the dielectric constant, electric conductivity, elasticity, and refractive index.

Let's take such characteristic of a crystal as hardness. Diamond scratches and cuts glass, but not vice versa. A less hard object can be scratched with a harder one, so it can be used for surface treatment. And how to process the diamond itself? To do this, you need a harder material than diamond. There are several suitable materials for this, but we learned about them not very long ago. How was it possible to cut the hardest of the known precious stones — a diamond?

The fact that came to the rescue is that the hardness of many crystals, including diamond, is anisotropic. Therefore, for centuries, the facets of a diamond have been processed with the help of another diamond, positioning them so that the faceting goes in a certain direction, in which the hardness of the diamond used for processing is more than average, and the hardness of the diamond under processing is less. That is, one diamond could cut another, but only in a certain position of each stone.

3.2.3. *Amorphous substances*

In addition to crystalline solids, there are also amorphous ones that do not have long-range order and a crystal lattice. They are isotropic and when heated, they either soften with increasing temperature, like glass or wax, or burn, decompose, or turn into a set of other substances, like plastic does.

There is no specific melting point for them. At high enough temperatures, they often behave like a liquid. Some of the amorphous substances can be made anisotropic if, during solidification, it was under the influence of anisotropic factors, for example, in a strong electric field.

Under external influences, amorphous substances simultaneously exhibit elastic properties, like crystalline solids, and fluidity, like a liquid; therefore they are sometimes called viscoelastic media. So, upon impact, they behave like solids and can split into pieces. But under very prolonged exposure (for example, stretching), amorphous substances flow.

Molten glass is a typical liquid, although quite viscous. As the temperature drops, the viscosity of glass increases dramatically. At room temperature, it increases so much that we can consider glass in a window or a glass vase to be a solid.

Glass can crystallize, which is extremely undesirable for window glass. This is a very rare event, but it does happen sometimes. Therefore, it is cooled quickly so that the crystallization process does not have time to occur before cooling. But, in principle, glass can also be considered a liquid with a huge, almost infinite viscosity. The glass panes, which have stood vertically for several centuries, thicken slightly at the bottom and are thinner at the top, showing that the glass is very slowly, but permanently changing its shape. Sometimes they talk about a special glassy state of matter.

If you have understood details about crystals and glasses, it's time to talk about liquids. Liquid water has no long-range order, and no elementary cell, and the position of its molecules is not fixed.

The main thing is that the structures of ice and water are completely different. Their density is also different. By a whim of nature, the density of ice is less than the density of water, so icebergs and pieces of ice float on the surface of water.

3.2.4. *Heat of fusion and freezing*

To melt ice, it is necessary to break the bonds in its crystal lattice. This requires additional energy. The greater the mass of melting ice, the more heat is required. To melt each kilogram of ice, like any other crystal, you need to spend a certain amount of heat, called the specific heat of fusion. For water, this is 330 kJ/kg. For another substance, the specific heat of fusion is different, but it is always positive.

When the temperature drops, the liquids freeze at the freezing point, which is equal to the melting point at the same pressure. In this, each kilogram of liquid when freezing releases heat equal to the specific heat of fusion.

What can be said about the specific heat of melting ice, which is equal to the specific heat of freezing water? To heat 1 kg of ice by 1 K, an energy of 2.1 kJ must be expended, which is approximately 150 times less than the specific heat of fusion. To heat 1 kg of water by 1 K, energy of 4.2 kJ must be expended. Therefore, melting ice requires approximately the same energy as heating water by 75 K or heating ice by 150 K. So, the transition of a solid phase into a liquid phase requires serious energy costs.

When we throw ice with a temperature below the phase transition into water with a temperature above 0°C, the ice heats up and the water cools. If the thermal energy released when the water is cooled to 0°C is not enough to heat the ice to the same temperature of 0°C, then the ice will continue to heat, and the energy will be taken up when part of the water freezes. As a result, both ice and water will acquire the same temperature, equal to the phase transition temperature, in our case 0°C, and part of the liquid will freeze.

If the heat energy released when the water is cooled to 0°C is enough with excess, then this excess will go to the melting of part of the ice. In any case, the mixture of water and ice will have a temperature of 0°C.

If it is in a warm room, then heat from the environment flows into it. This does not change its temperature, but causes some of the ice to melt over time and the mass of the remaining ice will be less and less. If the environment is colder than 0°C, then part of the water begins to freeze and the weight of ice will increase over time.

As long as there are both phases in water with ice, it serves as an excellent thermostat, that is, a device that maintains a constant temperature. If all the ice melts away or all the water has been frozen, then the temperature, of course, begins to change.

3.2.5. *Phase transitions of the first and second orders*

The turning of water into ice or vice versa is a first-order phase transition. This term was coined by the Austrian and Dutch theoretical physicist Paul Ehrenfest, who proposed his now generally accepted classification of phase transitions. A first-order transition is always accompanied by the release or absorption of heat, and the structure of the two phases of a substance, the initial and the resulting one, is very different. Other characteristics are also different, for example, heat capacity.

There are other types of phase transitions. In physics, they also talk about second-order phase transitions, in which the symmetry of the structure of a substance or some other characteristics change. The densities of the substance before and after the second-order phase transition coincide. There is no heat release, but the heat capacity of the body changes abruptly. A solid can change the symmetry of its crystal lattice or turn from a paramagnet to a ferro-magnetic, which I will discuss in Section 4.1.

Ordinary iron at room temperature is capable of strong magne-tization, but at a temperature above a certain one, called the Curie point (for iron, this is 770°C), this ability disappears. When the tem-perature passes through this point, a second-order phase transition occurs. More exotic examples are the transition of liquid helium to a superfluid state, as well as a metal or alloy into a superconducting state.

3.2.6. *Metastable states*

Everything seems to be simple, but a first-order phase transition can give an unexpected surprise. I'm talking about the possibility of the existence of so-called metastable states. These are unstable states, in which substances can nevertheless remain for quite a long time. Let a certain state of matter be stable under some external parameters, for example, temperature and pressure, but this substance keeps changing into another phase.

Yes, it is beneficial for a substance to make a phase transition and to reduce its energy. Nevertheless, it needs some kind of "initial impulse" or "magic kick" in the form of the appearance of nuclei of a more energetically favourable phase. However, if the substance is very pure and there are no impurities and other centres of formation of nuclei of a more stable phase in it, then this process of transition to a stable state can be greatly slowed down. When centres appear, the phase transition process starts to proceed very quickly.

Such metastable states also exist near the liquid–solid phase transition boundary. If the water is purified from impurities, then it can be cooled down to a temperature of −48.3°C with the help of some tricks; this is the current record for supercooling of water.

A supercooled liquid can freeze almost instantly after external exposure and even without it. A body thrown into a supercooled liquid will immediately find itself inside a layer of suddenly formed ice. By falling, it will violate the metastable state, which will be replaced by the state with the lowest energy, that is, ice.[b]

It is believed that ordinary glass is an example of a supercooled, metastable state of matter. Amorphous glass crystallizes or, as they say, devitrifies, and becomes opaque after prolonged and strong heating. A lollipop is supercooled sugar syrup. It can crystallize.

As a fun fact, let me mention a supercooled liquid on which you can write with a stick or stylus. It was created by an international group of chemical scientists led by Kyeongwoon Chung from the University of Michigan at Ann Arbor (USA).[c] The liquid is strongly supercooled: at a crystallization temperature of 134°C, it can be cooled to 5°C. Inscriptions or drawings on the surface are formed by crystals that appear on it after touching, they glow yellow in ultraviolet light.

They have been compared to the Midas touch that turns everything into gold. Connoisseurs of the Russian language can remember the idiom "it is written with a pitchfork on water" (it means "It's still all up in the air"), which becomes a reality for this liquid.

3.2.7. *Freezing point shifts with pressure change*

Now it's time to draw the so-called phase diagram. The temperature is plotted on the abscissa, and the pressure is plotted on the ordinate. Recall that pressure is the force acting on a unit of body surface. The situation with temperature is somewhat more complicated.

In physics, there are several ways to determine the temperature, but if a system is in a state of thermal equilibrium, they all give the same value for the temperature of that system. It is clear that when we look at the weather forecast for tomorrow, we care little about

---

[b]There are more than one video on YouTube with transparent plastic bottles with supercooled water. After shaking, it turns into ice in 10 seconds, right before our eyes. To repeat this experiment, it is good to use distilled water, which is more prone to hypothermia than ordinary tap water, because it does not contain impurities and air bubbles. However, you can get on the video with another substance, such as sodium acetate aka "hot ice", which is passed off as supercooled water.

[c]For details, see http://www.dailymail.co.uk/sciencetech/article-3081449.

how physicists determine the value called temperature, but more about whether to dress warmer or lighter. Nevertheless, here is the simplest definition.

The temperature of a substance is 1.5 times less than the average kinetic energy of the thermal motion of its molecules. For historical reasons, temperature is not measured in units of energy, such as joules, but in degrees. One Kelvin or 1 K equals $1.38 \cdot 10^{-23}$ J. One degree Celsius equals 1 K, but these scales have different zero temperatures.

The first thermoscope, a forerunner of the thermometer, was invented by Galileo Galilei in 1597. The photo of a device called Galileo's thermometer is shown in Fig. 8.2 at the end of the book. It has a completely different design and was invented at the *Accademia del Cimento* of Florence by a group of scientists including Galileo's pupil, Torricelli, and Torricelli's pupil, Viviani.

The Celsius and Fahrenheit temperature scales were proposed in 1742 and 1724, respectively. They were widely used in different countries long before scientists found the physical meaning of temperature in the middle of the 19th century. By this time, everyone was already accustomed to degrees and there was no good reason to abandon these units.

The temperature of the phase transition changes when changing the external pressure. If the pressure is increased, does the freezing point rise or fall? This question is difficult to answer without a serious study of physics, although you can try to just guess. It can be formulated more formally. The freezing point of the liquid, T, changes with increasing pressure, P. We need to find the $T(P)$ dependence or, in the simplest case, determine whether the freezing point increases or decreases.

In physics, there is the Clapeyron–Clausius equation that determines this dependence, but I do not use it, but apply simple reasoning. When water freezes, its volume increases, because ordinary ice is one of the few solids that have a lower density compared to the liquid phase.

Therefore, during the phase transformation of water into ice, it performs work against external pressure. If we increase the pressure, then the work will increase. Let's assume that water freezes at pressure P and temperature $T(P)$. Then, at a pressure greater than P, it is energetically favourable for water to remain liquid at temperature $T(P)$, so as not to do this extra work.

To freeze, it must be additionally cooled a little. Thus, the values of $T(P)$ decrease with an increase in external pressure for the phase transition of water into ice. For those substances in which the solid phase is denser than the liquid, these values, on the contrary, increase.

Plotting the obtained dependence $P(T)$, we determine the boundary between liquid and solid phases, in our case, between water and ice. The ice region is to the left of this boundary at lower temperatures. The region of water is located to the right. Metastable states infiltrate into a foreign region in the phase diagram. They are somewhat similar to sabotage groups that penetrated the rear of the enemy during trench warfare.

3.2.8. *Gas*

Recall that the phase diagram also contains a region related to the gaseous state. The term *gas* was coined in the early 17th century by the Flemish naturalist Jan Baptista van Helmont to denote the "dead air" he received (today we call it carbon dioxide).

According to the most common version, this is a phonetic transcription of the Greek word *chaos* ($\chi\acute{\alpha}o\varsigma$). Some argue that van Helmont wrote: "I named such a vapour as gas, because it hardly differs from the chaos of the ancients."

Others believe that the term came from alchemy, where the Swiss Paracelsus (Philippus Aureolus Theophrastus Bombastus von Hohenheim) meant something like super-thin water. There is also an option that the word comes from *gahst* or *geist*, that is, a ghost or spirit.

Let's think about the fact that the existence of air and wind was known long before van Helmont. Ancient philosophers willingly talked about air as one of the four cardinal elements, the lightest element in the sublunary world. But none of them considered air as a part of the more general concept of gas.

Maybe this was due to the fact that in ancient times, they did not know other gases, and so, when faced with fetid vapours or the same carbon dioxide, they considered it simply dead or spoiled air? Be that as it may, in ancient times, everyone knew about the air, because they breathed it and sometimes smelled the odour it brought, either fragrant or stinky. But no one saw the air.

In everyday life, water in a gaseous state is called water vapour. It's invisible. Many people believe that they see steam above a pond, steam coming out of the spout of a kettle, or steam from a locomotive pipe. In fact, this is fog, that is, condensed steam, which has formed a cloud of tiny water droplets in the air. It is white because light is partly scattered, partly reflected, and partly absorbed on the surface of these clear water droplets.

Water vapour fills the top of a boiling kettle above the water. However, coming out of the spout, it immediately mixes with the air that fills the Earth's atmosphere. For this reason, we do not try to extract steam in its pure form, but do our experiments with easily available air. It represents to us the properties of a typical gas. At the same time, we can conduct our thought experiments at a more comfortable temperature.

So, let's take out a glass and fill it halfway with air. Why did it not work out? Naturally, because the gas fills the entire vessel in which it is placed. It has no definite shape or volume and it has no surface. The gas is not solid. You can easily verify this by twisting your fingers in the air. The gas consists of individual molecules. It has no long-range order, no exotics. It is always isotropic, that is, all gas properties are the same in all directions.

3.2.9. *Carbon dioxide and other heavy gases in the atmosphere*

Now let us clarify a little all these properties of gases. We can fill the vessel with gas to half, at least from a practical point of view. For example, carbon dioxide or another gas heavier than air can fill the bottom half of the glass. Worse, it can accumulate in a basement, well or other depressions, poisoning a living being trapped in this gas trap.

The so-called Cave of Dogs (Grotta del Cane) near Naples, even more specifically, near Lake Agnano, which dried up in 1870, is famous for emissions of carbon dioxide. This gas spreads like a liquid at the bottom of the Cave of Dogs. A person with his or her height can walk in an upright position over a dangerous layer, but a dog running in it suffocates, as shown in Fig. 3.4. Other countries also have caverns that emit carbon dioxide, for example, in Germany, near Bad Pyrmont in Westphalia.

However, this trick is possible because there is a layer of air above the heavy gas. If it were not there, the gas would expand and fill the entire volume available to it. The air of the Earth's atmosphere is held close to the planet by its gravity, so it replaces the missing upper wall of the vessel. Accordingly, a surface or a gas boundary can exist only in the "light over heavy" option.

If the dimensions of the vessel are significantly less than a hundred metres, then the gas not only fills the entire container, but also does it almost uniformly and its pressure can be considered the same anywhere inside the vessel. In the atmosphere with its many kilometres thickness, air pressure decreases with increasing altitude.

In early 2020, a young Instagram blogger celebrated her birthday at a Moscow bath complex. As a surprise, 25 kg of dry ice was dropped into the pool where the guests were swimming, "for a beautiful fume". This is frozen carbon dioxide, which, when heated

Figure 3.4.   Engraving depicts the Dog's Grotto near Naples, the bottom of which is filled with carbon dioxide. This work is in the public domain.

above $-78.5°C$, sublimes and becomes a gas. It is easy to estimate that more than 12 cubic metres of gas is formed from 25 kg of dry ice at standard pressure and temperature, which is quite enough for those who were swimming in the pool to feel what it is like to be a dog in the Cave of Dogs. Some could not be saved.

3.2.10. *Boiling, condensation, supercooled vapour, and superheated liquid*

Now, having familiarized ourselves with the general properties of gases, let us return to water vapour. It is easily obtained by heating water to the boiling point, at normal atmospheric pressure it is $100°C$. The water–vapour phase transition curve passes through this point. This is the first-order phase transition with a change in the state of aggregation.

The reverse transition is called vapour condensation. It occurs at the same temperature at a fixed pressure. When liquid boils, a certain amount of heat is spent on this per kilogram of liquid, which is called the specific heat of vaporization (or boiling); when the gas condenses into a liquid, it is released. It is equal to $2258\,kJ/kg$ for water at normal atmospheric pressure.

Naturally, metastable states exist near the phase boundary. These include supersaturated (supercooled) steam and superheated liquid. They exist at such pressures and temperatures at which another state of aggregation will be stable, namely, a liquid for a supersaturated vapour and steam for a superheated liquid.

If a metastable substance is shaken or something is thrown into it, the process of phase transition will proceed very quickly, violently, often with an explosion. Water allows sustained superheating up to $200°C$. Water heated to $300°C$ can exist in a liquid state at atmospheric pressure for a time of the order of microseconds.

Recently, in Finland, water overheated when boiling in a microwave oven and exploded upon further manipulations with it, severely scalding a person who just wanted to make coffee. The media regularly report on similar incidents in different countries.

To avoid such surprises, chemists often add quartz sand to their flasks. It does not participate in the reactions, but is used as a source of nucleation centres for the stable phase. Simply put, it secures the liquid from overheating, and chemists from burns, reprimands from the bosses, and other troubles due to an explosion.

Note that not only pure water but also coffee and even low-fat milk can overheat in the microwave. So, dear readers, be careful when heating liquids in the microwave.

Superheated liquid is actively used by physicists. The Nobel Prize in Physics for 1960 was received by the American Donald Arthur Glaser, who invented a bubble chamber — a device for recording traces or tracks of fast charged ionizing particles. It is filled with superheated liquid, usually liquid hydrogen. During the flight of a particle, the latter acts as a boiling centre, leaving behind hydrogen bubbles (see Fig. 3.5). From their photograph, you can determine the parameters of the passing particle. The camera is

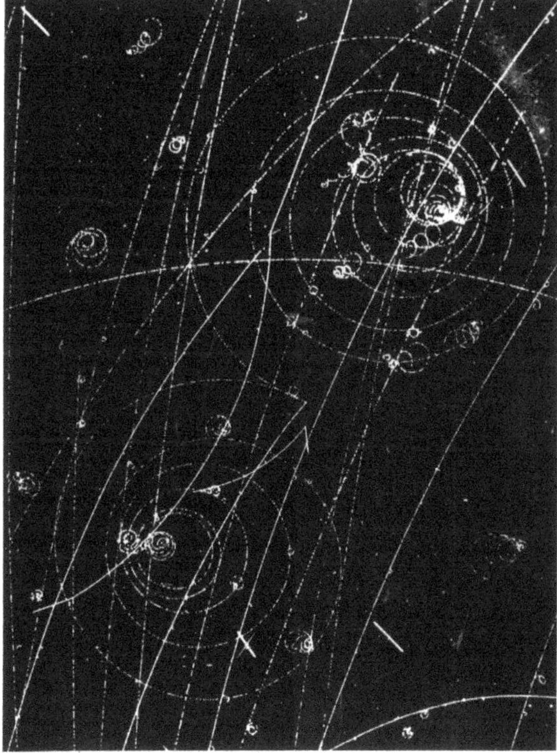

Figure 3.5. Tracks of charged particles in the 81-cm liquid hydrogen bubble chamber. Reprinted with permission from CERN. Original source: https://cds. cern.ch/record/2790874.

placed in an external magnetic field. Therefore, the particles in it move in a circle or a helical line. Due to energy losses, the speed of the particles decreases and the radius of the spiral decreases too.

Such chambers compete successfully with the older chamber invented by the Scotsman Charles Thomson Rees Wilson, who was awarded the Nobel Prize in 1927. The Wilson cloud chamber uses condensation of supersaturated steam. Thus, both metastable states near the gas–liquid phase transition curve have contributed to the physics of elementary particles.

3.2.11. *Sublimation and desublimation: Triple point*

There is also a transition from a solid to a gaseous state, called sublimation, and the reverse transition — desublimation. In our case, ice is sublimated into steam, and steam is desublimated into ice. The sublimation process allows you to dry your laundry that was hung outdoors in winter in frosty dry weather. Wet laundry quickly freezes, so water turns into ice, which sublimates slowly. Heat is absorbed during sublimation and released during desublimation.

So, two new boundaries between the phases appear in the phase diagram. They are corresponding to two phase transitions: steam–water and steam–ice. Which way are they tilted? Let us repeat the reasoning given in connection with the slope of the water freezing line and take into account that the vapour density is significantly less than the density of both water and ice.

It turns out that with increasing pressure, the boiling point of water rises, as does the sublimation temperature. Both curves of dependences $P(T)$, which define the boundaries between liquid and gas and ice and vapour, have a positive slope. As a result, we get a rough version of the phase diagram of water, shown in Fig. 3.6.

It can be seen that all three curves are connected at one point, called triple. It corresponds to the pressure and temperature at which a substance can exist in any of the three phases. For water, it corresponds to a pressure of 611.657 Pa $\approx$ 0.006 atm and a temperature of 0.01°C or 273.16 K. The phase diagram of water is quite typical; similar diagrams have many substances. But the boundary between the liquid and solid phases is usually tilted in the other direction.

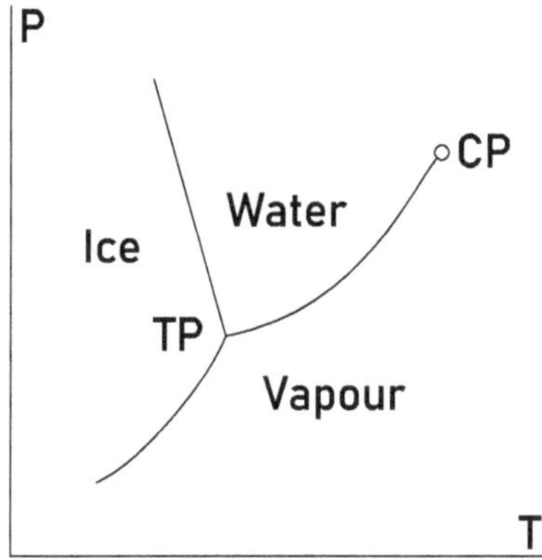

Figure 3.6.   A part of the phase diagram of water near the triple point.

3.2.12. *Boiling point of water at different pressures*

Due to the large difference in the density of steam and water, the boiling point of water changes when pressure changes significantly more than the freezing point. We can observe the consequences of this in different manifestations. Air pressure at high altitudes drops and with it the boiling point of water.

Therefore, climbers who climb especially high mountains complain that they cannot brew delicious tea. There is nothing strange about this, given that it is brewed at an inappropriate temperature. At 4000 m above sea level, where the pressure drops to 60 kPa, the water boils at about 85°C, and it takes more time to make a soup in the mountains. At the height of Everest 8848 m, water boils at a temperature of about 70°C. It is unlikely that the climbers would have had a tea party on the summit of Everest, but the tea certainly would be very badly brewed.

This effect can be used for good as well. To evaporate a liquid, that is, to transfer it to a gaseous state, it is not at all necessary to heat it to the standard boiling point. You can at first reduce the atmospheric pressure by applying vacuum pumping. This possibility was realized in the preparation of condensed milk, from which water is evaporated at a temperature of about 60°C under a slight vacuum.

The benefits can also be obtained by raising the boiling point. Food is cooked in a pressure cooker at increased pressure and temperatures above 100°C, so the cooking process takes less time. The older brother of the pressure cooker is called the autoclave. In such devices, doctors and chefs reliably sterilize what requires sterilization, from surgical instruments to canned food.

Sublimation was not ignored either. The frozen crystals of the coffee extract could be dehydrated by vacuum sublimation. This "freeze-drying" process better preserves the constituents of the extract, but due to the more energy-intensive technology, freeze-dried coffee is more expensive than other types of instant coffee.

3.2.13. *Dynamic equilibrium during phase transition*

From coffee and bad tea, let's go back to the diagram in Fig. 3.6. Surprisingly, it has not yet revealed all her secrets to us. We remember that water at a pressure of 1 atm boils at a temperature of 100°C. But why do the puddle and the water left in the cup evaporate even at normal room temperature? Can this process be considered boiling? In principle, yes, but the question is how to define the concept of boiling. When water evaporates, it undergoes a phase transition, spending on this corresponding energy, but the process turns out to be not violent, but strongly extended in time.

What is causing it? Let the water temperature be 20°C. From the phase diagram (or better from the tables of saturated water vapour pressure), we see that the pressure of the water–vapour phase transition at this temperature is 2.3388 kPa or 17.54 mm Hg (millimetres of mercury). What does this mean?

If we evacuate air from a vessel, filling it halfway with water, and place it in a room with the desired temperature, then after establishing thermal equilibrium, we see a vessel with water in the lower part and water vapours in the upper part. The temperature of both water and steam will be 20°C, and the vapour pressure will be 17.54 mm Hg. This is the so-called saturated steam pressure. It is equal to the atmospheric pressure at which water would boil at a given temperature, in our case at 20°C.

After this state is established, water and steam will be in a state of dynamic equilibrium. Evaporation and condensation processes completely balance each other. For us, the system looks static, not changing over time. But water molecules from time to time pass from a liquid to a gaseous phase and vice versa. In one second, a huge number of molecules pass from water to vapour, but as many make the reverse transition. Therefore, the mass of water in each phase remains unchanged. The phase transition energy released during vapour condensation completely transforms into the energy required for the evaporation of the liquid.

But everything will change if we heat or cool our vessel. The saturated vapour pressure is highly dependent on the ambient temperature. If we heat the water to 25°C, it rises to 3.169 kPa or 23.77 mm Hg. Therefore, some of the liquid water will evaporate to provide this pressure increase. And if we cool the room to 15°C, then the saturated vapour pressure drops to 1.7 kPa or 12.79 mm Hg and some of the vapour will condense into a liquid.

What happens if we make a hole in this sealed vessel and ambient air with a standard pressure of 100 kPa rushes into it? Until the water vapour from the upper part has time to escape out of the vessel through the hole (this is due to diffusion, and this process is much slower than the inflow of air due to the pressure difference inside and outside the vessel), it still provides a pressure of 2.3388 kPa.

This is called the partial pressure of water vapour. But the remaining 97.6612 kPa, required to equal the total atmospheric pressure, will be provided by nitrogen, oxygen, and other atmospheric gases. From the point of view of physics, this will be the sum of the partial pressures of these other gases under the described conditions.

If we quickly plug the hole in the vessel, then nothing else in it will change over time, naturally if the temperature is constant. Water will still be in dynamic equilibrium: the number of water molecules passing from liquid to gas will be balanced by the same number of molecules passing from gas to liquid. All other gases trapped in the vessel will remain in the gaseous phase.

Strictly speaking, a small fraction of them will dissolve in water. This process is of little importance for the discussed physical processes, but it is very important for fish living in water, because it is oxygen dissolved in water that enables them to breathe.

But what happens if you do not plug the hole or, conversely, completely remove the upper part of the vessel, allowing the surrounding air to reach the surface of the water? This changes everything. The dynamic equilibrium between evaporation and condensation of water will disappear. The fact is that atmospheric air as a whole is not in a state of equilibrium and the partial pressure of water vapour in it is usually less than the pressure of saturated vapour.

As a result, the number of vapour molecules entering the water decreases compared to the case of a sealed vessel, while the number of vaporized molecules remains the same. The water evaporates and eventually completely turns into steam. This is accompanied by the loss of thermal energy, which turns into the heat energy from water evaporation, and by the cooling of the evaporating liquid.

3.2.14. *Air humidity*

Physicists have introduced a quantitative indicator characterizing the evaporation process. This is the humidity of the air. Absolute humidity is the density of water vapour in the air. And relative humidity is the ratio of this density to the density of saturated steam at a given temperature. Or, equivalently, the ratio of the partial pressure of water vapour in air to the pressure of saturated vapour.

Naturally, this value cannot exceed 100%. More precisely, an absolute humidity of more than 100% corresponds to a supersaturated steam. It can be found in Wilson's chamber, but is very difficult to find in the Earth's atmosphere. However, airplane contrails visible in the sky appear when supersaturated water vapour condenses on soot particles from engine exhaust and air turbulence after the flight of the aircraft.

For a given partial vapour pressure, there is a dew point concept. This is the temperature at which air's relative humidity is 100%. At a lower temperature, some of the vapour condenses in the form of a liquid or desublimates in the form of ice. When humid air cools, rain clouds, dew, or fog appears. If the air temperature is below 0°C, then ice forms, often in the form of frost, icing or snow. Changes in air temperature and humidity provide all weather phenomena known to us.

For those who prefer numbers rather than words, consider a simple example. Let's say that the air had a temperature of 30°C during the day, and at night it cooled down to 25°C. Will dew fall out? We look at the table of saturated water vapour pressure and see that at 25°C it is 3.17 kPa or 23.8 mm Hg. This is the maximum possible water vapour pressure, there may be less moisture in the air, but there cannot be more. And at a temperature of 30°C, the pressure of saturated water vapour is 4.24 kPa or 31.8 mm Hg.

In the atmosphere, it will be equal to 4.24 kPa multiplied by the relative humidity of the air. And if this humidity exceeds 3.17 kPa / 4.24 kPa = 74.8%, then the air after cooling is no longer able to hold all the water in the form of vapour and some of it will turn into a liquid, more precisely, into water.

If we look at the directory, then we learn that the air is considered dry with up to 55% moisture, moderately dry with 56% to 70% moisture, moderately humid with 71% to 85% moisture, and very wet with 86% moisture or higher. We see that in our example dew will fall out even from moderately humid air. I add that the air humidity in the room according to the norms should not exceed 65%, but in areas with high humidity, it is allowed to exceed this indicator up to 75%. It is clear that a comfortable sleep at 75% humidity is impossible without an air conditioner.

Note that we considered cooling by only 5°C. And if it cools more strongly, for example, from the same 30°C to 15°C with the corresponding saturated water vapour pressure of 1.70 kPa or 12.79 mm Hg? Then dew will fall out at a relative humidity of more than 1.70 kPa / 4.24 kPa = 40%, i.e. even from dry air.

If the hotter air is cooled by 15°C, then the maximum humidity without dew loss increases slightly. If the air cooled from 40°C with a saturated vapour pressure of 7.37 kPa to 25°C, then dew will fall out at a relative humidity of more than 3.17 kPa / 7.37 kPa = 43%.

The readers may wonder if everything is that simple. And they would be right. When evaluating, I did not take into account one factor. If the atmospheric pressure did not change, then everything written above is absolutely correct. Note that atmospheric pressure changes rather weakly. If special accuracy is important to us, then we could get the correct result by taking into account the ratio of atmospheric pressure at day and at night. This can slightly change the result.

Seeing the morning dew, you can run through it barefoot, but you can also think about how many factors have combined to manifest together in these droplets. To begin with, you need a night-time temperature drop. The period of rotation of the planet around its axis must differ from the period of its revolution around the star (suppose that it is single, like the Sun); otherwise it would always be turned to the star by one side.

There should be a lot of water in the atmosphere (you can discuss methane or carbon dioxide dew, but I do not go over the exotic). This requires the existence of reservoirs, into which liquid water enters in order to evaporate again in the form of water vapour.

If we restrict ourselves to our planet, then dew falls more abundantly where there is a greater difference between day and night temperatures. That's in areas with a continental climate, far from the oceans and seas. In summer, this difference is greater than in winter. It decreases as you approach the poles. It is greater in clear weather than in cloudy weather because clouds, like vegetation or snow cover, prevent night cooling of the soil.

Daily temperature variations are greater if the surface has low thermal conductivity and acts as a heat insulator. For example, on a sandy surface the difference between day and night temperatures is one and a half times greater than on a granite surface.

On a dark surface, it is greater than on a light one, because it emits and absorbs thermal radiation better. And the relative humidity of air can differ in different places, even close ones. In a meadow near a river or lake, it can be greater than in a neighbouring hillock. It is on a meadow that dew falls primarily when the air cools.

Dew can save your life if you find yourself without water in a rocky desert. It falls in the morning and can be collected on the surface of stones, rubble, or pebbles. In sandy deserts, for example in the Sahara, they learned to extract water in the sand with the help of the so-called solar condenser. The Sun's rays evaporate water from a pit in the sand, and the steam is condensed on a special film.

The rate of evaporation of water from a puddle in natural conditions is greater in a hot desert. Why? Let's remember our open vessel with water. Water evaporates and the evaporation rate increases with a decrease in the relative humidity of the air and with an increase in its temperature; the latter leads to an increase in the saturated vapour pressure and the rate of transition of liquid molecules into a gaseous state.

Wind will also help speed up evaporation. When evaporating near the surface of the liquid, the concentration of vapour increases. The stream of air carries this vapour away, preventing its molecules from returning to the water.

3.2.15. *Critical point*

Let's go back to the draft diagram in Fig. 3.6. It would seem that it is completed. But this is not entirely true. There are some nuances that spoil this rosy picture. For example, the fact that the curve corresponding to the boundary of the liquid and the gas regions breaks off before reaching very high pressures. It ends at some point called critical. Water has the critical temperature $T = 647\,\mathrm{K} = 374°\mathrm{C}$ and the critical pressure $P = 218.3$ atm.

How is this even possible? What happened to the curve? We have already introduced the concept of specific heat of vaporization. The value of this quantity changes with a change in temperature or pressure, i.e. with moving along the phase transition curve. However, in contrast to the specific heat of freezing and sublimation, which are always positive, the specific heat of vaporization decreases when approaching the critical point and reaches a zero value in this very point.

The intersection of the liquid–gas phase equilibrium curve corresponds to a first-order phase transition. The intersection of the curve exactly at the critical point is a second-order phase transition. If we try to continue this curve further, then there would be no phase transition in this region. This behaviour is typical for all substances except for those that simply cannot exist in a given range of temperatures and pressures, for example, due to the fragmentation of their molecules as a result of heating.

Even a transparent substance becomes cloudy near the critical point. This effect is called critical opalescence. Added to the word "opal" is the Latin word *escentia*, glow. If a substance near the critical point, or rather at a temperature slightly less than the critical one, is placed in a vessel, then, due to its own weight, the pressure in the lower part of the vessel is greater than at the top. So a boundary of phases in dynamic equilibrium appears in the vessel. There is gas at the top, liquid at the bottom.

Random fluctuations, which I talked about in Section 1.1, arise in each of the phases. They cause the appearance and rapid disappearance of microscopic regions with an "alien" phase near the border. The less energy needs to be spent on the birth of such a region, the more of them are there. And this energy is proportional to the heat of the phase transition, which vanishes when approaching the critical point.

Therefore, many such regions arise in matter near it. They strongly scatter the light passing through them and the matter near the critical point becomes opaque, somewhat similar to the opal stone, about which Pliny the Elder wrote that it contains "the living fire of ruby, the glorious purple of the amethyst, the sea-green of the emerald, all glittering together in an incredible mixture of light".

But the very existence of a critical point means that the gas and liquid regions are not completely separated. You can take a liquid, squeeze it strongly, and then heat it, then reduce the pressure to normal and get a gas. In this case, during the entire process there would be no boiling, but the liquid turns into gas, continuously changing its temperature and pressure. There is no clear border separating one from the other domains of gas and liquid.

Therefore, the very division into gas and liquid becomes rather arbitrary because we can transform one state of aggregation into another through a slow, continuous process. From what moment can we assume that we no longer have a liquid in a vessel, but a real gas? Shouldn't liquid and gas be combined into one state? But then boiling would be considered a transition from this unified state to the same one, which looks clearly unreasonable.

Let us now give the final version of the phase diagram of water, shown in Fig. 3.7. With its help, we are once again convinced that it is possible to turn water into steam, or vice versa, without any phase transition. You just need to go around the critical point and the phase transition curve.

Some physicists propose to introduce a new phase state called supercritical fluid. In the diagram, this is the area above and to the

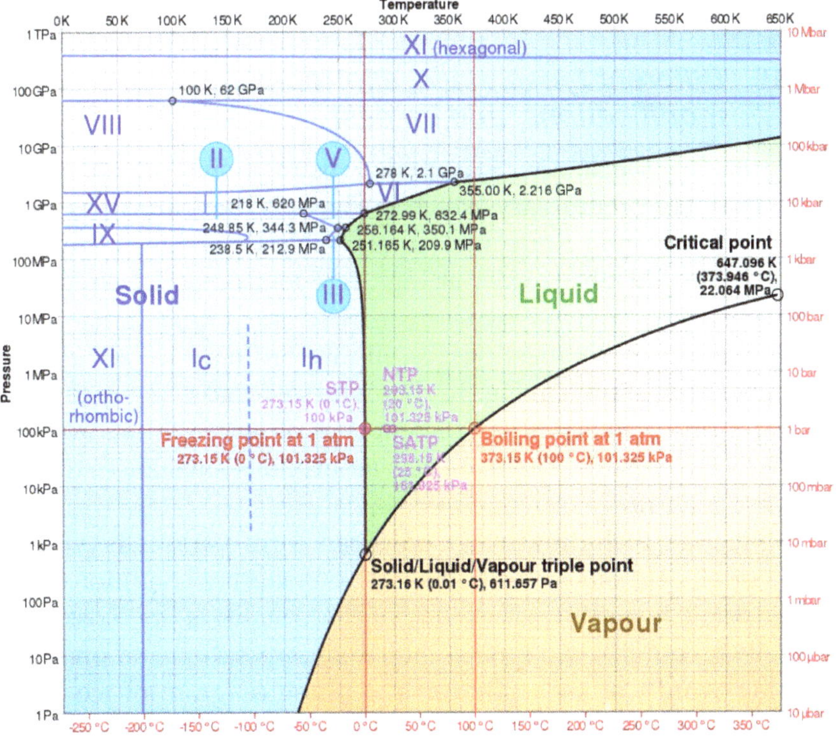

Figure 3.7. The phase diagram of water. The pressure is shown in a logarithmic scale. The Roman numerals indicate various ice phases. From Wikipedia. This file is licensed under the Creative Commons Attribution-Share Alike 3.0 Unported license by user Cmglee.

right of the critical point, bounded from above by the area of ice. Then one could argue that we can convert water to steam either by boiling, i.e. phase transition, or through an intermediate state of a supercritical fluid.

It is clear that the question of whether this new concept is useful or not is not so obvious. If we introduce it, it will somewhat mask the problem of the possibility of a continuous transition from liquid to gas, or vice versa. And in science it is not customary to mask problems. Scientists do not have a consensus on whether it is worth separating the supercritical fluid region into a separate state of matter, but this question is more of a methodological nature. It is connected not so much with the essence of the phenomenon, but with how to explain it to pupils and students.

3.2.16. *Different types of ice, allotropic modifications, and tin pest*

In Fig. 3.7, we see another feature that distinguishes the phase diagram from the preliminary sketch in Fig. 3.6. There are areas of ice marked by different Roman numerals. What does this mean? And how can a solid phase exist at high temperatures and enormous pressures?

To answer these questions, recall that ordinary ice is a crystal with the lattice shown in Fig. 3.2. But who said that water molecules cannot form crystals with a different type of lattice? Moreover, not all solids are crystals. There are also amorphous substances that do not have long-range order and crystal lattice.

Molecules including water ones can form various crystalline forms or amorphous states depending on external conditions such as temperature and pressure. One of these states is the most energetically favourable at a certain choice of temperature and pressure. In other words, at each point of the phase diagram of ice, there is the most stable type of ice. Other types of ice with a different lattice either transform into this selected type of ice or exist as a metastable state.

And if in everyday life, we do not encounter other modifications of ice,[d] some substances are found in various forms, called allotropic modifications. For example, carbon in crystalline form is found in the form of graphite, diamond, and also carbyne, lonsdaleite, nanodiamond, fullerenes, fullerite, carbon fibres, nanofibres, nanotubes, and graphene, and many of these modifications were discovered or created very recently.

There are also amorphous modifications: soot and carbon nanofoam. The stable state under normal conditions is graphite, and diamond is a metastable state.

Sulphur has the second-largest number of allotropes among the chemical elements. There are 11 known allotropes of phosphorus. Oxygen has only two of them — oxygen itself O_2 and ozone O_3, which shows that modifications can differ not only in the type of lattice, but also in other properties, for example, the number of atoms in a molecule.

Modifications of tin are curious. At high temperatures of 161–232°C, the so-called γ-tin is stable, but at room temperatures other allotropes compete. The α-tin (grey tin) with a cubic diamond-type lattice is stable at temperatures below 13.2°C and β-tin (white tin) with a tetragonal crystal lattice is stable at temperatures above 13.2°C. At temperatures below 13.2°C, white tin becomes metastable and turns into grey tin, and the faster, the lower the ambient temperature. Conversion rate is maximum at −33°C. This phenomenon is called the tin pest, tin disease, tin blight, or tin leprosy because it frequently makes tin objects decompose into powder.

It has caused the death of many tin articles, from collections of tin soldiers to buttons on antique military uniforms. Tin leprosy is one of the reasons for the death of Robert Scott's expedition to the South Pole in 1912, which was left without kerosene because it had

---

[d]In Kurt Vonnegut's satirical postmodern novel *Cat's Cradle*, the plot is based on the discovery (naturally fictitious) of a modification of ice that is more stable than ordinary one.

leaked from sealed tin cans struck by the tin disease. There were no problems with the cold in the Terra Nova Antarctic expedition.

Let's go back to water. I repeat once again that under the conditions we are accustomed to, we deal with only one modification of ice, called I_h. Only it can get into your juice or cocktail, sink the *Titanic* or be under your skates. But when the temperature and pressure change, other modifications enter the stage.

Currently, three amorphous varieties and 17 crystalline modifications of ice are known. Let us give the structure of crystals of some phases in Fig. 3.8.

Their densities are significantly different. Ice XVI, obtained in a complex way in 2014, has the smallest density of $0.81\,\text{g/cm}^3$, ice XII has a density of $1.3\,\text{g/cm}^3$, and the density of ice X is $2.51\,\text{g/cm}^3$. Naturally, this particularly dense modification of ice is stable at ultrahigh pressures of more than $70\,\text{GPa}$. The transition from one modification to another is a first-order phase transition and is accompanied by the release or absorption of heat.

Add to this that we can easily distinguish between ice and snow. But both are one substance in the same state of aggregation. And in Fig. 3.7, there is no area with the label "snow". So what is it? When we talk about crystals, we usually imagine something like cut diamonds or crystals from a mineralogical museum, that is, a body filled with a solid phase of some substance. But it's not always the case.

As an example, take a look at crystals of scolecite, a mineral from the silicate class. This is a hydrated calcium silicate $CaAl_2Si_3O_{10} \cdot 3H_2O$. It got its name from the Greek *skolex*, meaning

Figure 3.8.   The structures of ice modifications III, IV, VI, and XII. Published with the kind permission of Grigori Malenkov, the author of the paper "Liquid water and ices: understanding the structure and physical properties."

Figure 3.9.   Fluffy scolecite crystals.

"worm", due to its property to wriggle slightly under the influence of an open flame. It is more often found in the form of radiant prism-shaped filaments, shown in Fig. 3.9, but could have a shape similar to the crystals we are accustomed to.

And this is a crystal, even with air between the threads! I think that after such a "fluffy crystal" you will no longer be surprised by a patterned snowflake consisting of ice. Snowflakes always have the shape of a six-pointed star or hexagon, as Johannes Kepler wrote in his treatise *Strena, seu de nive sexangula* (On hexagonal snowflakes).

We observe snow in a different state: falling, loose, trampled, melting and with varying degrees of freezing and glaciation. After the publication in 1911 of an article by the American anthropologist Franz Boas, a myth arose that in the language of the Inuit living among the snow, there are a huge number of words for its different types. In subsequent studies, it was refuted. However, there is an article claiming that in the Sami languages, which are spoken by part of the inhabitants of Norway, Sweden, and Finland, there are about 180 words related to snow and ice, as well as up to a thousand words for deer. I do not undertake to judge how true it is.

Let us recall again that all the states depicted in the phase diagram of water consist of the same molecules. This is water, albeit in different guises. So if you walk in winter along the surface of a frozen pond, you can safely tell everyone that you have walked on water. There will not be a drop of lies in this, only the omission of some details.

3.2.17. *Plasma*

It's time to talk about the state of matter, which I have repeatedly mentioned. It was discovered by William Crookes in 1879 and named "plasma"[e] by Irving Langmuir in 1928. It looks like an ionized gas. Ordinary gas is composed of electrically neutral molecules. In plasma, electrons of some of the molecules are detached and move separately. Therefore, plasma consists of a mixture of negatively charged electrons, positively charged ions, and neutral molecules.

The percentage of ions determines the degree of ionization of plasma, which varies greatly for different processes and conditions. To see plasma, it is enough to look at the gas discharge tube, the electric spark, or the Sun. If you light a match, then the flame is a very weakly ionized plasma, whereas inside the Sun it is highly ionized.

In general, plasma remains electrically neutral, which distinguishes it from electron or ion beams. Strongly ionized plasma conducts electric current very well. Therefore, if electric charges are separated in generally neutral plasma, currents appear that will quickly restore the neutrality of its parts. Plasma is isotropic, but it becomes anisotropic in a strong magnetic field.

High electrical conductivity and a number of other features force us to consider plasma as a specific state of aggregation. The transformation of a gas into plasma is called ionization and is not accompanied by an abrupt phase transition; it is a continuous change. It is characterized by a shift in the degree of ionization, that

---

[e]The Greek word $\pi\lambda\acute{\alpha}\sigma\mu\alpha$ means something fashioned, designed, invented, and is derived from $\pi\lambda\sigma\acute{\alpha}\sigma\sigma\varepsilon\iota\nu$ — to form, to model.

is, the ratio of the number of free electrons to the total number of neutral atoms and charged ions.

If you increase the gas temperature, i.e. move to the right in the phase diagram, then after reaching the characteristic ionization temperature, the gas will gradually turn into plasma. Therefore, the plasma region in the diagram lies to the right of the gas region. There is no clear border between them. The reverse process of converting plasma into a gas is called recombination.

There is no plasma region on the phase diagram of water shown in Fig. 3.7. The so-called thermal dissociation occurs in it. Water vapour after strong heating turns into a mixture of hydrogen and oxygen. And already these gases become plasma with a significant degree of ionization at even greater heating.

3.2.18. *Liquid crystals*

Liquid crystals are anisotropic liquids with molecules oriented in space in a certain way. So, it is not the position of individual molecules that is ordered, but their orientation. Liquid crystals have all the properties of liquids, except for the anisotropy of optical properties. Moreover, these properties can be changed by the temperature or the electric current passing through liquid crystals. This is used in the manufacture of smartphone screens, monitors, indicators, thermometers, and other familiar items that have entered our daily life.

My story was mainly about the phases of a substance consisting of molecules of the same type. It is clear that new possibilities appear for a mixture of different molecules. If the gases simply mix, forming gas mixtures like air, then the rest of the phases are capable of more.

There are simple solutions and alloys, but there are also dispersed systems, made from two or more phases, which practically do not mix and do not chemically react with each other. We are talking about suspensions, emulsions, gels, aerosols, foams, pastes, and composite materials. It is clear that I am only mentioning all these systems; their description would increase the volume of the book to an absurdly large extent.

3.3. Earth: Subsoil, Surface, and Atmosphere

3.3.1. *Earth's core*

What does the Earth consist of? What areas does modern science divide it into? Let's walk through them, starting from the centre of the Earth. There is a core in the very centre, which is divided into internal and external. It is quite dense. We know this because the Earth's average density of $5.515\,\mathrm{g/cm^3}$ is significantly higher than the average density of layers near the surface, which is about $3\,\mathrm{g/cm^3}$, so there must be areas with increased density inside.

The core is hot. We know this because the temperature rises as we go deeper into the bowels of the Earth. In addition, it spews molten lava and hot geothermal waters to the surface.

But what is the source of heat? There are two of them. The first is a decay of radioactive elements inside the Earth. They are primarily uranium, thorium, and the potassium isotope ^{40}K. Half-lives of the isotopes ^{238}U and ^{40}K are comparable with the lifetime of the Solar System, while the half-life of the ^{232}Th is comparable with the lifetime of the Universe, so these elements simply do not have enough time to fully decay.

Another source of energy is the lowering of heavier rocks into the depths of the Earth with the rise of lighter ones upward, that is, the use of the potential energy of the Earth's gravitational field. It is clear that this mechanism worked more efficiently immediately after the formation of the Earth than in our time. But the heat released a long time ago is stored in the bowels and will warm up the Earth's surface from the inside for a long time.

The mechanism of separation of rocks by density with the sinking of heavy ones and the lifting of the lightweight ones is called gravitational differentiation. It works not only on the Earth, but also on other planets. Naturally, on the gaseous planets of the Solar System, where matter at great depths could be in a state of supercritical fluid, this process is faster due to the lower viscosity of the matter.

The radius of Uranus is approximately four times larger than that of the Earth, and Neptune is not much smaller than Uranus, so we understand that the process of gravitational differentiation there can provide quite a decent energy release. Neptune, for example, emits 2.6 times more energy than it receives from the Sun. And Jupiter and Saturn are significantly larger in size.

At the same time, in the depths of the gas giants, the pressure and temperature are so high that the methane that got there breaks down into hydrogen and carbon. Hydrogen floats up, but carbon is released in a solid state and falls towards the centre having the allotropic modification in which it is supposed to be at such temperatures and pressures. And it stands out in the form of diamonds! So some scientists argue that diamond rains are falling inside Uranus and Neptune.

Surprisingly, this statement was recently verified in a terrestrial laboratory, more precisely, at the SLAC National Accelerator Laboratory at Stanford University. Honestly, instead of methane CH_4, they used styrene C_8H_8, heated to 5000 K by laser and compressed with a pressure of $1.5 \cdot 10^{11}$ Pa. As a result, it turned into diamonds, releasing hydrogen.

Let's return to the Earth's core, the existence of which we know from the analysis of the propagation of seismic waves, that is, displacement waves caused by various sources inside the Earth and on its surface, for example, earthquakes.

It is believed that the outer core is liquid, has a radius of about 3500 km, which corresponds to a depth of 2900 km from the surface, and a density of 9.9–12.2 g/cm^3. Inside it is a solid core with a radius of 1220 ± 10 km (70% of the lunar radius), a density of 12.8–13.1 g/cm^3 and a temperature of 5700 K (5400°C), comparable to the temperature of the Sun's surface. The outer core is colder; its temperature drops from 4000–8000 K (3730–7730°C) at the border of two cores[f] to 3000–4500 K (2730–4230°C) on its outer surface bordering the mantle.

---

[f]The temperature of the inner core and the boundaries between the cores are taken from reference books. The study of scientific articles showed that there is no inconsistency in the estimates; just an estimate of 4000–8000 K is given as an interval between two boundary values, and 5700 K is a very rough estimate with incomprehensible accuracy. But it was not easy to get it either.

According to the most common view, the Earth's core is composed primarily of iron. Many people calculate the composition of other components, but their estimates differ. In any case, it is similar in composition to nickel–iron meteorites.

There are also some extravagant theories, like a nucleus made of metallic hydrogen. This is the state of hydrogen predicted by theorists at ultrahigh pressures when it becomes a conductor. In 2016, it was reported that it was obtained in the laboratory at a pressure of 495 GPa, and in 2019, the results were confirmed. This theory does not describe well the properties of the Earth's core, but some suggest a similar core to be inside Jupiter.

3.3.2. *Mantle and crust*

The core is surrounded by a mantle — a silicate shell, which accounts for 67% of the Earth's mass and about 83% of its volume. It extends from the border with the outer core to the border with the Earth's crust at a depth of 5–70 km. However, in some places under the oceans, the mantle comes to the surface, more precisely to the ocean bed. Tectonic processes have moved some rocks formed from the mantle to the surface of the Earth, for example in Newfoundland and Labrador in Canada.

The density of the mantle rocks is 3.4–5.6 g/cm^3. The pressure at the bottom of the mantle is about 140 GPa; it does not allow the substance to melt, despite the high temperature, which reaches 500–900°C at the upper boundary of the mantle and more than 4000°C near the lower boundary.

The substance in a viscous amorphous form is located in the asthenosphere, a layer located in the upper mantle, at depths from 100 to 700 km. The temperature is lower there, but the pressure is also lower. In the rest of the mantle, the substance is in a crystalline state.

The Earth's crust differs from the mantle in the composition of its rocks. For the most part, it consists of basalts. When a seismic wave passes from the crust to the mantle, its velocity rapidly increases from 6.7–7.6 km/s to 7.9–8.2 km/s. Seismologists call this boundary the Mohorovich layer. There is a similar crust on the

Moon and Mars. There are two types of crust on the Earth — continental and oceanic ones. The first includes granites and sedimentary rocks.

The mass of the Earth's crust is estimated at $2.8 \cdot 10^{19}$ tons (of which 21% is oceanic crust and 79% is the continental one). This is only 0.47% of the total mass of the Earth. The area between the surface and the Earth's asthenosphere is called the lithosphere, it reaches depths from 5 to 200 km, on average about 60 km, and consists of the crust and the upper part of the mantle.

3.3.3. *The motion of lithospheric plates*

The discovery of the motion of large sections of the crust, called tectonic plates, along with the continents and their parts located on these plates was a very important achievement of the geosciences. It not only explained the movement of the continents relative to each other, but also allowed to find out how exactly the continents looked in the distant past, how they split up, moved, and joined in new configurations.

The theory of global tectonics has linked the movement of lithospheric plates with zones of seismic activity and active volcanism. Some of the new predicted phenomena were subsequently discovered, in particular by underwater research.

You can find not only maps, but also videos showing the movement of the continents at a very accelerated pace on the Internet. Links to them can be found, in particular, at the end of the article "Plate Tectonics" in Wikipedia. We will return to the movement of the continents when discussing the global climate. The movement of continents also makes it possible to understand exactly how the ancestors of today's animals expanded their habitats, moving from one part of the world to another. Therefore, this information is important for palaeontologists as well.

By the way, scientists predict that now a new ocean is forming before our eyes, which in the future may catch up in size with its older brothers. It passes through Ethiopia and the Red Sea. Two plates on its sides, African and Arabian, move away from each other at a rate of about 1 cm per year, i.e. fast enough by geological standards. The formation of the ocean began about 25 million years ago and will last for hundreds of millions of years.

3.3.4. *Water on Earth*

But let's digress from the future ocean and look at the already existing World Ocean. It occupies approximately 70.8% of the Earth's surface and, together with rivers and lakes, constitutes the so-called hydrosphere. The rest of the planet's surface is occupied by continents and islands. The variety of climates, types of surfaces, and landscapes make it an excellent place to live, both for work and leisure.

I am sure that readers are well aware of this even without this book. If not, then instead of reading, they should watch exciting films about the beauty of nature, the romance of sea travel, underwater exploration, or just booklets of travel companies.

There is no hydrosphere on other planets. Where did the water that forms it come from? Scientists suggest that it was brought to our planet from the outer regions of the Solar System, from the ice lying outside the freezing line.

There, ice and even some simple organic molecules froze onto the dust particles as they moved inside the molecular cloud. And together with dust, comets, and other small bodies, they got to us, leaving their contents after heating.

This idea has been worked out on a qualitative level rather than on a strictly quantitative one. However, it has one important confirmation — the ratio of hydrogen isotopes in water. The fact is that the ratio of the number of deuterium and protium atoms in the protoplanetary cloud varied depending on the distance to the Sun.

In the hydrosphere, it is approximately the same as in the outer belt of asteroids or on comets arriving from it. But the water that came from the Earth's mantle has a lower percentage of deuterium. Naturally, the ratio of oxygen isotopes is also being studied.

It seems surprising that such a huge mass of water could have been brought in from other regions of the Solar System. I confess that I am amazed along with everyone. Nevertheless, if we recall that 4 billion years is a huge period of time, and in the young

Solar System there were much more asteroids and comets than now, then the hypothesis no longer looks deliberately absurd.

3.3.5. *Water on Mars and Venus*

"But well, why then there is no water on other terrestrial planets, on Venus and on Mars?" the reader could ask rightly. If water was brought to the Earth from the outskirts of the Solar System, then it must also fall on the neighbouring planets. The question is good and the answer to it was received quite recently. More precisely, not answer, but answers, because the situation is completely different on Venus and on Mars.

Let us start with Venus. There was water on it, but only billions of years ago. To be completely honest, there is water on Venus even now. There is some water vapour in the Venusian atmosphere. This atmosphere is heated to 460°C and consists of sulphurous and carbon dioxide gases with clouds of sulphuric acid drops. But water vapour, being condensed in a liquid state, would form a puddle 3 cm deep on the entire surface of the neighbouring planet. Venus has lost a lion's share of its water. Scientists came to this conclusion after the data of the devices in the European spacecraft Venus Express appeared in April 2006.

Under the influence of solar radiation, water molecules disintegrated into hydrogen and oxygen, which were carried away into space. Instruments aboard the Venus Express have detected noticeable losses of hydrogen and oxygen on the night side of Venus. And since the heavier deuterium evaporates slightly more slowly than protium, its concentration is increased in the upper layers of the Venusian atmosphere.

It is much colder and solar radiation is weaker on Mars. But the gravitation is also weaker and molecules of hydrogen and oxygen escape from the atmosphere. However, there is some water on Mars. And it lies mostly underground or rather sub-Martian. Ice is concentrated in the so-called cryosphere — a layer of permafrost tens or hundreds of metres thick. On the planet's surface, only polar caps contain ice mixed with dust.

The results of studies of the interior of Mars to a depth of 5 km were announced in 2018. They were obtained using the MARSIS (Mars Advanced Radar for Subsurface and Ionosphere Sounding) radar for sounding the ionosphere and deep layers of the Martian surface, installed on the Mars Express spacecraft of the European Space Agency. It discovered a subglacial lake of liquid water at a depth of 1.5 km under the South Polar Cap. Scientists have figured out by 2020 that it is a system of at least four communicating lakes.

So there is water on Mars, but it cannot release water vapour into the atmosphere. Nevertheless, there was a lot of water on Mars, including deep oceans, 2.5–3.5 billion years ago. Then a strong climate change began and about a billion years ago the planet took its present shape.

So, the water went not only to the Earth, but also to its neighbours, who, however, quickly squandered their wealth. But stop discussing other planets. It's time to return home.

3.3.6. *Atmosphere*

There is an atmosphere of air above the Earth's surface. Now its main component is nitrogen gas. The composition of the atmosphere has changed greatly over the history of the Earth's existence. The primary atmosphere, naturally, contained much more hydrogen and helium than modern ones. These gases have partially leaked into space.

Lighter molecules of hydrogen and helium have higher thermal velocities than nitrogen and oxygen. Some molecules have velocities exceeding the escape speed and can leave the Earth's atmosphere. We discuss this mechanism a little further. But escape velocities are much higher for more massive planets and gases almost do not leave their atmospheres. In the summer of 2019, astronomers reported the results of a study on the exoplanet Gliese 3470b in the red dwarf system Gliese 3470 in the constellation Cancer at a distance of about 100 light-years from Earth. Its mass is 12.6 times that of the Earth, which allowed it to preserve the primary atmosphere rich in hydrogen and helium.

In addition to leakage, hydrogen on our planet was partially bound by chemical reactions, and volcanic gases came from the depths of the Earth: carbon dioxide, ammonia, and water vapour, which also actively changed in chemical reactions.

The emergence of life after a while led to the greatest ecological catastrophe for the entire existence of life on the Earth. Some living things began to emit a very toxic substance as waste products, which accumulated in the atmosphere and led to the death of the vast majority of all living things. It was oxygen. New generations of living organisms have not only adapted to the presence of oxygen, but can no longer exist without it, you and I included.

For obvious reasons, we know much more about the atmosphere than about the interior of the Earth. Specialists in atmospheric physics have long ago broken it into layers, and layers into sublayers. But we will not go into details, but simply list the layers from bottom to top. The troposphere or the lower part of the atmosphere contains about 80% of its total mass and 99% of all water vapour. Its upper boundary is located at an altitude of 8–10 km in polar latitudes, 10–12 km in middle latitudes, and 16–18 km in tropical latitudes; in winter it is lower than in summer. Clouds, cyclones, and anticyclones, all this is about the troposphere.

Above is the so-called tropopause, in which the decrease in temperature with height, observed in the lower layers (by about 0.5–0.7°C per 100 m), stops. The stratosphere is located at an altitude of 11–50 km. It contains the ozone layer.

Even higher are the mesosphere (up to heights of 80–90 km) and the thermosphere up to 800 km. However, the air density in the thermosphere is low. According to the definition of the International Aeronautical Federation, everything that is above 100 km, the so-called Karman line, already belongs to space. It was named after the Jewish-Hungarian American scientist and engineer Theodore von Kármán.

In 1944, the German V–2 rocket, reaching an altitude of 188 km, became the first man-made object in history to cross the Karman line. The first creatures that overcame it and returned to the Earth alive were the fruit flies, sent by the United States on a similar rocket in 1947. The first mammal was the rhesus monkey Albert–2, launched by the United States in 1949. Well, about the first cosmonaut Yuri Gagarin you know for sure.

3.3.7. *Air leak*

Some molecules of atmospheric gases are capable of leaving the atmosphere and leaking into space. There are several ways to do this.[g] I start with one of the atmospheric leakage mechanisms, first described by the British physicist James Hopwood Jeans in 1931. A molecule needs for this space flight a speed not less than the escape velocity, described in Section 2.2. In addition, it should not collide with other molecules on its way. The latter can be expressed more simply: the molecule must fly out of the exosphere,[h] i.e. start from a height greater than the height of the exosphere lower bound, called exobase. The altitude of the Earth exobase ranges from about 500 to 1000 km (310 to 620 mi) depending on solar activity. So air leaks in space at very high altitudes.

But how can molecules achieve escape velocity? I have already said that temperature is proportional to the average kinetic energy of gas molecules. Knowing the temperature of the gas and the mass of the molecules, it is possible to calculate the thermal velocity of the molecules at which they have the corresponding kinetic energy. Naturally, the lighter the molecule, the higher is this speed at the same temperature. Therefore, only lighter molecules can have the high speeds required for escape from the atmosphere.

But the thermal velocity even for hydrogen in the exosphere is still significantly lower than the escape velocity. Recall that during thermal motion the velocities of molecules between their collisions with each other differ in magnitude. Some are almost at rest; others show record speeds. To some extent, this can be considered as random fluctuations of this quantity. So some of them may have

---

[g]They are well described in D.C. Catling and K.J. Zahnle, The Planetary Air Leak // Scientific American — May 2009, pp. 36–43.

[h]The atmospheric layer where the air density is so low that the molecules are essentially collisionless.

speeds higher than the escape velocity and fly off into space. The velocity distribution of molecules was found by Maxwell back in the 19th century.

Physicists speak in this case about the high tail of Maxwell's distribution. Therefore, only a small fraction of the molecules fly away. The likelihood of escape drops very dramatically if the thermal velocities are low or the escape velocity is high. So molecules are reliably held in the atmosphere if they are heavy or it is the atmosphere of a massive planet.

As a result, the atmosphere of the Moon, Mercury and small celestial bodies of the Solar System is practically absent. The atmosphere of not very massive Mars is very rarefied. Jupiter reliably holds its atmosphere. In addition, atmospheric molecules can be knocked out by the solar wind. Since Jupiter is much farther from the Sun than Mercury, this effect is stronger for Mercury, not Jupiter.

What about the Earth? Its atmosphere loses about 3 kg of hydrogen and 50 g of helium every second. Let us estimate how significant the mass loss of air can be for 4.5 billion years, multiplying its current rate 3 kg/s by time. The resulting estimate $4 \cdot 10^{17}$ kg is many orders of magnitude less than the mass of the atmosphere. Naturally, this estimate is too primitive. Indeed, during this period, the composition of the atmosphere, its density and temperature changed. We know little about the parameters of the early Earth's atmosphere, in particular the content of hydrogen and helium in it. However, it is quite enough to conclude that the Earth's atmosphere, like the atmosphere of Venus, leaks into space very weakly.

More precisely, the atmosphere of Venus is sufficiently resistant to the escape of molecules by the process considered by Jeans. A more essential leakage mechanism for the Venusian atmosphere is associated with the knocking out of molecules under the influence of the solar wind.

3.3.8. *Ionosphere and long-distance radio transmissions*

The ionosphere is a layer of the atmosphere, partially ionized due to exposure to cosmic rays and solar radiation. It is a mixture of common gases and plasma and is generally electrically neutral. The degree of ionization is rather small, but increases with altitude, reaching 50% at 1000 km, which is higher than the low orbits of some satellites. It is considered a special layer or area when discussing electric currents and the characteristics of the propagation of electromagnetic waves within it.

Having plasma in its composition, the ionosphere is to some extent a conductor. If an electromagnetic wave meets a conductor, then it can pass through it if its frequency is greater than a certain frequency, called plasma frequency, or it can be reflected if the frequency is less than it. Plasma frequency for the ionosphere corresponds to short radio waves with a wavelength of 10–20 m.

Therefore, long, medium, and most short radio waves are reflected off the ionosphere like from a mirror, allowing radio receivers to receive radio transmission from the far lands. Waves are repeatedly reflected from the ionosphere and from the Earth's surface on the way to receiver. As a result, long wave signals, emitted by a radio station somewhere in Brazil, can reach listeners in Boston or Amsterdam.

The plasma frequency depends on the concentration of free electrons in the ionosphere, so it varies depending on the time of day. It is maximum in the evening and minimum in the morning. The fact is that these very free electrons arise when the air is ionized by cosmic rays and solar radiation. Cosmic rays act constantly, and the Sun shines only during the day, but not at night. Many citizens of the USSR knew that broadcasts of the Voice of America could be caught in the shortwave range, say, 16 metres, in the evening, but not in the morning.

But ultrashort UHF or VHF waves for TV and FM radio, not to mention visible light, cellular signals, and GPS signals, pass through the ionosphere and are not reflected from it. Therefore, they can only be caught within the line of sight of the transmitter. You cannot receive an FM signal in Paris from Los Angeles and even from Warsaw. But you can see the light from the Sun which has passed through the ionosphere.

3.3.9. *Global atmospheric circulation*

Air in the atmosphere does not stay in one place. It heats up and rises up in warm near-equatorial regions and goes down in cold sub-polar regions, driving the entire global atmospheric circulation system. This is the name of the planetary system of air currents, which combines trade winds, monsoons and winds associated with cyclones and anticyclones.

So, at the equator the air rises, whereas at the pole it falls. It could move towards the equator at low altitudes and from the equator at high altitudes, forming a single closed path. But now the global circulation system consists of three cycles rotating like three intermeshing gears transmitting rotational motion to one another. These are three convective cells called the Hadley cell, the mid-latitude cell, and the polar cell (see Fig. 3.10). In each of them there is closed air circulation, and its directions in the adjacent cells are opposite.

In this case, the Coriolis forces, arising from the rotation of the Earth and described in Section 1.7, deflect the air moving in the

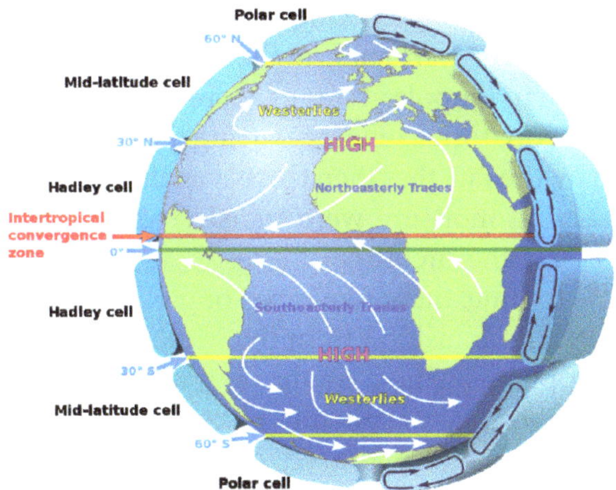

Figure 3.10. Scheme of the global atmospheric circulation in the modern era (from Wikipedia; drawn by Kaidor), based on NASA's depiction of the Earth's global atmospheric circulation. This file is licensed under the Creative Commons Attribution-Share Alike 3.0 Unported license by user Kaidor.

meridional direction and it shifts to the west or east. As a result, the picture of the global circulation explains all the known features of the winds in different parts of the globe and their changes over time.

But this picture determines not only the direction of the winds, but also the transfer of moisture. In one half of the cell, ascending currents are dominating and descending currents prevail in the other half. The moisture evaporating in the first half falls out mainly in the second. Thermal energy is carried mostly by water vapour. The heat spent on the evaporation of water is released where the vapour carried by the air currents falls out in the form of precipitation, condensing back into the liquid.

For example, in the equatorial cell of the Northern Hemisphere, low-altitude winds blow from north to south, so that tropical rainforests appear in its southern half and arid savannas, in the northern half. In mid-latitudes, where the wind direction is reversed, deserts appear in the south, and subtropical and deciduous forests in the north. The hydrosphere also has its own global circulation and these two circulations are connected with each other. Thus, it is very difficult to understand the general phenomena that determine the climate.

Such a picture of global circulation exists now. However, in the history of the Earth (in particular, in the Mesozoic), one can also find eras, when the circumpolar regions were very warm, and plants growing in the equatorial zone, preferring an arid climate. This can be explained by the fact that then there was the only convective cell in which heat and moisture were transferred from the equatorial regions directly to the circumpolar ones.

From time to time, these two types of global circulation replace each other (e.g. the transition from the Mesozoic to the Cenozoic). Why is this happening? There is a theory that this is caused by continental drift. The continents split into pieces, ran into each other, and merged into a single whole, moved, and did not sit still in general. Climatologists believe that if the poles and equator of the Earth are free of land, and the continents are located in mid-latitudes and, in addition, are deployed along the meridians, as it was in the Mesozoic, then this leads to the formation of the single convective cell.

But time passed and the continents drifted. India, moving northward, stumbled upon Asia, and the Himalayas emerged at the site of their collision. A little later, Africa collided into Eurasia from the southwest. As a result, the mountain ranges of Southern Europe and the Iranian Highlands arose. Only isolated basins (Mediterranean, Black and part of the Caspian Sea) remained from the once huge ancient ocean of Tethys.

Antarctica broke away from Australia and South America and came to the South Pole, and a closed Antarctic circulation was formed around it, that is, a system of winds blowing in a circle. The Isthmus of Panama arose, blocking the channel from the Pacific Ocean to the Atlantic Ocean and, accordingly, the former equatorial circulation in the hydrosphere. The result is climate change, the beginning of the era of glaciation and mass extinctions of species. But remember that this is only a theory that requires confirmation.

3.4. Earth's Magnetic Field

3.4.1. *Magnetic and geomagnetic poles*

The Earth is surrounded by its own magnetic field. It is similar to the field of a magnetized bar or current coil, especially over long distances. Such a field is called dipole. The axis of the magnetic dipole passing through the centre of the Earth and creating a field close to the observed one makes an angle of about 9.5 degrees with the axis of rotation of the Earth.

The points of intersection of this magnetic axis with the Earth's surface are called geomagnetic poles. The coordinates of the geomagnetic poles in 2017 were as follows: in the Northern Hemisphere, 80.5° north latitude and 72.8° west longitude; in the Southern Hemisphere, 80.5° south latitude and 107.2° east longitude. The two geomagnetic poles are antipodal, that is, their latitudes coincide, and their longitudes add up to 180°. The poles move along the surface of the Earth, shifting from 0.05° to 0.1° per year.

Now the geomagnetic poles are not very far from the geographic ones, so on most of the planet the compass shows

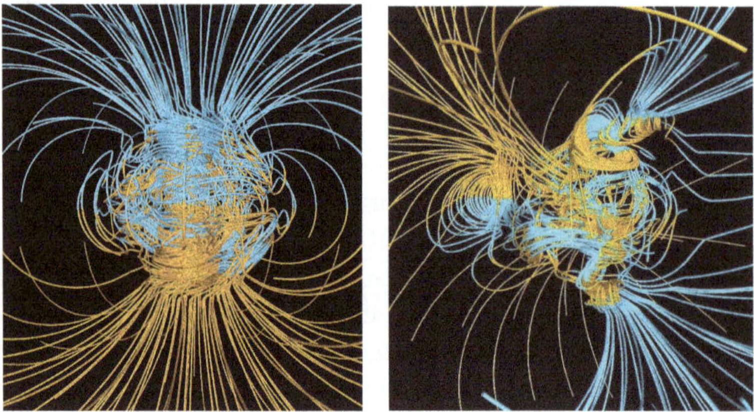

Figure 3.11. Model of the magnetic field of the Earth now (left panel) and during polarity reversal (right panel). This file is in the public domain. Original source: NASA.

a direction close to the meridian. Problems arise in circumpolar regions and near magnetic anomalies.

The dipole field only roughly describes the Earth's magnetic field. The left panel of Fig. 3.11 shows the true magnetic field lines plotted from measurements and simulations. The differences from a simple dipole field are noticeable. They are especially strong in the Southern Hemisphere, where the South Atlantic anomaly is located, which I discuss below.

Because of the complex picture of the Earth's magnetic field, the concepts of the Earth's magnetic (not geomagnetic!) poles have been introduced. These are the points at which the magnetic field is directed vertically. They do not have to be antipodal.

The North Magnetic Pole was discovered on June 1, 1831, by the English polar explorer James Clark Ross. It was located in the Canadian archipelago, on the Boothia Peninsula, off the coast of Cape Adelaide, with coordinates of $70°05'$ north latitude and $96°47'$ west longitude. The next explorer to visit the North Magnetic Pole was Roald Amundsen in 1903, but the pole moved slightly. In 2009, it moved to a point with coordinates $84.9°$ north latitude and $131.0°$ west longitude.

It was on the territory of Canada, but was moving towards Russia. Now the pole has left the 200-mile zone belonging to

Canada and is moving at a speed of 55 km per year towards Siberia. This speed is so high that the change in the model of the Earth's magnetic field, planned for 2020, had to be carried out a year earlier.[i]

James Ross determined in 1841 the location of the magnetic pole of the Earth's Southern Hemisphere (75°05′ south latitude and 154°08′ east longitude) in Antarctica, when he passed 250 km from it. The South Magnetic Pole was first reached on January 15, 1909, by David, Mawson, and McKay from the expedition of Sir Ernest Shackleton. According to them, it was located at a point with coordinates 72°25′ S. and 155°16′ E. However, this is being questioned, because after a century the pole was at a point far from it. In 2015, the coordinates of the pole were 64.28° S. and 136.59° E., and it moved rather slowly, 10–15 km per year.

Naturally, the magnetic and geomagnetic poles were moving permanently. Where were they in ancient times? The answer to this question is given by palaeomagnetism. Some ancient geological rocks retained information about the direction and strength of the magnetic field during their formation. The processing of data from many rocks allows not only reconstructing the position of the poles and the magnitude of the magnetic field, but also the position of the rocks at the time of formation, which makes it possible to trace the drift of the continents.

What is the source of the Earth's magnetic field? Most likely these are electric currents in the liquid core of the Earth, caused by its rotation. The process of generating a field is called a dynamo mechanism and is similar to the mechanism of changing the solar magnetic field.

Generally speaking, generating a field requires a liquid core and fast rotation. This explains the weak magnetic fields of the Moon, Mercury (no or small core), and Venus (rotates slowly). The magnetic field of Mars is about 500 times less than that of the Earth

---

[i]You can see and download a picture of the pole movement and magnetic meridians, sometimes very different from the seemingly natural arcs of the great circle, on the site http://www.esa.int/spaceinimages/Images/2019/02/Magnetic_north_on_the_move.

due to the lack of differential rotation of its core. So, of the terrestrial planets, only the Earth can boast of a magnetic field. The gas giants have a sufficiently strong magnetic field. Jupiter's magnetic field is fourteen times stronger than Earth's. The magnetic fields of Uranus and Neptune are not dipole, so these planets formally have two north and two south magnetic poles each. Jupiter's moon, Ganymede, is the only known satellite that has an active dynamo process inside and creates a magnetic field.

3.4.2. *Magnetosphere*

The area where the Earth's magnetic field prevails is called the magnetosphere. Its shape and size are determined by the field parameters and the influence of the solar wind. A kind of tail appears near the Earth's magnetosphere due to the solar wind flow.

The inner magnetosphere extends somewhere around 6–7 Earth radii. Stars, planets, and satellites that have any significant magnetic field have their own magnetospheres. The magnetic field strongly affects the movement of electrons and ions in the plasma inside the magnetosphere.

The Earth's magnetosphere serves as a shield that protects us from charged particles of cosmic radiation. Once in the region of the Earth's magnetic field, they begin to rotate along helical lines around its lines of force.

But there is a weak point in this shield. This is the South Atlantic Magnetic Anomaly, which is sometimes divided into the Brazilian and Cape Town Anomalies. The magnetic field strength is less in this region and the protective properties of the magnetosphere are weakened. Therefore, flying over it, all spacecraft once suspended their work due to an increase in radiation from cosmic rays.

This applies to the Hubble Space Telescope and other scientific equipment. Additional radiation protection measures are being taken at the ISS. A number of satellites were lost as they flew over the area. In October 2012, the SpaceX CRS–1 Dragon spacecraft docked to the ISS had problems caused by radiation. Some laptops inside the shuttle spaceships were damaged when they flew through the area of the anomaly.

The story of the discovery of the South Atlantic Magnetic Anomaly is curious. In 1958, the United States conducted a secret test called Operation Argus, during which three nuclear explosions were carried out in space.[j] A lot of charged particles were created by the explosion. Analysis of their movement in the Earth's magnetic field showed that something was wrong with this field in the area of the explosions, coinciding with the region of the anomaly. This random selection of test sites has attracted increased interest in the magnetic field in the area, which soon led to the discovery of the anomaly.

Figure 3.12 shows the distribution of the intensity of the Earth's magnetic field over its surface according to the data of the International Association of Geomagnetism and Aeronomy (IAGA). There are four special regions on the surface. These are the two magnetic poles where the field is stronger and the blue-coloured region of the South Atlantic Magnetic Anomaly where it is weakest. The area in Siberia, located approximately antipodal to the site of the anomaly, also has a stronger magnetic field. The distribution

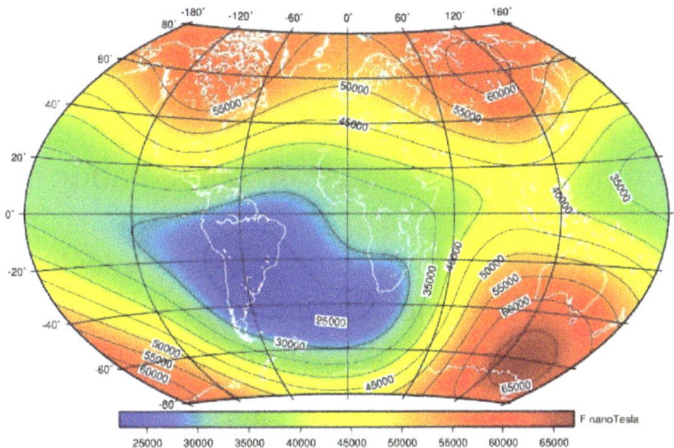

Figure 3.12. Distribution of the intensity of the geomagnetic field over the surface of the Earth according to IAGA data. Image provided courtesy of the British Geological Survey. Details could be found in Alken, P., Thébault, E., Beggan, C.D. *et al.* International Geomagnetic Reference Field: the thirteenth generation. Earth Planets Space 73, 49 (2021).

---

[j]The well-known scientist James Alfred Van Allen told in detail about the scientific objectives of the project in the declassified movie available on https://www.youtube.com/watch?v=QUM3PDVnk7M.

of other parameters can be viewed at https://geomag.bgs.ac.uk/research/modelling/IGRF.html.

In addition, our shield from time to time becomes completely permeable and cannot protect the Earth at all. This happens when the polarity of the Earth's magnetic field is reversed, so the north and south magnetic poles change places and the magnetic field weakens sharply. We are lucky on the Earth and the shield for terrestrial inhabitants disappears much less often than on the Sun.

According to palaeomagnetism, this polarity reversal or geomagnetic reversal occurs at intervals of 500 thousand to 50 million years. This process is not at all periodic, but rather random. The last polarity reversal took place 780 thousand years ago. In Fig. 3.11, in the left panel, you can see what the Earth's magnetic field looks like in the normal state, and in the right one, it is shown during the geomagnetic reversal.

3.5. Space Weather

Let us recall that, in addition to light and heat, the Sun also supplies the Earth with a stream of particles from flares and occasionally bombards it with CMEs, which reach us some time after the flight. What protects all life on the Earth from the destructive radiation of the Sun and, to a lesser extent, from the radiation of other space objects?

Electromagnetic radiation, namely X-rays and hard ultraviolet light, is absorbed in the atmosphere. Ozone and oxygen do it. The boundaries of the regions of absorption of ozone and oxygen are quite close. Not surprisingly, this is because their molecules are composed of different numbers of the same oxygen atoms, O_2 for oxygen and O_3 for ozone.

We are protected from the flow of charged particles by the Earth's magnetic field, which deflects these particles and forces them to move along the Earth's magnetic lines of force. In the polar regions, the magnetic lines of force enter the atmosphere. It slows down the charged particles orbiting these lines. These particles excite molecules of atmospheric gases, causing colourful auroras.

Nature does not protect us against the harmful effects of solar radiation on the Moon and Mars, where there is no magnetic field and there is no or almost no atmosphere, as well as inside space-ships during interplanetary flights. Here, the issues of protection from radiation must be solved by a person. But on the Earth, one should not forget about the influence of the described factors, which are now being combined into the concept of space weather.

Space weather is a branch of space physics and heliophysics concerned with the time-varying conditions within the Solar System, including the solar wind, with a focus on the space surrounding the Earth, including conditions in the magnetosphere, ionosphere, thermosphere, and exosphere. It takes into account the impact of all factors associated with space, both permanent and emerging from time to time, such as CMEs. After the CME arrives on the Earth, a geomagnetic storm can occur. A violent storm can cause serious trouble.

Even simple geomagnetic storms are troublesome. They contribute to the interruption of radio communications in high latitudes, cause increased corrosion of pipelines, interfere with the readings of GPS navigators, and prevent accurate positioning when drilling wells, which is necessary for the production of shale oil and gas. But a particularly strong one can cause a real disaster.

3.5.1. *Carrington superstorm of 1859*

A classic example of a strong magnetic storm was the events of 1859, called the Carrington-Hodgson event or the Carrington storm. British astronomers Richard C. Carrington and Richard Hodgson on September 1 observed a solar flare so strong that it could be seen without any telescope. The flare caused large coronal mass ejections. They rushed to the Earth and reached it after 18 hours, and three or more CMEs came to the Earth in succession.

As a result, the largest known geomagnetic storm occurred on Earth. It was not by chance that I wrote "known". It is clear that over the billions of years of the existence of the Earth, storms and stronger ones have happened on it. But since the beginning of regular observations of solar flares and the appearance of magnetometer instruments that measure the Earth's magnetic field with great accuracy, this storm has been a record one.

Auroras were observed even above the equator, and they were bright. Some people in the northern part of the United States (not Alaska) have managed to read newspapers in the light of the northern lights. Gold miners in the Rocky Mountains, waking up at night from the light, decided that it was dawn and started preparing breakfast. Telegraph wires have become a source of induced EMF. Telegraphers on the New York–Boston line disconnected the batteries and transmitted telegrams for several hours using this free voltage.

Many telegraph lines were simply turned off, and some were burned down. For many years, it was believed that the frequency of geomagnetic storms of comparable strength is one storm in 500 years.

What if the Carrington storm happened now? Since 1859, humanity has had much more sophisticated equipment and infrastructure, especially vulnerable to geomagnetic storms. Weaker storms have repeatedly caused failure of energy generation and transmission systems, for example, in Quebec in 1989.

A Carrington-level storm would simply burn many transformers on Earth. Pilots and passengers on planes high in the air would receive a large dose of radiation. Computer networks and the Internet could also be affected.

The consequences of a superstorm now would be much more serious than 150 years ago. This is an inevitable price to pay for progress. After all, we are increasingly dependent on technology and its infrastructure, which means we are more vulnerable to its failures. Some insurance companies have estimated losses from superstorms, but US government experts have concluded that they will be so great that there are not adequate methods for estimating losses, both human and material.

In terms of catastrophic consequences, a Carrington-class superstorm is comparable to a world thermonuclear war, but much more likely. In fact, the question is not whether there will be a superstorm, but how soon it will happen.

We almost checked the apocalyptic scenario when, on July 23, 2012, a CME flew past Earth, capable of causing a storm like Carrington's. NASA only reported this event on April 28, 2014. After that, the estimates of the likelihood of a catastrophic event were revised. It is now believed by some scientists that the probability of a superstorm in the next decade is about 10%.

The reader may think that his or her guide, that is, I, is exaggerating, like many guides in different countries of the world. So I am ready to tell you in more detail about the disaster that did not take place in 2012.[k] People learned about it later and a little by accident, because no special flares on the Sun were observed on the Earth and a powerful CME was not noticed.

But at that time, two spaceships were in their orbits to observe the Sun. The NASA space mission STEREO (Solar Terrestrial Relations Observatory), launched in 2006, served science for a long time, supplying information, for example, snapshots of the Sun's photosphere. Contact with the STEREO-B spaceship was lost in 2014, but STEREO-A is still in operation.

STEREO-A equipment was turned on in July 2012 for completely different observations and suddenly scientists found that the devices went off scale, showing something incredible. At first, this was attributed to a failure, but, as it turned out, apparatus recorded the passage of two CMEs with an interval of 10–15 minutes. If they had been launched 9 days earlier, they could hit the Earth.

Judging by the speed of the solar wind recorded during the flight, which, according to various estimates, was from 2000 to 3500 km/s (about one-hundredth of the speed of light), the arrival of the CME on the Earth would lead to a storm no less strong than Carrington's. For the latter, measuring the speed of the solar wind was impossible, but modern estimates suggest that it was about 2000 km/s.

---

[k]Further details can be found on the NASA website at https://science.nasa.gov/science-news/science-at-nasa/2014/23jul_superstorm under "Near Miss: The Solar Superstorm of July 2012".

If these CMEs staged a superstorm, then, most likely, some of the pilots and passengers of planes in flight would quickly die from radiation and power transmission networks would be disconnected, affecting all electrical appliances, water supply, and other infrastructure facilities, communications, and much more that distinguish our civilization from that of 200 years ago. Everyone can fantasize about the further course of events.

"If it had hit, we would still be picking up the pieces," says Daniel Baker of the University of Colorado, who published a detailed study of the event. "In my view the July 2012 storm was in all respects at least as strong as the 1859 Carrington event," he added. "The only difference is, it missed."

However, one can disagree with the estimate of the likelihood of such a superstorm on Earth. Scientists' opinions and assessments differ greatly. I have already cited Baker's estimate, but in a 2019 article, Spanish scientists concluded that he overestimated the probability ten times. I will not take sides in this dispute, just note that it is extremely difficult to assess the likelihood of any very rare event.

To complete the story about the influence of the Sun on life on Earth more optimistically, we note that in the United States and a number of other countries, structures are being created to monitor space weather and make its predictions. Governments develop superstorm action plans and measures to reduce the damage caused. Just as houses are designed with the expectation of an earthquake, infrastructure facilities must be designed to limit losses from geomagnetic storms.

On the other hand, it is easy to imagine that, in the future, a network of satellite- and ground-based observations and measurements will not allow any CME to fly up to the Earth unexpectedly, but it is difficult to figure out how we could fight the emergence of a superstorm, even if we were warned in advance of the possibility of its development. As you know, the Latin proverb states, *Praemonitus, praemunitus*, or "Forewarned is forearmed".

3.6. The Moon, Our Satellite

Formally, the Moon does not belong to the globe, as I explain in the second chapter of the book. However, other satellites are good, but we have our own one, illuminating the Earth at night and influencing the sublunary terrestrial world. It is also the only celestial body that humanity has visited. So, in this chapter I talk about the Moon, including where it came from.

3.6.1. *Moon motion*

I start with general information. The Moon orbits the Earth in an elliptical orbit with a major semiaxis of 384,399 km and a small eccentricity equal to 0.055. The orbit is close to a circle and the distance to the Moon does not change much, from 405,000 km at apogee to 363,000 km at perigee. So, the apparent size of the Moon changes insignificantly, which does not prevent the tabloid press from periodically frightening us with an allegedly terrible "Super Moon".

The plane of the Moon's orbit is close to the ecliptic, i.e. to the plane of the Earth's orbit, but does not coincide with it. The period of the Moon's revolution is approximately equal to one month, more precisely it lasts 27 d 7 h 43 min 11.5 s. In fact, this is the reason for the appearance of this unit of measurement of periods of time. The mass of the Moon is 81 times less than that of the Earth.

The Moon rotates on its axis for the same period equal to a month; therefore, it is always directed to the Earth with one side and hides the other, called the dark side. If you do not quite understand why this is happening, I can quote a piece from Jules Verne's book *From the Earth to the Moon: A Direct Route in 97 Hours, 20 Minutes*: "Go into your dining-room, and walk round the table in such a way as always to keep your face turned towards the centre; by the time you will have achieved one complete round you will have completed one turn round yourself, since your eye will have traversed successively every point of the room. Well, then, the room is the heavens, the table is the Earth, and the Moon is yourself."

Note that the equality of the periods of rotation and revolution around the Earth is not at all accidental. However, a very long time ago, these periods could be different, and from the Earth, it was possible to observe different sides of the Moon. But later they became equal due to a phenomenon that we will discuss a little further. It works very effectively, and not only for the Earth–Moon system. All large satellites of Jupiter are also always turned to it with one side.

The plane of the Moon's orbit rotates with a period of 18.6 years and its perigee, in turn, rotates with a period of 8.85 years. Both effects are caused by the influence of the gravitational field of the most massive object in our planetary system — the Sun. The second one was discovered in the second century BC by the great Greek astronomer Hipparchus of Nicea ("Ἵππαρχος).

It's time to be surprised. When calculating the motion of the Earth and Moon around their centre of mass, astronomers consider the influence of the huge Sun as a less significant factor compared to the attraction of the Earth and the Moon. It is associated with the impact of solar tidal forces, which quickly fall with distance from their source, inversely proportional to the cube of the distance to it.

3.6.2. *Moon phases, solar and lunar eclipses*

The Moon shines by reflecting sunlight and does it very badly. The proportion of reflected light is only one-eighth (for comparison, the Earth has more than a third), and it is directed in all directions. So at night you cannot read a newspaper in the light of the Moon, even on a full moon. By the way, what is it? The Moon is illuminated only from the side where the Sun is. During a full moon, the Earth is between the Sun and the Moon, close to the straight line connecting them. Therefore, the entire visible side of the Moon is illuminated and we see a bright lunar disc in the sky.

Moreover, if the Earth is very close to the Sun–Moon line, then its shadow would fall on the Moon and it ceases to shine, because the light of the Sun is blocked and there is nothing to reflect. And people on half of the Earth's surface who can see our satellite at this moment witness a lunar eclipse. So lunar eclipses are only observed on a full moon.

In about a week, the Moon will move to a quarter of its orbit, and the Sun to 1/52 of its orbit (there are 52 weeks in a year). Let's wait a little longer and the angle between the directions from the centre of the Earth towards the Moon and the Sun will reach exactly 90°. Congratulations, the Moon is now in the third (or last) quarter and half of the circular lunar disc is shining in the sky. It is the left one, if you look from the Northern Hemisphere. The Moon in the third quarter is visible in the morning.

In another week, the Sun, Moon, and Earth will again be approximately on the same straight line, but the Moon will be between the Earth and the Sun, illuminating its dark side. Therefore, we will not see the Moon and will say that the new moon has come. Before him, in the sky, only a narrow crescent of the waning crescent Moon or the "old" moon in the shape of the letter c will be visible. You can also see a narrow waxing crescent Moon or young "growing" moon after the new moon in the shape of the arch of the letter p.

Sometimes, during the new moon, the shadow from the Moon falls on the Earth. This shadow does not cover the entire Earth's surface, since the Moon is much smaller than the Earth, but the shadow moves along the surface as a dark spot. The moment this spot passes through a certain place, people in this place can observe a solar eclipse. In the very middle of the spot, it is a total or annular eclipse, and further from the centre, it is a partial eclipse. In the more distant regions, no eclipse is visible. So a solar eclipse is a spectacle not for all.

Astronomers calculate in advance where a solar eclipse is expected and arrive at a place convenient for observation with the necessary equipment. And the richest hire a plane that accompanies the path of the lunar shadow on the surface, gaining an opportunity to observe a solar eclipse not for a couple of minutes, but for a couple of hours.

A week after the new moon, the first quarter will come and once again we will see the illuminated half of the disc in the sky. But it will happen in the evening and the right side of the disc will be illuminated. In another week, the next full moon will come and everything will return to where I started. However, if you observe the phases of the moon from the Southern Hemisphere, say from Melbourne, then feel free to change right to left in this description and vice versa.

3.6.3. *Ebb and flow*

The Moon is the source of not only light, but also the gravitational field. Since it is not very massive, but is located close to the Earth; it manifests itself primarily in the form of tidal forces, which cause the ebb and flow. Tidal forces decrease in inverse proportion to the cube of the distance to their source and are proportional to the mass of the body that causes them. The corresponding formula can be found in textbooks and reference books, but it can be obtained at the elementary level if you remember that the tidal acceleration is the difference in the free fall accelerations for two close points, and that the free fall acceleration decreases as the inverse square of the distance.

Because of the rapid drop in tidal forces with distance, lunar tides on Earth are stronger than solar tides. Indeed, the Sun weighs $2.0 \cdot 10^{30}$ kg and is located at a distance of $1.5 \cdot 10^8$ km from the Earth. The Moon weighs $7.3 \cdot 10^{22}$ kg and is at a distance of $3.8 \cdot 10^5$ km. Thus, the Sun is $2.7 \cdot 10^7$ times heavier and 395 times farther than the Moon. If we cube the ratio of distances, we get $6.2 \cdot 10^7$, which is 2.2 times more than the ratio of masses. So, lunar tides are 2.2 times stronger than solar tides.

Tidal forces raise the water level on the line connecting the centre of the Earth with the massive body that is their source. The uplift occurs both at the site directed towards the body and in the opposite one. The water level drops slightly in perpendicular directions, as shown in Fig. 1.4.

The tides caused by the action of the Moon and the Sun act independently of each other. Sometimes their effect is added and the tides become especially high, sometimes the solar tide is subtracted from the lunar tide. The first option is realized on a full moon or new moon, when the Sun, the Moon, and the Earth are located approximately on the same straight line and the tidal uplifts caused by the Sun and the Moon coincide. Astronomers call this arrangement of the three celestial bodies a *syzygy*, from Ancient Greek σύζυγος. It causes the bimonthly phenomena of spring tides when lunar and solar tidal forces act together to reinforce each other. During this time period, the ocean or the sea level rises higher and falls lower than the average.

At the first and third quarter, when the directions from the Earth to the Sun and the Moon are perpendicular (this arrangement is called quadrature), their tidal forces counteract each other, and the tidal range is smaller than average. This is time for neap tides. A month passes from one new moon to another, so the time interval between two successive spring tides (or neap ones) is approximately 2 weeks, and a week passes from spring to neap tides, and vice versa.

Due to the rotation of the Earth, areas of high and low water levels move across our planet. And a person rotating with the Earth observes that a complex change of ebb and flow occurs in one place. Sailors, fishermen, and coastal residents are perfectly familiar with it.

Similar ebbs and flows exist in the Earth's crust. It is solid, so deformation, i.e. changes in shape and size, are much less noticeable. However, they are not only known to scientists, but have also been studied and measured since the middle of the 19th century. Some enthusiasts regularly propose hypotheses about the relationship of earthquakes with the phases and position of the Moon, but when checked, they invariably turn out to be incorrect, even statistically.

3.6.4. *The Moon is moving away from the Earth, slowing down its rotation*

A physicist or astronomer can see behind all this routine a slowly but surely operating mechanism for stealing something from the Earth. The tidal force slows down the rotation of our planet, so it steals angular momentum. Meanwhile, the Moon moves away from the Earth.

How does this robbery happen? In Fig. 1.4, a not quite correct but simplified picture is drawn. Tidal humps in it are located right on the line connecting the centres of the Earth and the Moon. But the Earth rotates together with the water in the oceans and seas. Therefore, the location of the tidal humps is slightly turned in the direction of the Earth's rotation relative to the direction to the Moon. The shift is associated with the friction of water against the surface of our planet. The force of attraction of these humps to the Moon provides the moment of force, which reduces the angular momentum of the Earth and transfers it to the Moon. It is the result of gravity attraction between the Moon and the tidal hump.

Accurate time standards claim that the length of the day increases by 23 microseconds per year. That is, the rotation of the Earth is slowing down, albeit very slowly. And the Moon is just as slowly moving away from the Earth. The distance between them, averaged over the period of the Moon's revolution, increases by 38 mm over the year. This is clearly only a little on an astronomical scale. But over billions of years, the small changes could become very large, so they should not be treated with disdain.

These changes were predicted at the end of the 19th century by the English astronomer George Howard Darwin, the son of the famous Charles Darwin. Naturally, the calculation of the angular momentum decelerating the planet is very complicated, and it depends not only on the change in the distance from the Earth to the Moon over time, but also on the configuration of the continents on the Earth.

However, it is not necessary to calculate in order to answer many questions. It is enough to use the law of conservation of angular momentum in the Earth–Moon system. Naturally, we neglect its possible change due to the effect of some external factor, for example, the tidal forces of the Sun or planets. We also do not take into account the evolution of the Sun, which, in about 7.7 billion years, will turn into a red giant, significantly increasing its size and, possibly, absorbing both the Earth and the Moon. Consider a simple model with only the Earth and the Moon.

If we calculate the angular momentum of the Earth–Moon system relative to the centre of the Earth, it includes two terms. The first is associated with the Earth's rotation, and the second with the movement of the Moon in its orbit. The described effect, which can be called a lunar brake, leads to a redistribution of the angular momentum between these terms, but their sum is constant.

Now the angular momentum of the Moon's motion in its orbit is four times more than the one associated with the rotation of the Earth. Consider the motion of the Moon in an almost circular orbit of radius r. I wrote almost because this radius is slowly increasing, so the trajectory of the Moon is an expanding spiral very close to a circle. It can be calculated that if the radius of the Moon's orbit changes, then with constant masses of the Earth and the Moon, the angular momentum of the Moon is proportional to $\sqrt{r}$. Over time, the angular momentum of the Earth decreases and the Earth rotates more and more slowly, but the orbital angular momentum of the Moon increases. The latter requires an increase in the distance between the Moon and the Earth.

An interesting confirmation of this theory was provided by the study of fossil corals with traces of the diurnal and monthly cycles of changing illumination on them. Scientists were able to find out how many times the length of the month exceeded the length of the day at the time when these corals grew. It turned out that for very old corals, both the month and the day were less than now. Therefore, a couple of billion years ago, the Moon was closer to the Earth. Note that this means that in the very distant past, for example, when animals left the ocean to live on land, the height of the tides was greater than now. This can be important for biologists and palaeontologists.

3.6.5. *The origin of the Moon*

George Darwin even suggested that the Moon was torn away from the still liquid Earth by centrifugal forces. Let's test this hypothesis. At the time when this event was supposed to occur, almost the entire angular momentum of the Earth–Moon system fell on the rotation of the Earth, because its mass is 81 times greater than that of the Moon, and the latter was very close to the planet. As we have already estimated, it was then approximately five times more than the angular momentum of the Earth now, because the total angular momentum of the system is conserved.

And since the size and mass of the Earth did not change much, this gives rise to the conclusion that the Earth then rotated almost five times faster than it now does. The day at that time was five times less than now and lasted about 5 hours.

Okay, but would that rotation be enough for a piece to come off the planet? It is easy to see that it is not. If we put a stone somewhere on the equator, then it will be able to fly into space only if the speed of its rotation exceeds the first cosmic speed. When a satellite moves in a particularly low orbit, i.e. it is with this minimum speed that it makes a complete revolution around the Earth in about one and a half to two hours. And the speed of even the rapid rotation of the young Earth would provide one revolution in 5 hours, which is not enough to tear off part of the Earth's matter.

Here is the first conclusion: the Moon was not torn off from the Earth by centrifugal forces. In our time, the hypothesis that the Moon broke away from the Earth has again become relevant, but with an important addition. The Moon did not come off, but was knocked out of the Earth during a gigantic collision with another planet, which was named after the Titanide Theia, mother of Helios, Eos, and the goddess of the Moon Selena. Theia's blow fell almost tangentially along the edge of the Earth and was able to tear off a piece, from which our satellite was formed. This hypothesis is supported by the results of geological and isotopic studies of lunar rocks.

It is clear that the collision hypothesis is just one of many, including very strange hypotheses of the formation of the Moon. The option of a centrifugal separation can be ruled out after the simplest estimates. But there are also hypotheses of the joint formation of the Earth and the Moon, the capture by the Earth of the Moon, which supposedly was once a minor planet, the merger of several smaller satellites into one large one, and many more exotic options.

3.6.6. *The motion of the Moon and the rotation of the Earth in the distant future*

So, with the help of simple estimates, we were able to shed light on the beginning of the process of increasing the radius of the Moon's orbit, in particular, to find the length of the day at this time. Now let's figure out how it will end. The angular velocities of the Moon's revolution and the Earth's rotation eventually will be equal to each other and the Moon will always be above a certain side of the Earth, as now the Moon itself is always turned to the Earth with one of its sides. The day and month will be equal in duration.

Let's estimate the length of a month in those distant times. This can be done from the same angular momentum. If the angular velocity of the Earth's rotation would be equal to the angular velocity of the Moon's rotation, then a negligible fraction of the total angular momentum of the system would be associated with the rotation of the planet. Almost all of the total angular momentum will be caused by the motion of the Moon. So, in comparison with its current situation, the Moon's angular momentum will have to increase by a quarter.

Since this value is proportional to $r^{1/2}$, the average radius of the Moon's orbit should increase by $1.25^2 \approx 1.5$ times. Moreover, according to Kepler's third law, the period of the Moon's revolution, i.e. the length of the month (and the length of the day which will be equal to it) will increase in comparison with the current month by $1.25^3 \approx 2$ times. So, in those distant times, a month and a day will last for about two modern months.

Note that the "tidal brake" in the past successfully affected the Moon as well. It was the tidal force that has led to the current situation when the period of rotation of the Moon is equal to the period of its revolution and our satellite is always turned to the Earth on one side. This equality will be maintained in spite of the increase in the length of the month due to the theft of the angular momentum from the Earth, as described here.

3.6.7. *Colonization of the Moon*

Let us now try to look into the not so distant future. How about colonizing the Moon? There are several factors preventing this from happening. This is the absence of an atmosphere and magnetosphere to protect against cosmic rays. Most likely, solid and deep underground shelters will be needed for all living things. Gigantic temperature differences between the lunar day and night, each lasting two Earth weeks, also require the transfer of all mechanisms deep below the surface.

But there is water on the Moon. In the fall of 2020, its discovery was announced with the help of SOFIA (Stratospheric Observatory for Infrared Astronomy). This is the only observatory on board an aircraft which allows scientists to take pictures with almost cosmic quality. It was created by NASA and the German Centre for Aviation and Astronautics. It found about 300 grams of water per cubic metre of soil on the surface of the Claudius crater illuminated by the Sun.

Oxygen and hydrogen can be extracted from water. Solar panels can be used to generate energy. Food can be grown on the spot.

I don't know about mining, but there are places on the Moon that scientists dream of getting to. It is a crevasse that the Sun's rays have never looked into, and where frozen samples of the original material from which the Solar System was formed have been stored for possibly billions of years.

But there is a place right in the centre of the dark side of the Moon, where you can put a radio telescope, so it is protected from radio interference from Earth caused by radio, television, and other transmissions by the entire mass of the Moon. Moreover, lunar rocks, called regolith, contain many metal particles, meteorites, so they have good electrical conductivity and, for this reason, can screen radio signals well.

Engineers dream of installing solar panels on one lunar hillock always illuminated by the Sun. So, space explorers from different countries know which areas they would like to capture in the first place and for what.

3.6.8. *Helium-3 and lunar mining*

Extraterrestrial mining is often mentioned by people. Since the Moon is the closest celestial body, they suggest starting with lunar mines. What minerals could be mined there to be transported to the Earth? There are many there, but it is unprofitable economically. So, for now, the talks are limited to the mining of the isotope helium-3.

I am sure that most of the readers have guessed that this is the isotope of helium, ^{3}He, whose nucleus consists of two protons and a neutron. It is stable, but it is much less common than the isotope ^{4}He, which has two neutrons. In the Earth's atmosphere, the proportion of ^{3}He is 1.37 ppm relative to ^{4}He. The physical properties of these isotopes are noticeably different. The critical point of ^{3}He is observed at a temperature of 3.35 K (5.19 K for ^{4}He), at atmospheric pressure it boils at a temperature of 3.19 K (helium-4 at 4.23 K). The density of liquid ^{4}He is 124.73 g/l is twice that of 59 g/l for ^{3}He.

Like any primordial element, helium has existed on the Earth since its formation. As a noble gas, it could not be chemically bound in any compounds, but it can exist inside the Earth's rocks. Until now, the Earth's mantle excretes a little helium, including in the composition of natural gas and during volcanic eruptions. A little ^{3}He enters the atmosphere from space or is formed after interaction with cosmic rays. But much more ^{3}He leaves the atmosphere, escaping into space. It can be estimated that the entire atmosphere of the Earth contains about 3800 tons of this isotope. Production of ^{3}He is very low. It is also produced during the radioactive decay of tritium and as a by-product of some nuclear reactions. It is quite expensive; a litre of liquid ^{3}He costs about \$1000.

It is mainly used for cryogenic purposes, scientific experiments, and neutron radiation detectors that employ the reaction $n + {}^3\text{He} \rightarrow {}^3\text{H} + p + 0.764\,\text{MeV}$. This isotope is indispensable for obtaining ultralow temperatures. Its use in medicine began. The use of ^{3}H for controlled thermonuclear fusion is being actively discussed.

It is assumed that the ^{3}H isotope is formed inside the lunar regolith under the influence of the solar wind and cosmic rays, so it can be mined from the lunar soil. Unfortunately, it is estimated that more than 150 tons of regolith must be processed to produce one gram of ^3H. So the stories about the production of ^3H on the Moon are more of a fascinating idea than something really expected in the coming decades.

4 THE SCALE OF DIVERSITY: THE WORLD AROUND US

We continue our journey. The road expands, forks, winds; the path along it must be chosen regularly. This is understandable: we are approaching our human scale. Roughly speaking, this is what is smaller than a continent and larger than a grain of salt or pepper. What to look at, what to write about?

Since this is the scale on which humanity lives, we know a lot here and a lot is interesting to us. Physics, since its inception, has precisely studied the phenomena of these dimensions, because even the simplest telescopes and microscopes have not yet been invented. Even an ordinary magnifying glass was inaccessible, to say nothing of a microscope.

The Universe was the abode of gods and monsters, for many people the neighbouring town was its edge, and if the stars were visible in the sky, then nobody knew how far these strange lanterns were hung up. But there was rain, lightning, a rainbow, fire; water boiled in the cauldron over the fire; everything was moving around; in temperate latitudes, winter followed summer, and vice versa. Beautiful, strange crystals were found under the ground, and pieces of amber attracted any specks of dust if rubbed with a cat's skin.

As time went on, the possibilities of science grew. But this scale remained the scale of the laboratory, the scale of the nature around us, of what we directly saw, heard, smelled, held in our hands, could touch or climb to the top.

It's also the scale of diversity. We live in a world filled with very different objects, and each baby must solve an incredibly difficult task: to learn to distinguish between hundreds and thousands of different objects, between living and non-living things.

Idioms related to this distinction can be found in all languages. An Englishman would compare chalk and cheese, a Russian would remember God's gift and scrambled eggs, and a Pole would ask, *"Co ma piernik do wiatraka?"* Let's agree with the Poles. Indeed, *piernik* (a gingerbread) is not *wiatrak* (a mill); it is difficult to confuse these so different objects. Even the most absent-minded person will be convinced of this by trying to drink tea with a mouthful of a mill. And all these objects are on our scale.

The wealth of choice does not arise by chance, despite taking into account the fact that we must confine ourselves to earthly phenomena. We have one Universe. There are many galaxies and stars, but it would be strange to write about the Tau Ceti star, the Betelgeuse star system and the planet Uranus instead of the Sun, Solar System, and Earth. We are not that cosmopolitan. But still the choice is huge.

On a small scale, there is also a choice, but one must understand that there are fewer types of atoms of chemical elements and elementary particles in nature than there are species of insects in one beet field. Diversity also extends to the sciences related to the comprehension of the world. The microcosm is the world of pure physics. Even the chemistry of the microcosm does not exist, not to mention biology or geology. It is impossible to imagine the economics or anthropology of the microworld.

The Universe as a whole is studied by physics, as well as astronomy and cosmology, which are also based on physics. But unlike the microcosm, we can also talk about chemistry, for example, about how chemical elements are formed or spread across galaxies. There are elements of biology: we can search for extraterrestrial life, including intelligent life, in the depths of space.

But even the craziest astronomer has no hope of ever getting to know an intelligent galaxy, or at least a living galaxy. However, the spectacle of a galaxy wagging its tail as the neighbouring one approaches would not leave any astronomer indifferent.

For centuries, scientists have tried to explore everything that can be investigated and reveal as many secrets of the Universe as possible. But not only scientists have curiosity. Physics and other sciences are so connected with the life of an ordinary person that it is difficult to do without them, even if science and technology are not interesting to you at all.

Life sometimes reminds us of this when the electricity suddenly goes out in the house. It's hard not to react, especially on a late winter evening, when your favourite TV series or an interesting football match is on TV, and you additionally have to work on your computer and send a couple of emails. And to top it off, you didn't make it to the house, but got stuck in a stopped elevator.

Many phenomena, processes and events deserve attention and a place in the book. But alas, our journey is like taking a tourist bus. We drove along a boring freeway, lazily observing the surrounding landscapes, and entered the city of Paris, with its buildings, people, museums, cabarets, restaurants, boutiques, monuments, catacombs, memories of past events, and current everyday life around us. And there are only a couple of hours to inspect all this. Tourists would like to have time to look at the most interesting; the rest will be left for later.

So this chapter, devoted to the scale of our life, cannot cover even a percentage of what could be written about. One has to choose. And I, as your guide, will take you not throughout the museum of nature, but directly to the hall with the inscription "Electricity, magnetism, optics".

Why exactly in there? Yes, simply because if you don't visit this hall, then insufficient attention will be undeservedly paid to the electromagnetic interaction. And I will be ashamed before the computer on which I am typing this text now, before the outlet into which the plug of this computer is stuck, before the power station that provides voltage to the outlet, before each transformer in the path of electric current, and of course, before many wonderful scientists and engineers who have turned a piece of amber that attracts specks of dust into all of the above.

Finally, I will tell you about the very basics of kitchen physics related to heating and heat transfer. First, we will look at the physics of boiling water in a kettle, and then we will add scientific notes to the TV competition of crazy potato cooking.

4.1. A Magnet and Its Cronies

Why am I beginning my story with the magnetic properties of substances? After all, this is the second part of the word *electromagnetism*. In addition, an electric charge exists, and magnetic charges have not yet been discovered. But there are two well-known things that put magnetism first in our excursion: a permanent magnet, at least a souvenir fridge magnet, and a compass, which is the best friend of sailors and tourists. Each of them has its counterparts in the world of electric charges, but you are hardly familiar with them. So let's start with familiar things, gradually moving on to more exotic ones.

There have been many prejudices and superstitions regarding magnets for a long time. Some of them are fun to mention, and I will tell you about them, but not in this section, where scientific results are discussed, but in Section 8.2, dedicated to the emergence of modern science. It includes the story of William Gilbert's discovery of the simplest properties of a magnet and an explanation of the basics of the Earth's magnetism, which was discussed in Section 3.4. Now it's time to take a closer look at what we know about magnets and magnetism today.

4.1.1. *Compass*

But first, let's give credit to the compass, which has faithfully served humanity for centuries. Albert Einstein often recalled how his father showed him the compass. "I can still remember — or at least believe I can remember — that this experience made a deep and lasting impression upon me. Something deeply hidden had to be behind things."

Invented in China over 2000 years ago during the Han Dynasty, it was initially used not for navigation, but for divination and magical rites. Only during the Song dynasty, in the 11th century or a little earlier, it began to be used to indicate the direction of movement in the deserts. One of the ancient Chinese wrote that "the seers rub the end of the needle with a magnetic stone so that it points to the south".

Sailors began to use it at the very beginning of the 12th century. The nautical compass appeared in Europe in the 12th and 13th centuries. In European literature, the first mentions of the compass are found in the works of the Englishman Alexander Neckam, *De Untensilibus* and *De Rerum*. In none of them did he describe the compass as a novelty; at that time, it was already quite common.

The main thing in a compass is a magnetic needle, that is, a small magnet that can rotate freely on the needle. Everyone knows that a magnet reacts to the presence of another magnet. It attracts, repels, or turns. The second magnet is the Earth, which creates a geomagnetic field. It makes the compass needle turn so that it is oriented along the direction of the external magnetic field. Usually, this direction is along the magnetic meridian, but the compass can be deceived by placing a magnet or an iron object next to it.[a]

Note that the Earth's magnetic field has not only a horizontal but also a vertical component. The angle that the direction of the field makes with the horizon is called magnetic inclination. To prevent the compass needle from skewing because of this, the arrow is made slightly asymmetrical so that the two skews compensate each other.

Since the vertical component of the geomagnetic field has different signs in different hemispheres of the Earth, a compass that works perfectly in Reykjavik can skew the needle in New Zealand. It is said that companies that produce compasses for the whole world divide the Earth into five zones — the equator, two

---

[a]Jules Verne has used this idea in his novel *Dick Sand, A Captain at Fifteen*, published in 1878. In the areas of magnetic anomalies, nature has "planted" rich deposits of iron ores that deflect the compass needle from the magnetic meridian.

circumpolar zones, and two mid-latitude zones — and make compasses for each of these zones.

From the point of view of physics, the compass needle acts as a magnetic dipole. If it can rotate freely, the dipole always orientates in an external magnetic field so that its direction coincides with one of the fields in the place where it is located. And since it is directed along the elongated and symmetrical needle, its orientation shows the direction of the magnetic field.

4.1.2. *Magnetic dipole moment*

It's time to explain the new words and concepts that we used in the previous paragraph. A dipole is an important element of physics, especially in electromagnetism. There is an electric dipole, for example, a pair of closely spaced positive and negative charges of equal magnitude. Such an electric dipole is characterized by its electric dipole moment. The direction of this moment is the direction from negative charge to positive, and the modulus is equal to the product of the modulus of charges by the distance between them.

Any body or particle that has a nonzero magnetic dipole moment can be considered a magnet, large or small. Such a magnet in an external magnetic field tries to turn so that its moment is directed along the field. A force acts on a loose magnet, which tries to move it to where the magnetic field is stronger. Place two strong magnets next to each other and they will get as close as possible.

But the magnet can be fixed or held in any position. And the force acting on the fixed magnet can be directed anywhere. Try to bring two magnets with the same poles closer to each other with your hands and clearly feel the force of their mutual repulsion.

There are no magnetic charges in nature, but there are analogues of them located at the poles of a magnet, including at the ends of the compass needle. They are formed due to the magnetic properties of the substance from which the arrow or magnet is made, usually iron.

In the theory of magnetism, the reason for their appearance is sometimes called as microcurrents or molecular currents. This name not only explains nothing, but also misleads, because no currents flow either in the magnet or in the compass needle.

In the so-called Gilbert model a small magnet is modelled by a pair of magnetic poles of equal magnitude but opposite polarity. We call the poles of a magnet the place where these mock magnetic charges, "north" and "south", are concentrated. The magnetic dipole moment is directed from south to north pole. Its magnitude is the product of the so-called strength of its poles and the distance separating them.

Each magnet has at least two poles, but there may be more. The sum of these ersatz magnetic charges (strengths of poles) is zero. If the magnet is broken in half, then new poles appear in the place of the break, and each half will have all the properties of a magnet of excellent behaviour: at least two poles and zero total magnetic charge.

Ordinary electric current has a magnetic moment (the adjective "dipole" is often skipped). If you make a ring, turn or loop from a wire and pass a direct current through it, then this current has a magnetic moment directed perpendicular to the plane of the turn. This was shown by André-Marie Ampère.

So, the compass needle is a magnetic dipole. This makes the compass the simplest device that once helped physicists understand what a magnetic field is. I mentioned this in Section 2.3. In the same place, I gave an example of embarrassment with the definition of a magnetic field and a permanent magnet in a certain reference book on physics. After talking about the dipoles, I can continue to discuss it.

The magnetic field manifests itself as an additional force acting on moving electrically charged bodies when this field is present. This force is called the Lorentz force and its formula adorns any textbook on electromagnetism. The field arises when electric charges move and because of the magnetic properties of substances, which ultimately boils down to the fact that atoms, molecules, and elementary particles have their own magnetic moments, that is, they are somewhat similar to small magnets.

Electrons, protons, and even electrically neutral neutrons have their own magnetic dipole moment. The magnetic moments of nuclei, atoms, and molecules are the sum of both the moments of individual particles that form these more complex objects, and the moments caused by the movement of charged particles, such as the motion of electrons around nuclei.

Let me remind you that the magnetic moment is a vector quantity. Therefore, if its directions of two particles are opposite, they could be mutually compensated when added, and the system consisting of this pair would have a zero magnetic moment. We are primarily interested in only one property of the total magnetic moments of molecules: whether they are equal to zero or not. It is the answer to this question that determines the magnetic properties of the substance.

Let's move on from the simplest reasoning to practice. If we bring a small magnet closer to different objects, we quickly realize that it is attracted to some, but not to others, such as wood, stone, copper, or plastic. Let's leave the first ones for last and start by taking a closer look at those substances that do not seem to react to a magnet and a magnetic field. And here is the first surprise: all substances, except vacuum, react to the magnetic field. But the force arising from the action of a magnetic field on most substances is very small and we simply do not notice it. However, this force reminds of itself in a very strong magnetic field.

4.1.3. *Diamagnets*

If the molecules of some substance have zero magnetic moment, then according to its magnetic properties it belongs to diamagnets. There are many of them in the world around us: water, nitrogen, bismuth, table salt, all noble gases from helium to radon, gold, silver, copper, mercury, hydrogen, carbon dioxide, silicon, zinc, and many other substances. All of them have their own magnetic field, which slightly weakens the external field. It is very weak: record magnetic field weakening does not exceed 0.1%, while for nitrogen it is only 0.003%. The percentage is clearly small, but even a small percentage of profit or loss from a huge amount of money can

Figure 4.1.   Levitation of diamagnets. Photo of a plate of pyrolytic carbon hovering over a strong neodymium magnet and a frog in a strong magnetic field. (Left) In the public domain. (Right) This file is licensed under the Creative Commons Attribution-Share Alike 3.0 Unported license by user Lijnis Nelemans.

turn into a good amount, and the effect, small in a weak field, will become large in a strong one.

All diamagnetic materials experience a force that tries to push them out of the field to where the magnetic field is weaker. In super-strong magnetic fields, this force can become quite tangible. In the left panel of Fig. 4.1, you see a photograph of a pyrolytic carbon plate hovering over a strong neodymium magnet. The weight of the plate is compensated for by the force of pushing the diamagnetic out of the magnetic field.

Here's an even funnier example in the right panel of Fig. 4.1. It shows a frog levitating in the strongest magnetic field, which is about 300,000 times stronger than the Earth's magnetic field. This frog, like any living creature, consists almost entirely of water, and water belongs to diamagnetics. In 2000, Sir Andre Geim and Sir Michael Berry of the University of Bristol received the Ig Nobel Prize for the study of magnetic levitation, frogs in particular. Geim received the Nobel Prize in Physics for the creation of graphene in 2010, becoming the only person to be awarded both the Nobel Prize and the Ig Nobel Prize.

Water droplets or frogs are not the only ones capable of levitating in a magnetic field. Geim once sent his hamster Tisha on a flight. This pet also distinguished itself by becoming co-author of one of Geim's articles under the name H.A.M.S. ter Tisha. So if someone boasts to you about their scientific publications, it is easy to laugh it off by remarking that even a hamster is capable of this. The flying frog, on the other hand, remained unknown.

If we make a compass with an arrow from a diamagnetic material, such as bismuth, then in superstrong magnetic fields, it will be oriented perpendicular to the direction of the magnetic field. But which end will be "western" and which "eastern" is not known in advance. Such an arrow can be turned 180° with some effort and it will freeze in this position.

4.1.4. *Paramagnets*

We got acquainted with the main features of diamagnets. It's time to move on to describing paramagnets — substances whose molecules have their own magnetic dipole moment and therefore behave like a compass needle. These include oxygen, aluminium, tungsten, tin, platinum, magnesium, sodium, and many other substances.

In all of them, the same phenomena act as in diamagnets, but they are overpowered by a stronger effect. It is associated with the alignment of the magnetic moments of the molecules of the paramagnetic material along the external magnetic field. At zero absolute temperature, all molecules in the external magnetic field would be oriented so that their magnetic moments were directed along the field. At temperatures familiar to us, thermal vibrations force molecules to change their orientation in space, but the magnetic dipoles are more often directed along the field than against it.

These magnets create an additional magnetic field that amplifies the external one. However, this gain is small, about thousandths of a percent. But for strong fields this gives noticeable effects. For example, in superstrong magnetic fields, we could navigate using a compass with an aluminium needle. It would be oriented along the direction of the external magnetic field, but either of its ends could become both north and south.

Paramagnets tend to be drawn into the region of a strong magnetic field, so they are attracted to the magnet. But the force of attraction is many thousand times less than that of iron, so we do not feel it when playing with ordinary magnets that can be found at home.

4.1.5. *Ferromagnets*

Iron is strongly attracted to the magnet. Such substances are called ferromagnets. They include cobalt, nickel, some rare earth metals, and alloys. What can one say about the magnetic moment of the molecules of these substances? Is it zero or not? At this point, the reader should experience some bewilderment: if molecules have a magnetic moment, then the substance belongs to paramagnets, and if not, then to diamagnets. Where do ferromagnets come from? The point is, they are paramagnets, but very special paramagnets which are werewolves or shape shifters.

For every ferromagnetic material, there is a temperature called the Curie point after the French physicist Pierre Curie. It is 770°C for iron, 360°C for nickel, and 1130°C for cobalt. And it is only 17°C for the rare earth metal gadolinium. At higher temperatures, all ferromagnetic substances behave like ordinary paramagnets.

It is known that at metallurgical plants, when remelting scrap iron, powerful electromagnets are used to move it to the right place, but no one moves the molten metal with their help. Simply molten iron is paramagnetic in its magnetic properties and the magnetic field of an ordinary industrial electromagnet is not enough to raise it.

At the Curie temperature, a second-order phase transition occurs (I mentioned second-order phase transitions in Section 3.2) and on a full moon ... sorry, at lower temperatures, matter changes its magnetic properties, showing its ferromagnetic nature. Therefore, gadolinium is a paramagnet in your fist or in a heated room and an obvious ferromagnet on a frosty street. So on the street it can be magnetized, and in the room you do not notice any effect, although it is, but extremely weak. Fans of mysticism say that a werewolf can be recognized by looking at its reflection in the mirror. "Werewolves" — ferromagnets are easily identified with a magnet.

So what happens inside a ferromagnet? Let us recall that the neighbouring molecules of a substance interact with each other not only by magnetic forces between two dipoles. It is the quantum interaction, which is called the exchange interaction, which makes all the molecules of a solid (and ferromagnets are solids) stay together and interact, and not scatter in corners, as is customary in gases.

The exchange interaction in ferromagnets at temperatures below the Curie point is arranged in such a way that the energy of the interaction of neighbouring molecules is minimal when their magnetic moments are parallel to each other. The ferromagnet wants to reduce its energy and turns the molecules in a certain area so that their magnetic moments are oriented in one direction, no matter which exactly.

4.1.6. *Domain structure of ferromagnets*

The region of space in which all magnetic dipoles are oriented in the same direction is called a magnetic domain. Below the Curie temperature, a ferromagnet spontaneously breaks up into many such domains. The size of the domain itself can reach 0.1 mm, so it can be seen with the naked eye. The boundaries between them are called domain walls. Typical thicknesses of these walls are 10–100 nm, which is less than the wavelength of visible light, but they contain about a hundred layers of atoms.

Iron is usually polycrystalline, i.e. an agglomerate of small crystals. Then the domains are separate crystals. But there are also single crystals of a ferromagnet. Domains in them come in different forms, sometimes quite specific.

The left panel of Fig. 4.2 shows domains that are clearly visible when viewed through a metallographic microscope. The surface of the ferromagnet is first polished and then coated with a suspension of fine iron powder in mineral oil. The powder collects on the domain walls, showing where they are located. Arrows are drawn on the photo showing the directions of magnetization in individual domains. Domains can form a kind of labyrinth in thin films.

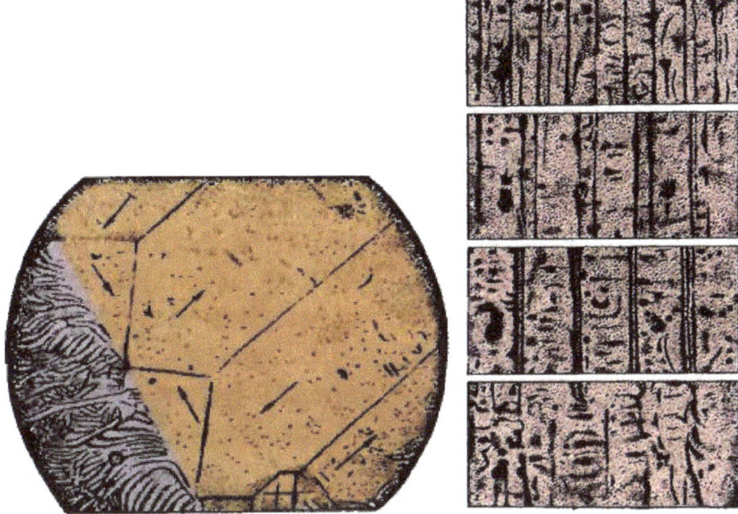

Figure 4.2. Domains in ferromagnets and the motion of domain walls with an increase in the intensity of the external vertical magnetic field. Extracted from: http://worldofschool.ru/fizika/el-dinamika/yavleniya/em/magnet/ferro magnetiki.-domeny.

Let's discuss what determines the sizes of domains in ferromagnetic single crystals. This question was resolved by Lev Landau and Evgeny Lifshitz in 1938, but no one bothers us to discuss it now. Let's start with thinking about why the whole crystal does not become one large domain. In this case, the interaction energy between its molecules would be minimal. But at the same time, it would be a rather strong magnet, and the magnetic field also has an energy density. The sum of the field energy and interaction would not be minimal.

To decrease the energy of the magnetic field and the total energy, it is advantageous for the crystal to turn into a set of domains with the opposite orientation of the magnetization, as in Fig. 4.2. In polycrystals, the magnetization is oriented in different directions, as in Fig. 4.3.

But why then don't they break down into even smaller domains? It's all due to the additional energy of the domain walls. The right panel of Fig. 4.3 schematically shows how the orientation of the magnetic dipoles of individual molecules changes near the

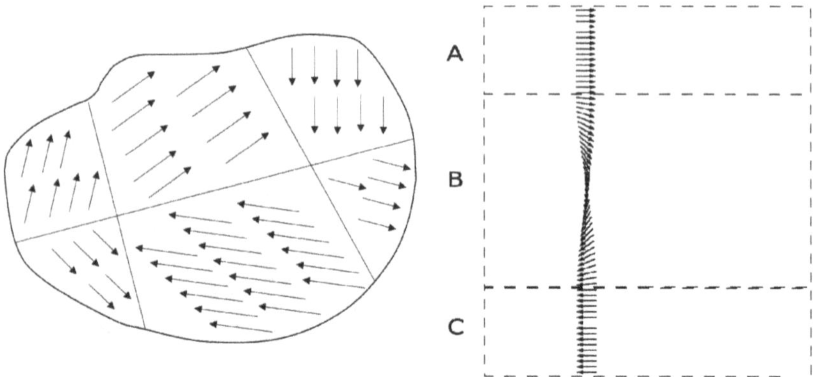

Figure 4.3. In polycrystals, the magnetic dipole moment in domains is oriented in different directions (left). The direction of the magnetization vector turns near a domain wall at the boundary of two domains in a monocrystal. Here the A and C areas are domains, and B is a domain wall (right). (Left) This file is licensed under the Creative Commons Attribution-Share Alike 4.0 International license by user Cipintina. (Right) In the public domain.

domain wall separating two domains with opposite orientations of magnetization.[b] It can be seen that near the domain wall, the magnetic moments of neighbouring molecules are not quite parallel. Therefore, the energy of exchange interaction between them is slightly larger than the minimum.

As a result, the wall region has an additional energy of exchange interaction proportional to the area of this wall. If the dimensions of the domains became too small, the crystal would decrease the energy of its magnetic field, but it would lose in total energy due to the increase in the additional energy of the walls. Minimum energy condition determines the size of the domains.

But can its sizes change? Can domain walls move? In the best traditions of jurisprudence, I will prove that they have both a motive and an opportunity for this. I will not go too far and reproduce the speech of the prosecutor at the hypothetical trial "The people against the domain walls", starting with the opening address to the judge and the jury, but will give the main reasoning.

---

[b]It shows one of two possible alternative versions of this change.

The motive for the domain walls, which are not capable of revenge, envy or jealousy, is simple. This is self-interest. They are not interested in money, stocks, or jewellery. But a ferromagnetic crystal, like any physical object, wants to lower its energy within the framework of the laws of physics. But why is it necessary to move the walls for this, changing the width of the domains? The point is that in an external magnetic field, the optimal domain width is already different.

Suppose that we direct the external magnetic field along the direction of magnetization of large domains in Fig. 4.3. Then the magnetic moments in one domain are directed along the field, so the energy of magnetic dipoles decreases. In the neighbouring one, they are directed against the field, and the energy of these magnetic dipoles increases.

It is clear that if the first type of domain becomes larger in size and the second smaller, then the total energy would decrease. There are optimal sizes for domains with magnetization along and against the external field for any field strength.

Let's take another look at Fig. 4.3. If we add an external magnetic field directed horizontally from left to right, then molecules from domain A will decrease their energy, and on the contrary, molecules from domain C will increase it. Therefore, if the domain wall B descends, then the size of domain A will increase, and domain C will decrease. This will reduce the total energy of the system.

It remains to be understood how exactly the boundaries between domains move. A domain wall is not a material object like the Great Wall of China, but a border, a place where neighbouring atoms have slightly different directions of magnetic moments. To shift it, it is necessary to turn a couple of hundred layers of atoms by a small angle. But rotating atoms inside a solid is not so easy.

Therefore, the domain wall can be moved only by applying a force greater than a certain minimum. For example, if we place a piece of iron in a strong magnetic field, then the boundaries of the domains would be moved, but after removing the field, they may not return to their original places. Iron will acquire a nonzero dipole magnetic moment and be simply magnetized, becoming a permanent magnet.

The process of moving the boundaries can be seen in the right photo in Fig. 4.2, which was taken with an increasing external magnetic field. It can be seen that, at an especially strong field, the magnetic dipole moments of all atoms of the single crystal are directed in one direction and the sample transforms into one large domain. This is called saturation magnetization. You can see this process in motion, but not in the book, but in Wikipedia, in the illustrations for the articles "Domain (magnetism)" and "Ferromagnetism". Moreover, a lot of videos on this topic can be found on the Internet.

Note that you can facilitate the process of moving the domain walls with the help of a "magic kick". I mean any blow that provides a wave of deformation inside the ferromagnet. It is easier for molecules to turn around when this wave passes through. If a piece of iron of some varieties is struck hard, such as with a sledgehammer, it can sometimes become magnetized because of this blow and a rather weak geomagnetic field.

It is clear that ferromagnets have a memory of what happened in the past. Not all physical objects are capable of this. The force of attraction of two small bodies is determined by their masses and the distance between them at the very moment and does not depend on where they were before. But look at two identical pieces of iron lying side by side. One of them is magnetized, while the other is not. The point is that their pasts differ: one was once magnetized and this magnetization remained, and the second was not magnetized at all.

Naturally, people could not help but use this memory for their own purposes. In addition to old magnetic tapes and floppy disks, we know modern storage devices in which information is recorded when a small area of the plate is magnetized.

4.1.7. *Hysteresis loop*

The memory effect causes the magnetization of the sample to change with a change in the external magnetic field, as shown in Fig. 4.4. You can see the typical behaviour of the ferromagnetic

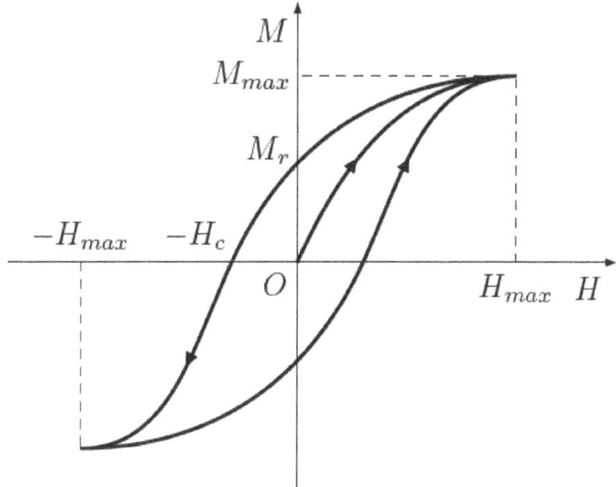

Figure 4.4.   "Hysteresis loop" — a characteristic change in the magnetization M of a ferromagnetic material with a change in the external magnetic field H. This file is licensed under the Creative Commons Attribution-Share Alike 4.0 International license by user TFra6.

magnetization M (this is the total magnetic dipole moment of unit volume of the sample) as a function of the external magnetic field intensity H. This particular curve is called a hysteresis loop or simply hysteresis from the Greek word ὑστέρησις — lagging. It was proposed around 1890 by Sir James Alfred Ewing to describe the behaviour of magnetic materials.

Let's start at a point in the upper right corner of the loop. It corresponds to a magnetic field in the direction which we will conventionally consider positive, say northward. The external magnetic field is so strong that the magnetization of the substance, having the same direction, reaches saturation and does not increase with further strengthening of the field.[c]

Now let's try to decrease the external field. In this case, the point on the graph corresponding to the state of the system will shift to the left and down along the left side of the hysteresis loop. It will reach the ordinate axis when the external magnetic field disappears

---

[c]When the external field is enhanced, the point on the graph will move to the right, but not up.

($H = 0$). But the magnetization of the ferromagnet will not disappear. The ordinate of the intersection point is called remanence or residual magnetism. It characterizes the strength of a permanent magnet, into which the ferromagnet will turn after the external field disappears. This value is often indicated as a percentage of saturation magnetization.

Further, there is a section of the loop from the intersection with the ordinate axis to the abscissa axis. On it, the magnetization is still directed to the north, i.e. it is positive, and the field is already directed south, in the opposite direction. The movement along the curve describes the demagnetization of a ferromagnet by a magnetic field directed against the magnetization.

The point of intersection with the abscissa axis ($M = 0$) corresponds to complete demagnetization of the sample. It requires a certain strength of the magnetic field, called coercive force, from the Latin word *coercitio* (holding). This value indicates how much opposing magnetic intensity must be applied to a magnetized material to remove the residual magnetism.

In real life, we are not dealing with pure iron, but with its alloys with carbon and other metals called steel, iron or cast iron. Different grades of steel differ in their coercive force. They are conventionally divided into soft magnetic materials with a low coercive force and a narrow hysteresis curve and hard magnetic materials with a high coercive force and a wide hysteresis curve.

The former are good for transformer cores; they are easy to re-magnetize. The latter are used to make permanent magnets. The conventional boundary between them is the value of the coercive force equal to 4 kA/m.

It corresponds to a field with intensity one hundred times greater than the Earth's magnetic field in the middle latitudes. For soft magnetic materials, the coercive force varies in the range 8–800 A/m, which corresponds to 0.2–20 of the geomagnetic field strengths. It is clear that the "softest" of the soft magnetic materials are already magnetized in the Earth's field.

Let's continue our traversal of the hysteresis curve. We move from right to left, which corresponds to an increase in the strength of the external magnetic field in the south, "negative" direction. Under its action, the magnetization increases in the same southern direction, eventually reaching saturation. We are at the bottom left of the hysteresis curve. A further increase in the field will not give any change in magnetization.

If we decide to start decreasing the field strength in a southerly direction, then we will start moving to the right on the graph. But we will move not along the left branch along which we got to the left point, but alongside the right border of the hysteresis curve. Removing the external magnetic field, we get a magnet with remnant magnetization in the south direction. Then we will demagnetize the sample by applying an increasing external field in the north direction. This will happen when the strength of this field becomes equal to the coercive force. At the end of the path, we will return to the starting point, completing the bypass of the hysteresis curve.

However, we are not obliged to change the strength of the external field in exactly this way: from saturation in one direction to saturation in the opposite direction. We can start to decrease the strength of the magnetic field before reaching saturation. As a result, we can reach any given point within the hysteresis loop. To determine the magnetization of the sample, it is not enough for us to know the strength of the external magnetic field (if there is no saturation), but the entire history of its change. This is the memory effect.

Domain structure, memory effect, and hysteresis are the main properties of ferromagnets at temperatures below the Curie point, when they are capable of demonstrating everything that made physicists classify them as a special class in terms of magnetic properties. Well, at temperatures above the Curie point, they simply disguise themselves as ordinary paramagnets.

Finally, I add a little exoticism. These are so-called ferrofluids, which are a suspension of solid ferromagnetic particles in a liquid because ferromagnets themselves cannot be liquid. In an external magnetic field, ferrofluids can form very beautiful structures, which also change in an alternating field.

4.2. Conductors, Dielectrics and Company

Having familiarized ourselves with the magnetic properties of various substances, it's time to look at electrical ones. Moreover, the zoo here is a little more colourful and exotic enough. The reason is clear: unlike magnetic charges that are absent in nature, electric charges perfectly exist and can move through matter. And this immediately divides all substances into two groups: conductors and dielectrics or insulators.

4.2.1. *Conductors*

Conductors are substances that contain electric charges that are not tied to a specific place and can move to any point in the body without the application of force. In other words, there is no force holding them in place. The most familiar to us are metal conductors, such as wires through which current flows. In metals, there are conduction electrons that are not associated with any one atom. These electrons, like cats, walk by themselves, wandering into all corners of a metal object. But there are other types of conductors: solutions of acids, alkalis, or salts, in which anions or cations move, i.e. negatively or positively charged ions, and plasma, in which electrons and positive ions move.

The time-independent electric field, which physicists call electrostatic, does not penetrate into the conductor. This happens because the charges inside the conductor shift under its action so that charges of different signs appear on the surface of the conductor. Their electric field completely compensates for the external electric fields inside the conductor.

So the electrostatic field is only observed outside the conductors. But the presence of conductors changes the picture of the electric field near them.

And what happens if an alternating electromagnetic field tries to penetrate into the conductor? This succeeds, but only partially. Due to the penetrated field, currents arise in the conductor, called eddy currents or Foucault currents in honour of the French physicist Jean Bernard Léon Foucault.

They not only heat the conductor, spending field energy on it, but also create their own magnetic field, which tries to compensate for the external one. As a result, both the magnitude of the field and the strength of the Foucault currents are attenuated inside the substance, especially in depth.

Therefore, the microwave field penetrates only to a shallow depth, heating the surface layer of the conductor. Often this is just a waste of energy for unnecessary and even harmful heating, but sometimes this phenomenon can be used for something useful.

For example, it is used to melt metal, harden it, or for physiotherapy procedures. Microwave radiation heats up a sore throat, but does not reach the brain, which we are not going to heat up. You can also consider the induction cooker. The microwave works on a slightly different principle, although it is similar in some ways.

But if an electromagnetic wave with an even higher frequency hits a conductor, then it can pass through it if this frequency is greater than a certain boundary value, called the plasma frequency. Or it can be reflected if this condition is not met. In Section 3.3, I have already mentioned the plasma frequency in the ionosphere. The plasma frequency increases with an increase in the concentration of conduction electrons.

In metals, the concentration of electrons is millions of times higher than in the ionosphere. As a result, the plasma frequency increases so much that metals reflect visible light. To be convinced of this, just look in the mirror. And remember that it is not glass at all that reflects the light in it, but a thin layer of metal amalgam on its back surface. Glass is needed for the strength of the mirror structure and to protect the amalgam metal from mechanical damage and oxidation.

When the properties of metals are listed, the characteristic metallic lustre of the surface has also been mentioned since time immemorial. Today, we understand that this is simply the reflection of light from a conductor with a high concentration of free electrons.

Due to the abundance of conduction electrons, metals have high electrical and thermal conductivity. Moreover, these two values are almost proportional to each other. What follows from this? If you stir very hot tea with a silver spoon, it burns your fingers. But with a spoon made of stainless steel or aluminium, you cannot be afraid of getting burned.

This means that the thermal conductivity of silver is greater than that of steel. It can be concluded that the electrical conductivity of silver is also higher than that of aluminium or steel. If it weren't for the cost of silver, it would be an excellent material for electrical wiring.

4.2.2. *Dielectrics*

So, conductors are what have free charge carriers. And if they are not, then the substance belongs to dielectrics. Each of the charged particles in the dielectric has its own place and cannot move far from it. But if the dielectric is placed in an external electric field, then the positive and negative charges shift slightly along the field and against the field, forming many electric dipoles.

In this case, the body will acquire an electric dipole moment, which is caused (physicists say "is induced") by the action of this external field. The field caused by the induced charges and dipoles is superimposed on the external field that generated these charges.

How strong is the field of induced charges compared to the external one? It depends on the properties of the dielectric. All dielectrics are usually divided into two groups: polar and non-polar. Like bears. Molecules of polar dielectrics have their own electric dipole moments, while non-polar ones do not.

Polar dielectrics include, for example, water. Its molecule forms an isosceles triangle, in two equal corners of which there is a positive hydrogen ion, and in the third — a negative oxygen ion.

The dipole moments of molecules of polar dielectrics prefer to line up along the external field. In this, the picture is similar to the action of a magnetic field on paramagnets. Both this effect and the displacement of charges in dielectrics always act in one direction, decreasing the external field. Naturally, polar dielectrics do this much more efficiently. Indeed, in this case, both mechanisms of field weakening work. Water weakens the electric field 81 times. But non-polar kerosene weakens it only 1.8 times.

4.2.3. *Ferroelectrics and electrets*

Are there electrical analogues of ferromagnets? Yes, they do exist, and they are called ferroelectrics. They have all the basic properties of ferromagnets: domain structure, memory, and hysteresis. Only, instead of permanent magnets, permanent sources of an electric field are obtained from them. They also have their own Curie temperature, above which they behave like ordinary polar dielectrics.

In this kind of electric zoo, in addition to the electrical analogy of paramagnets and ferromagnets, you can find a couple of exotic specimens that have no correspondences in the magnetic zoo. Meet: electrets and pyroelectrics.

Let's start with electrets. This word was proposed by the famous physicist Oliver Heaviside, connecting the beginning from electricity and the end from the word *magnet*. This is what looks like a magnet, but for electricity. We have already discussed ferroelectrics, but they are a completely different type of material.

I will describe the simplest type of electrets. An amorphous substance is taken, such as wax or resin, which has molecules with electric dipole moments. This polar dielectric is heated, causing its viscosity to greatly decrease, and then placed in an external electric field strong enough for all the dipole moments of the molecules to be oriented along it. Then it is cooled directly in the electric field, and only then is the field removed.

The electret becomes viscous and solid enough to prevent the polar molecules from turning after the field is removed and therefore retains its electrical polarization. The simplest electret can be prepared at home using paraffin wax or molten rosin in castor oil, heated, and then cooled in a strong electric field. But it does not have a domain structure and other specific properties of a ferromagnet or ferroelectric.

If you measure the voltage between the two ends of the electret with a voltmeter, it will be nonzero. Therefore, the electret can be used as a voltage source, especially if the current in a circuit is small. It is used in microphones, air filters, and sensors.

The Japanese used electrets in military equipment extensively during World War II. The Japanese physicist Mototaro Eguchi investigated their properties for the needs of the Japanese Ministry of Defence since 1922. This study was considered top secret.

Therefore, when the American fleet captured a Japanese destroyer in 1943, American sailors were puzzled by its communications system. It worked without batteries. Instead, something strange was used: a kind of electret pill made from a mixture of carnauba palm resin and wax. It could work as a power source for several hours. Its modern counterparts can keep the potential difference in an idle state for hundreds of years.

4.2.4. *Pyroelectrics*

Of the many substances related to pyroelectrics, tourmaline is the best-known. Depending on the colour and transparency, it belongs to precious or ornamental stones. Pyroelectrics are electric field sources. So how are they different from electrets or ferroelectrics? We prepared the electret recently, say, an hour or a century ago. It is a good insulator and, despite the voltage between its ends, there is practically no current flowing through it unless it is used as a battery.

And the crystals of tourmaline were prepared for us by nature, and it happened billions or hundreds of millions of years ago. Freshly prepared tourmaline has an electrical dipole moment and its own electric field. And no matter how small its conductivity was, since that moment a huge period of time has passed. So the electric charges managed to move under the action of the field, although this process was very slow. If you measure the voltage between the sides of the tourmaline crystal with a voltmeter, it would be zero. The field of free charges accumulated over the centuries at the edges of the tourmaline compensates for its polarization field.

But if you heat or cool the crystal a little, the balance will be disturbed. The field produced by displaced charge does not change, but the polarization field changes with temperature. The voltmeter will immediately detect the potential difference when the temperature of the tourmaline changes. Therefore, this class of substances is called pyroelectrics, from the Greek word *pyr* ($\pi \tilde{v} \rho$) which means fire.

4.2.5. *Electrical properties of crystals, piezoelectricity*

But we imperceptibly went from isotropic substances to anisotropic crystals. Are there any peculiarities in their electrical properties? Yes, some of them have, but not everyone. All crystals that exist in the world are of one of three types.

The first, called cubic crystals, includes crystals with the highest symmetry. In terms of their electrical and optical properties, they are completely isotropic. The other two classes are uniaxial and biaxial crystals.[d] Their electrical and optical properties are really anisotropic.

How to see it? Just with your own eyes. These crystals, due to their anisotropy, exhibit a phenomenon called birefringence. If you look through them at distant objects, then everything looks the usual way, but if you put them on something close, for example, a picture or text, then the image is doubled. This effect was first described by the Danish scientist Rasmus Bartholin in 1669, who observed it in calcite, a crystal having one of the strongest birefringences. I will tell you something about the cause of this phenomenon in Section 4.4.

Note that all pyroelectrics belong to biaxial crystals. In addition, all of them are also piezoelectrics at the same time (but not all piezoelectrics are pyroelectrics). And what is this new character? The first root "piezo" comes from the Greek word $\pi\iota\acute{\varepsilon}\zeta\omega$ — to squeeze or press. Indeed, a direct piezoelectric effect is observed in these crystals, in which compression or tension of the sample leads to the appearance of charges on its surface. They are sometimes strong enough to make a spark.

Piezo lighters and contact piezoelectric fuses work on this principle. So do pressure sensors, microphones, pickups in old electric turntables, in which sound is recorded on vinyl records, and dozens of other devices.

---

[d]I do not discuss the differences between them here, but, in short, biaxial crystals are crazier.

In addition to the direct piezoelectric effect, there is also a reverse one. By applying an electrical voltage to the piezoelectric, it can be ordered to shrink or expand. This effect is used in sound and ultrasound generators, and also for ultra-precise positioning, for example, in a scanning tunnelling microscope, in adaptive optics systems for mirrors of the best telescopes or in inkjet printers. The forward and reverse piezoelectric effects are used in combination in quartz watches and frequency standards.

4.2.6. *Superconductors and semiconductors*

I want to briefly mention two other special classes of substances that we missed on our first inspection: superconductors and semiconductors. Superconductors are substances that are capable of transmitting electric current without resistance and therefore without loss. For this, they must be well cooled, often to a temperature close to absolute zero. For the discovery of this phenomenon, the Nobel Prize in Physics for 1913 was awarded to the Dutchman Heike Kamerlingh Onnes.

For a long time, this mysterious phenomenon was inexplicable for physicists, but then the Americans John Bardeen, Leon Cooper and Robert Schrieffer proved that this is a quantum effect, a kind of gift to the macrocosm from the microcosm. For this, they received the Nobel Prize in Physics in 1972. Superconductivity is living according to the quantum laws.

But this is not the only Nobel Prize awarded for the study of this phenomenon. In 1962 the Soviet physicist Lev Landau received the prize "for his pioneering theories for condensed matter, especially liquid helium", while in 2003 the prize "for pioneering contributions to the theory of superconductors and superfluids" was shared by Vitaly Ginzburg, Alexei Abrikosov, and the British–American physicist Anthony Leggett. The first two physicists received a prize for their theories developed in the USSR.

Superconductors do not admit electric fields inside them. Some of their types also do not admit magnetic fields. A magnet can levitate over the surface of a superconductor because of the latter reason. I will discuss the causes of superconductivity in Section 7.4.

Some people are sure that semiconductors are called materials that conduct current so-so, something between conductors and insulators in conductivity. This is not entirely true. Semiconductors stand out as a special class due to the fact that they have free charges, but their concentration varies greatly with temperature. For example, an increase in temperature from 100 to 600 K increases the carrier concentration by 17 orders of magnitude in germanium, i.e. 10^{17} times. This concentration is constant in metals. Highly heated semiconductors become similar in their electrical properties to metals. The concentration of electrons in them ceases to change with further heating.

We went over the electrical and magnetic properties of substances, touching a little on the optical. And we found a real zoo there, worthy of its Gerald Durrell.[e] And this despite the fact that I prudently covered the cages with the signs "ferrimagnetics", "antiferromagnets", and "antiferroelectrics" (no, there is no antimatter in the last two). So the readers were once again convinced that we are dealing with the scale of diversity. This whole zoo not only surrounds us, but among its inhabitants there are many substances and objects with which we often deal.

4.3. Lightning

4.3.1. *Electric sparks and discharges*

It is clear that a couple of centuries or millennia ago, we did not know many of the inhabitants of our zoo. And what of the electrical and optical natural phenomena that most amazed an ordinary person at that time and continue to amaze many to this day? I think there are three great ones: lightning, rainbow, and mirages. It is difficult not to notice them, and having noticed them, it is difficult not to pay special attention to them. They certainly adorn our scale and deserve a special mention.

---

[e]Famous English animal catcher and zoo owner, author of many books about animals, zoos, etc.

Lightning is just a very big overgrown spark. Or, more scientifically speaking, it is a spark discharge in the air. Air is, of course, a good insulator, but far from ideal. It consists of electrically neutral molecules of different gases. However, as I will tell in Section 5.1 when describing the photoelectric effect, current can flow even in an ideal insulator — a vacuum, which has been confirmed for many decades by vacuum tubes in receivers and other electronic devices. So it is not surprising that the current can also flow in air, although the mechanism of this phenomenon is completely different than in a vacuum diode or triode.

We need a strong external electric field for a spark. If some molecule located in this field breaks up into a positively charged ion and an electron or a negatively charged ion, forming plasma, then these pieces would fly in opposite directions, increasing their speeds along the route in an electric field. Then pretty soon each of the ions will crash into a molecule of nitrogen, oxygen, or other air component.

If the electric field is weak or the ions have little time to accelerate, then the attempt ends there. But if the field is stronger than a certain threshold value and the pieces have had enough time to accelerate, then when they collide with the molecule, they will break it into two or more charged pieces, which, in turn, will begin to accelerate by the electric field. Each collision leads to an increase in the number of ions. Very soon there will be so many of them that they will turn the air into a weakly ionized plasma in which an electric current flows. Due to the appearance of plasma, the air conductivity is greatly increased.

The further fate of ions depends on how our field is organized. If a high voltage is applied to an elongated pointed object or wire, then a stream of ions will begin to drain from the end of it. This is called corona discharge and allows you to observe a beautiful glow of plasma in the dark. Corona discharge is used in technology as a method of coating an electric charge to a surface, for example, on paper in a laser printer or copier. It is also used in spark plugs. Mariners observe the so-called Saint Elmo's lights on the masts of ships, especially in the tropics. This is a corona discharge flowing down from the protruding parts. However, a discharge can be observed around objects without protruding parts.

Corona discharge is not the only option. You all observed the glow discharge in gases in neon letters on signs and advertising slogans. If you increase the current, you get an arc discharge, also known as electric arc or "voltaic arc", which is used in floodlights, projection devices, electric furnaces, and metal welding. Naturally, there is the same spark discharge that creates a plasma channel in the air through which current flows like a wire. In the case of lightning, this is a tremendous current, 10,000–500,000 amperes.

In June 2020, on the eve of the International Day for Protection against Lightning, the World Meteorological Organization recognized two new world records related to lightning. Satellite measurements showed that the lightning recorded on October 31, 2018, in the southern part of Brazil had a horizontal length of $709 \pm 8\,\mathrm{km}$, which is the same as the distance from Berlin to Antwerp. And in northern Argentina on March 4, 2019, a single discharge lasted 16.73 s.

On the same occasion, Indian media reported that 24 people in Uttar Pradesh and 83 people in Bihar were killed by lightning strikes in one day on June 25, 2020. These were peasants working in the fields and children walking in the street. In comparison, in the United States, for the entire 2019, 20 people became victims of lightning. However, these results were soon surpassed by lightnings recorded in 2020. A lightning over the south coast of the United States stretched for 768 kilometres and one in the border area of Uruguay and Argentina lasted 17.1 seconds. New records were recognized by WMO at the beginning of 2022.

4.3.2. *How the clouds are charging*

This is certainly curious, but physicists are primarily interested in the causes of the phenomenon itself. A spark requires a voltage between its ends, for lightning it is from tens of millions to billions of volts. Where are these ends located? Lightning can strike between storm clouds, from a cloud to the ground, from the ground to a cloud. Scientists learned about lightning striking from the clouds up into space after the appearance of artificial satellites. So it is the clouds that are the place where one of the ends of the spark is located, or even both. And these clouds are charging. Why?

First, let's ask ourselves a simpler question: why don't clouds fall to the ground? The answer is simple: in fact, they fall to the ground, but they do it very slowly.

After all, they consist of the smallest droplets of water that fall at a low speed due to the fact that it is at this speed that their low weight is compensated by the force of air resistance when falling. Therefore, they fall evenly, with a small constant velocity relative to the air. If the droplets collect in large droplets, then they fall down in the form of rain, hail, or snow, depending on their temperature. And they do it much faster.

And if there is friction between droplets in the air, then it is natural to expect their electrification. We are constantly faced with the fact that friction creates electric charges on the surface of bodies. We even sometimes water or spray clothes with a special antistatic agent to remove this charge. Here I have to admit that I, who for many years have been talking about processes near black holes or in atomic nuclei, cannot tell you in simple terms why bodies are electrified during friction. There are theories of this phenomenon, but they can hardly be called simple and understandable.

The friction of water droplets against air leads to the separation of positive and negative charges. Negative ones hit the ground, while positive ones accumulate in the ionosphere. The result is a vertical downward electric field. We all, to some extent, live in a huge charged capacitor on a planetary scale.

So, the reason for the potential difference between clouds or a cloud and the earth became clear. Lightning passes between them, leaving behind a temporary conductor — a channel filled with plasma. The plasma in the channel is very hot and, like any heated body, emits light, which we see as lightning. Well, the sound accompanying a lightning bolt is, accordingly, thunder.

The greater the initial air conductivity, the easier it is for the lightning spark to overcome the obstacle in the form of the atmosphere. Humid air conducts much better than dry air. Therefore, thunderstorms in mid-latitudes are rarely observed in winter, when cold air has low humidity, which is limited by the dew point, mentioned in Section 3.2. And clouds do not electrify as much when solid snowflakes or small ice floes fall.

4.3.3. *Fair-weather current*

Try to imagine the giant capacitor that I talked about, and most likely you will have several questions. Two of them I am ready to answer, having predicted them in advance. The first is: What about the air over the desert or other places where there are practically no clouds and no friction of droplets in the air? What charges the capacitor in this place?

The answer is simple: the capacitor is not only so large that deserts occupy an insignificant fraction of its surface, but charges can also move along the "plates" from one place to another. Soil or water acts as one plate, and the ionosphere as the second. All of these media conduct electricity rather well, so nothing interferes with the horizontal redistribution of charges from one place to another. So any droplets rubbing against the air anywhere on our planet can charge this capacitor.

And the second question immediately arises. Lightning discharges this peculiar capacitor, so its charge does not get bigger and bigger. But is there really no mechanism for discharging the capacitor in the same deserts where there are no thunderstorms, i.e. transfer of excess charge from the ionosphere to the ground and vice versa? Naturally, it exists and is called the fair-weather current. No matter how small the conductivity of air is, especially dry air over the desert, it is nonzero. As a result, even if there is no thunderstorm, a weak fair-weather current still flows from top to bottom. Its density is about 2 picoamperes per square metre. It is clear that we do not see it and do not notice it. So the discharge of the heaven-to-earth capacitor can do without thunder and flash of lightning.

It only remains to add that scientists record thunderstorms on other planets of the Solar System — Venus, Jupiter, Saturn, and Uranus. Flares are observed from spacecraft, and their radio emission is recorded by terrestrial radio telescopes.

Thunderstorms on Earth are sometimes unusual, caused by volcanic eruptions, dust storms, or tornadoes. However, the reason for the electrification remains the same — the friction of particles, not water droplets, but others, such as grains of sand. Photos of tornado lightning are easy to find on the Internet. It's good that we cannot observe them daily and with our own eyes.

4.3.4. *Lightning rod*

However, even ordinary lightning bolts are dangerous enough to be enjoyable to watch. However, harm from them has decreased significantly after Benjamin Franklin came up with a lightning rod. This is a high, grounded conductive stick, often sharpened from above, whose purpose is to take on lightning strikes and calmly forward current through itself into the ground without unnecessary fires and destruction. Fires following lightning strikes on buildings or trees were not uncommon during Franklin's time, especially in the plains of the US Midwest. The lightning rod helped to significantly reduce their number. Insurance companies quickly appreciated the benefits of the new invention, and the insurance fees for buildings equipped with a lightning rod were set lower than for buildings without it.

As Walter Isaacson wrote in his book *Benjamin Franklin: An American Life*: "*Eripuit cœlo fulmen sceptrumque tyrannis*, he snatched lightning from the sky and the sceptre from tyrants." Now, the lightning rod is an integral part of all high-rise buildings. You can find a lot of the photographs of lightning that struck the Eiffel Tower. But this lightning causes no harm precisely because of the powerful lightning rod.

4.4. Mirages and Gravitational Lenses

The thunderstorm is over, the lightning has stopped flashing, and none of the bolts have struck us; only the lightning rod has been struck. Life goes on! A rainbow has appeared in the sky after the rain. That is the second of natural phenomena that cannot be overlooked. We will talk about it, but in the next section. In this, I remind you of some of the things that underlie geometric optics and lead to the appearance of mirages.

The optical properties of each substance are characterized by the value of its refractive index. Here we will talk about transparent media such as air, water, or glass, but opaque substances also have a refractive index. For a physicist, this indicator is the ratio of the speed of propagation of light in a vacuum and in a substance we are discussing.

4.4.1. *Refraction of light*

It is clear that such a definition says little to a person who has not felt the intricacies of optics. Therefore, it is easier to remember another term that can be considered synonymous with refractive index: optical density.

If there are two media, then the one with a higher refractive index is considered optically denser, and the second, respectively, optically less dense. The refractive index of air is close to 1, for water it is 1.33, and for window glass it is somewhere around 1.5. Therefore, water is optically denser than air, but less dense than glass.

At the border of two media, a ray of light is deflected if their optical densities do not coincide. The closer the ratio of the two refractive indices to unity, the smaller this deviation is. This ratio, called the relative refractive index, completely determines the refraction at the boundary between the two media. Therefore, a piece of glass thrown into an aquarium with water is less noticeable than the same piece of glass, but in air. You may not even notice it.

In some adventure stories, the hero hides diamonds by simply throwing them into a transparent jug of water. Stupid enemies look at them, but do not see. Naturally, this can only be written by a person who has never seen diamonds in water. The refractive index of diamond is about 2.42, which is 1.82 times that of water.

Therefore, a diamond in water is much easier to see than glass in air. Even if it is a special optical glass or flint glass with an admixture of lead oxide and other metals, from which vases are made, which are sometimes called crystal in everyday life. Their refractive index could be up to 1.8. And only a very absent-minded person can fail to notice a crystal vase on the table among the surrounding air.

Let us assume that a single ray of light falls at some angle on the border between two media with different refractive indices. If the media are isotropic, then also a single ray goes into the second medium. The ray is refracted and partially reflected at the border. If the ray passes into an optically denser medium, then its direction becomes closer to the normal to the interface between the media; if into a less dense medium, then the direction becomes farther from the normal.

But if a ray hits a uniaxial or biaxial crystal, then two rays propagate in the crystal and for each of them the crystal has different refractive indices. This is what explains the phenomenon of birefringence. But we are not interested in it right now; we want to understand at first the simpler case of ordinary refraction of light. Despite its simplicity, it is exactly that which underlies the numerous mirages that each of us sees, at least occasionally.

In an optically homogeneous environment, i.e. a medium with constant optical density, the light beam propagates straight. If the optical density changes continuously, not abruptly, then the light beam is bent. However, our brain, when processing the image observed by the eyes, always proceeds from the fact that the light rays coming to our eyes are straight. This is almost always close to reality, but sometimes it leads to the appearance of mirages. They are caused by a combination of rays bending with the wrong algorithms used by our brains.

But why does air become optically inhomogeneous? If air is compressed or its density is increased, its optical density increases. Air pressure decreases with increasing altitude; this already somewhat bends the rays of light in the atmosphere. But mirages require substantially larger curvatures. They are provided by a change in air temperature.

4.4.2. *Inferior and superior mirages, Fata Morgana*

Let's start with the case of the so-called inferior mirage, which is often associated with the desert. It requires that the air temperature near a hot surface (say, sand in a desert) be significantly higher than at a low altitude above it. With increasing altitude, the air temperature drops, while its density and refractive index increases. The ray is bent. A light from the sky goes toward the hot surface, reaches a minimum height slightly above the level of the sand, and then, rising, falls into the eye of the observer.

He sees an image of the sky, palms, and other exotic things below in front of him and takes it for a reflection from some mirror surface, such as the water in an oasis. We are much more accustomed to observing the lower mirage over the hot asphalt. It seems that it is covered with puddles of water that disappear when approached.

Many readers have guessed that if there is an inferior mirage, then there must be a superior one. In this case, the light beam bends in the other direction, and the top of the trajectory looks up. Naturally, the optical density of air should thus decrease with height, and its temperature should increase. Warm air above cold water will perfectly provide the desired refractive index change. Sometimes the observer sees an inverted image in the sky.

In the case of the superior mirage, the layer of air with temperature varying with height is clearly thicker than in the case of the inferior one. Therefore, the observer easily notices the features of the mirage caused by the inhomogeneity of the rate of increase in the air temperature with height and especially by the jumps in this temperature. In the Arctic seas, mirages are common of a special kind, called *halgerndingar* in Icelandic. They are more familiar to us under the name Fata Morgana. Numerous videos and photos are not hard to find on the Internet. Distant objects appear as spires or towers rising from the sea surface with high vertical magnification in the case of Fata Morgana.

The change in temperature with height is the source of various types of mirages. But even if the temperature is constant, the density of the air changes with height, bending the rays. However, the radius of curvature of the rays is greater than the radius of the Earth. Therefore, all the usual explanations for phenomena that demonstrate the sphericity of the Earth work, for example, the fact that the observer sees the tops of the masts of an approaching ship, but not its deck.

Jupiter does not have a solid surface, but if it were, then the radius of curvature of the rays in the Jupiterian atmosphere would be less than the radius of this surface. Hypothetical Jupiterians would see themselves at the bottom of a kind of bowl, the bottom of which would move with the observer, and approaching objects, such as ships, proudly displaying both masts and decks, would move along the walls of this cup. This could be called a permanent superior mirage. Probably in Jupiterian schools there are a lot of losers who flunk jupiterography because they do not know that their planet has the shape of a ball.

4.4.3. *Gravitational lenses*

Could there be something like a mirage in outer space, where there is no air? At first, this seems impossible. But in fact, a similar phenomenon, called gravitational lensing, has long been known and used by astronomers.

In this case, the rays of light are bent not because of the different optical density of the medium in which they propagate, but because of the attraction of massive astronomical objects, past which this light passes. As a result, light from a very distant quasar or galaxy reaches the terrestrial observer not along a straight line, but along a slightly bent one, and we see an object in the sky slightly displaced compared to its true position.

In itself, this is not so interesting, because we cannot find out its true position. But if a particularly massive galaxy is located near the beam of light, then the light from a distant object can reach the Earth in several ways. This means that we see several images of this object in the sky. They will not only be close to one another and be at the same distance from us, but also change their brightness in the same way. The fact is that there are always some variations in the luminosity of quasars and other objects, caused by internal reasons. And no matter how the light comes to us, its brightness will repeat these oscillations, naturally with a delay associated with the different light propagation time along its way.

This time is not only huge by earthly standards, but also differs for each of the paths of light propagation. Therefore, all images of the object will not only demonstrate the same changes in brightness, but also do it synchronously, more precisely with a certain time delay. This property makes it possible to accurately distinguish all images of one gravitationally lensed object among numerous pairs of close but different objects.

You can find on the Internet a snapshot of the object Q2237+030 or QSO2237+0305, known as "Einstein's Cross". In the centre is a galaxy 400 million light-years distant from Earth, the attraction of which bends the rays of light from the distant quasar.

It is much further from us, 8 billion light-years away, i.e. 20 times farther than the galaxy. Four images of the quasar are visible around. Note that the number of images in gravitational lenses is always even. Most often there are two, less often four. In 2000, astronomers discovered the only currently known lens, CLASSB1359 + 154, which produces six images.

However, the conversation about earthly mirages led us far away, straight into the depths of the Universe. Your guide is clearly chattering instead of talking about the rainbow. It's time to keep the long-standing promise.

4.5. Rainbow

Now we can return to the rainbow. We talk about the colours of the rainbow: Red, Orange, Yellow, Green, Blue, Indigo, and Violet. The most common British mnemonic for this set is "Richard Of York Gave Battle In Vain". For us, monochromatic electromagnetic waves in the visible range correspond to a specific colour, from red (the longest waves with the lowest frequency) to violet (the shortest ones). These are just features of human vision.

In a vacuum, all photons move at the same speed, but in a different medium this is not entirely true. The speeds of light in it are slightly different for photons of different frequencies, or, as we perceive, for different colours. Accordingly, the refractive indices of the medium also differ slightly. Physics refers to this phenomenon as refractive index dispersion. In transparent media such as water, glass, air, and diamond, the refractive index of violet light is always higher than for red, and for the other colours of the rainbow it changes in accordance with their order in the spectrum. This is called normal dispersion. This dependence is not only confirmed experimentally, but also theoretically proven.

The dispersion of the refractive index is manifested in the decomposition of white light into a spectrum using a glass prism. Newton was the first to do so, in 1666. But long before that, water droplets were constantly doing this trick, showing us a rainbow. This physical phenomenon usually makes people happy and lifts their spirits.

Water droplets are in the air. Even if they are very small and are inhibited by air resistance, they still fall to the ground. Large drops can be found in the fountain when a jet of water falls down and turns into a swarm of drops, or in the falling drops of sun rain. One can find small ones in fog or fine water dust in the same fountain or in the clouds immediately after rain, before they have time to fall or evaporate. All of them are almost spherical in shape, being only slightly flattened due to the resistance of the air and friction against it when falling.

A ray of light enters the droplet and is refracted. Then it is reflected from the far surface of the droplet, comes back to the front surface, and exits the droplet, refracting again. As a result, it deflects by an angle of about $180° - 42° = 138°$ regardless of the size of the spherical drop.

But this happens if we forget about an important detail, which is provided precisely by the dispersion of the refractive index. The angle of deflection of light depends on the refractive index of water, which is slightly different for different wavelengths. As a result, red light is deflected at an angle of $137° \, 30'$, violet light at $139° \, 20'$, and the rest of the colours of the rainbow at angles between these values. White light from the source is decomposed into a spectrum, providing the usual alternation of colours.

This difference in the angles of deflection of light in less than $2°$ is the reason for the rainbow. The most common is the primary

Figure 4.5.   Photo of a double rainbow.

rainbow, which is part of multi-coloured concentric circles. Their centre lies in the direction opposite to the direction of the light source, which is usually the Sun.[f] Around this point, it is necessary to draw the outer red circle with an angular radius of 42.5°, the inner violet with a radius of 40.5°, and fill the gap between them with colour circles in accordance with the colours of the spectrum. In Fig. 4.5, you can see the primary rainbow. We will talk about the secondary one a bit later.

The rainbow can ideally be circular. Such a rainbow can be seen from an airplane or from a very high peak after sunrise or before sunset. But when you look at a rainbow on the plain, part of the circle lies below the horizon. There, underground, there are no water drops and there is nothing to reflect and refract the light that would come from this part of the rainbow. Therefore, the rainbow

---

[f]At night, this role could be played by the Moon, but its light is too weak. The human eye uses only part of the light-sensitive photoreceptor cells (rod cells, but not cone cells) for night vision, which cannot distinguish between colours. Therefore, at night, the colours of the rainbow can only be seen with much brighter light sources, such as spotlights.

below the horizon is not visible. You see it in the form of a multi-coloured arc.

Sometimes the entire circumference of a potential rainbow lies below the horizon and we cannot observe this phenomenon. This always happens when we are on a plain, and the sun is close to the zenith and it deviates from it by less than 42.5°. So, a rainbow can be seen only in the morning or evening in equatorial countries, but not close to noon, unless you see it from the top of a high mountain.

Let's go back to Fig. 4.5, in which you can see the secondary rainbow around the primary one. It has the reverse order of colours. The inner red arc has an angular radius of about 50°, the outer violet one about 53°. It occurs if a beam of light is twice reflected from its surface after entering the droplet and then goes out.

It is clear that higher-order rainbows are theoretically possible. They are not visible in nature, but physicists have learned how to get such rainbows in the laboratory using laser light. Between the primary and secondary rainbows is the Alexander's dark band, named after the ancient Greek philosopher Alexander of Aphrodisias ('Αλέξανδρος ὁ 'Αφροδισιεύς), who first described it in *circa* 200 AD.

On the Internet, it is not difficult to find photographs that show five to eight rainbows, with centres in different places. When explaining them, you have to choose between two options: the picture was taken on a planet illuminated by five to eight luminaries, or someone was mastering the basics of Photoshop or another image-editing program. The choice is yours.

4.6. Physics in the Kitchen: Trying to Boil the Water and Potatoes

Let's move closer to the stove and talk a little about physics in the kitchen. After all, the kitchen, a restaurant, or a cafe is a place where we usually forget about science, although it plays a crucial role in those processes that determine its purpose. But behind cooking, roasting, salting, smoking, wine, and steak, you can see hidden physics, not to mention chemistry and biology.

We will not discuss the proportions of saffron and cardamom, but we will look especially for the physical side of the processes. I hope that the aromas from pots, pans, and glasses will not distract us from this. But at the end we will have a double reward: a portion of new knowledge which underlie cooking and a portion of delicious food with appropriate drinks.[g]

We will not discuss in detail the things that are described in thousands of recipes and cookbooks. There are other books in which the physics of processes is set out in detail, making them tasty and even understandable. First of all, this is a chapter in the wonderful book[h] which covers boiling, frying, steaming, the physics of spaghetti, coffee, wine, and much more. My goal is much more modest — to discuss the simplest operations such as boiling water in a kettle. So the promised portion of delicious food is still for you or for the chef who is ready to cook it.

Let's simply try to boil water. Kettles were invented exactly for this, although no one bothers us to apply the basics of the heating process to a saucepan. They are all based on heat transfer and energy balance analysis.

In addition to the usual boiling of the kettle, we will discuss all sorts of non-standard options, such as those in which the water in the kettle does not boil or the water boils away completely. This is bad for those who just want to brew tea or to keep kitchen utensils safe and sound. But the same processes lie behind them as for ordinary boiling, albeit with slightly different quantitative indicators. For example, the kettle was put on a very low heat or the fire underneath was left to burn after the water has boiled away. So, let us also mention and analyse these undesirable offshoots of our speculations.

---

[g]The reader guesses that the guide is clearly luring him in with empty promises. It seems that he just wants to discuss heat transfer and related issues in a simple, understandable example, without mentioning the interior of the Sun or something equally distant and obscure.

[h]Andrey Varlamov and Lev Aslamazov, *Wonders of Physics* (Singapore: World Scientific), 2012.

4.6.1. *Energy balance when heating the kettle*

As corny as it may sound, the kettle heats up because it receives heat — either from an external heat source, such as a gas burner or a fire, or from an internal heater, such as an electric kettle, heated by an electric current. The main thing is that it receives more heat than it gives to the environment, spending the difference on increasing its own temperature. Everything is clear with the thermal energy inflow, provided by a fire or a heating element, but heat losses are a little more complicated.

A hot kettle can transfer heat to the environment by means of the three heat transfer mechanisms mentioned in Section 2.6, i.e. thermal conductivity, convection, and thermal radiation, and partly in the form of water–vapour phase transition energy. The thermal radiation of the kettle, even heated to 100°C, can be ignored. The other two mechanisms transfer heat much more efficiently.

The heat flux during heat transfer by thermal conductivity is roughly proportional to the temperature difference between the kettle surface and the air in the kitchen. This follows from the law discovered by Newton as a result of his experiments. But heat transfer by convection depends also on a number of details such as the shape of the kettle, its location relative to a heat source, or draughts in the kitchen. Naturally, it also depends on the temperature of the kettle. The more the kettle is heated in comparison with the environment, the more energy is carried away by convection at the same time.

Modern electric kettles greatly reduce heat dissipation into the environment. This greatly simplifies the study of the physics of the process of heating water in them. What is written next about boiling an ordinary kettle is trivially transferred to an electric kettle, and most of the exotic options discussed here are simply unattainable for it.

If water is boiling, then it also requires heat, proportional to the mass of the evaporated water. This is another mechanism for losing heat from the kettle. The difference between the supplied and removed heat is used for heating of the kettle. The rate of heating of the kettle is proportional to it. That's all we need to know to consider the process.

Let's add one formal point. In theory, heat losses due to thermal conductivity can become negative, i.e. directed in the other direction. This happens if the kettle is colder than the environment. In real life, this can be achieved not only by putting a kettle brought from the cold in the warm kitchen, but also by boiling the kettle in a sauna with an air temperature of more than 100°C.

In the latter case, we do not even need to put the kettle on fire: over time, the water will either boil by itself or will have enough time to simply evaporate from the heat of the surrounding air. Heat losses during convection can also be negative. Let's say we blow on the kettle with a powerful warm hair dryer. But let's not be so eccentric. Boiling the water with an industrial dryer in the sauna is not included in our plans. We boil it in an ordinary kitchen at normal air temperature. So, both conduction and convection reduce heat provided by a heater.

It's time to think about the heat source. The power of a conventional gas, induction, or electric stove can be changed, but within some limits. What can occur if we go beyond them and consider some non-standard oven?

Let's start with the case of a very weak heat source. The kettle just won't boil on it. It will heat up relative to the environment and its temperature will stabilize for a while. Indeed, as it heats up, the heat losses will also grow. They will become equal to the supplied heat after reaching a stable temperature.

But a stable state won't last forever. If you just pour water into the kettle and leave it indoors, sooner or later the water will evaporate. This might not happen in a day, and perhaps not even in a week. But it is clear that the water would not remain in the kettle for centuries.

If our weak heat source heats up the kettle, for example, to 60°C, then the evaporation process will be accelerated significantly. So if we are persistent idiots, then we will end up with a kettle without water, constantly heated by a weak heater to a moderate temperature. It remains only to understand why we did all this.

Let us increase the power of the source. Suppose that the temperature at which the heat input and output are equal, excluding the heat spent on evaporation, exceeds 100°C. In this case, we can make the kettle boil. It boils at 100°C (at normal atmospheric pressure, of course). At this temperature, the heat input exceeds the heat losses by thermal conductivity and convection.

Why? Simply because they would become equal at a higher temperature, and at 100°C with the same heat supply, we have smaller losses for both thermal conductivity and convection. Excess heat will be used for steam generation. If we increase the power of the oven that heats up a boiling kettle, then its temperature will not change, but the amount of water evaporated in one second will increase.

If we do not turn off the heating of the kettle and the water does not flood the fire, then the liquid in it will completely boil away and cease to stabilize the temperature, which was determined by the temperature of the phase transition between water and water vapour at normal atmospheric pressure. We will get an empty kettle, heated above 100°C, over a flame.

What will happen to it next? Will it melt or will it just stand like a monument to a careless owner? It depends on the power of the heater. It is difficult to melt a kettle itself. It is easier to melt the solder in a welded kettle. Some people are known to have done this trick quite successfully. However, few of them did it deliberately in the course of a planned physical experiment.

Theoretically, with a very powerful burner, we will get an empty kettle glowing in the dark and pretending to be a lighting device. So we would come to the conclusion that one should not always neglect thermal radiation when studying the boiling of a kettle. But it is much more important to turn off the heater in time and not be very distracted while boiling the water in the kettle.

4.6.2. *High-speed cooking of potatoes at the TV competition*

Now let's complicate the task a little. Why just boil water? Why not boil something in it, such as potatoes. Since potatoes stubbornly refuse to boil in cold water, the conclusion is obvious: if you want to eat boiled potatoes, you must first heat the water. Usually, to speed up the process, it is not just heated, but brought to a boil.

Everything is extremely simple. We will not even waste time trying to find out exactly how potatoes are cooked and what reactions occur during this process.[i] It is enough that they occur only in hot water, so you need to heat the water to a boil and give time for these reactions to complete their business, that is, turn raw potatoes into boiled ones.

Let's add a sports element. Once upon a time, I watched on TV a competition of young ladies called "Come On, Girls", which was something like an all-around. I add more details and let you know that it was Soviet television. Therefore, it was not the most beautiful, intelligent, or dexterous who had to win, but the most housewifely girl. One of the stages of this stupid all-around was to boil potatoes as quickly as possible. Each young lady received potatoes, a knife, a saucepan, water, a stove, salt, oil, and matches. At the command of the showman, everyone simultaneously started peeling and boiling potatoes. The winner was the girl whose potatoes were cooked the earliest. Readiness was determined by how easily the potato pierced with a fork.

Let's imagine that you also need to take part in a similar competition. But what is there to waste time on trifles: you have to win it. Therefore, all legal and semi-legal means must be used to win. Let's say that the organizers gave you the pan and it is no different from the rivals' pan. Therefore, everyone is on an equal footing here. How to win using your wits and the knowledge of the laws of physics? There are two important points: the amount of water and the start time of heating.

It is clear that the water should cover the potatoes in the pan, but no more. Indeed, the less content in a saucepan, the faster it will heat up to a boil. Potatoes are mostly water, so their heat capacity

---

[i]The polymer molecules of carbohydrate named protopectin are converted into molecules of the soluble pectin.

cannot be very different from the heat capacity of water. Therefore, the less water you pour, the faster it will boil.

By the way, the heat capacity of water (about $4200\,\text{J}/(\text{kg}\cdot\text{deg})$) significantly exceeds the heat capacity of aluminium ($920\,\text{J}/(\text{kg}\cdot\text{deg})$), cast iron ($500\,\text{J}/(\text{kg}\cdot\text{deg})$), steel 460 ($\text{J}/(\text{kg}\cdot\text{deg})$), copper ($385\,\text{J}/(\text{kg}\cdot\text{deg})$), silver ($235\,\text{J}/(\text{kg}\cdot\text{deg})$), gold ($129\,\text{J}/(\text{kg}\cdot\text{deg})$), that is, all the metals from which, even theoretically, a pan can be made. Well, if there is a choice of several options for pots, then it should not be very large and massive. In any case, it will have to be heated to $100°\text{C}$, so let less heat go to it. A gold pan will be better than an aluminium one, but you are unlikely to be offered one.

Moreover, the sooner the pot of water starts to heat up over the fire, the faster it will boil. So let's start the competition by putting a pot of water on the fire. If possible, the water should be poured as hot as possible. In ordinary life, we will not pour water from a hot tap into a saucepan, but then our goal is to eat delicious potatoes. Here they needs to be cooked as soon as possible, but no one will eat them.

Do I need to add salt? How much? I think this is not important. The effect of salt on the rate of heating of water is small compared to what we are discussing. When preparing real dishes, this issue is worth taking into account. For example, if we cook meat or chicken and add salt immediately, we get a delicious rich broth, and if we add it at the end of cooking, we get delicious meat. But you can think about more than just salt. When potatoes are boiled, calcium and magnesium cations, which are part of protopectin, are transferred into the water.

The cooking process slows down because of an acidic environment, as well as in hard water, that is, in water with an increased concentration of those same calcium and magnesium salts. Experienced cooks know that in sour soups, for example, in pickle, hodgepodge, or borscht, potatoes should be put first, and sour components of the soup, such as tomatoes or pickles, should be added later, otherwise the potatoes will remain firm. Considering that in competitions the readiness of a potato is checked by poking a fork into it, it is better to provide not an acidic but an alkaline environment.

For example, it would be nice to put soda in the water instead of salt. I don't know if potatoes boiled with soda are tasty. But we don't care about its taste; we need to defeat competitors! We mentally participate in an exciting competition where there are no trifles. The famous tea clipper races are replete with stories of how captains tried by all means to achieve even the slightest advantage: if you win, then the name of your ship becomes history, and if you lose, then woe to the vanquished.

So, we poured water so that it should later slightly cover the potatoes, which we had not yet begun to peel (so we need to estimate the required amount of water in advance), and set the pan on maximum heating. Now let's start peeling and cutting the potatoes. I advise you not only to peel the potatoes, but also to cut them into small pieces. They can be packaged more compactly in a small volume of water. This makes it possible to reduce the amount of water required ever further.

But the main point of cutting is different. If you cook whole large potatoes, then the heat must get to the centre of the potato to cook it. Therefore, the potatoes near the surface may already be cooked, but the insides may still be hard. Cutting eliminates this problem. In real life, large potatoes are cut into pieces so that they cook more evenly in volume, but now our goal is different.

We immediately throw pieces of potatoes into water so the cooking process begins faster. We do not use cheap tricks, trying to peel only one potato, but when peeling, no one bothers us to cut off a thicker layer, mentally grinning wickedly.

We finish peeling quickly and immediately close the pan tightly with a lid, leaving no gap or hole. A slit or a lid slightly shifted to the side helps to protect water from boiling violently and flooding the stove, but this danger does not bother us, because we are close by and carefully monitor the process.

Why is it important? The water in the pot is still far from the boiling point of 100°C, but vaporization is already taking place.

Let the initial water temperature be 25°C. The saturated vapour pressure at this temperature is 3.17 kPa or 23.8 mm Hg. This is the partial pressure that steam has just above the surface of the water.

Now the water in the pan has warmed up to 50°C. The steam pressure above the water has increased to 12.34 kPa or 92.6 mm Hg, that is, more than three times. Where did this steam come from? Naturally out of the water. And it has the temperature of water, just like the air in a saucepan over water. It is also heated by water. But the air in the room where the competition is taking place probably has a temperature of less than 50°C. If the lid has a gap or hole, then warmer air and water vapour would escape into the room, partially replaced by colder ambient air. This is loss of precious heat.

In addition, the ambient air has a lower absolute humidity and this leads to a faster rate of water evaporation. Let me remind you that water vapour is a hidden reservoir of heat, because it has the additional heat already spent on evaporation. This steam will condense on the colder lid, releasing energy, and the condensed hot water will drain back into the pot.

If we heat water in an open pan, then we will have not only a leak of steam and warm air, but also increased convection, which will increase the heat removal from the pan. By the way, if a breeze or just a draft is blowing in the room where the competition is held, you need to cover the pan from it with any improvised screen, at least with your own body.

But still, some of the warm air will definitely come out into the room. Why? The pressure of the gas above the water in the pan is equal to the atmospheric pressure. Of this pressure, which is approximately 101.3 kPa or 760 mm Hg, part is provided by steam, and the rest is provided by all gases other than water vapour. When heated from 25°C to 50°C, the vapour pressure increased by $12.34 - 3.17 = 9.17$ kPa. If the pan were hermetically sealed, then the total pressure inside would increase by this additive and would exceed the pressure in the room by almost an eleventh of the atmospheric pressure. It is on this principle that the pressure cooker works. But our saucepan is not a pressure cooker and the air will come out, reducing its partial pressure.

It seems that all measures have been taken to win? Not yet! We also have a secret trick that brought victory in the television competitions in question. In them, the winner threw a large piece of butter into a pot of boiling potatoes. Any other oil, such as olive oil, would work as well. Yes, even machine oil, because after the competition, as I have already written, no one will eat potatoes.

What is the essence of the trick? Why does this "ace up the sleeve" work? Oil does not dissolve in water and is lighter than it, so it forms a thin film on the surface of the water. It has little or no effect on the heat required to heat the pan, but it greatly reduces the rate of evaporation of the water.

As mentioned in Section 3.2, molecules are constantly leaving water when evaporated and returning to it when condensed. Thermal equilibrium requires the equality of the speed of these processes. However, many phenomena in the world are too fleeting for equilibrium to be established. If you pour a glass of water on the floor, then the only possible kind of steady balance under normal European summer conditions is complete evaporation of water. So it will be, but the process at low air temperatures will last a long time. We risk seeing a wet puddle a few hours later.

The film on the surface slows down this already slow process. Water molecules, falling on the surface separating them from the surrounding air, as a rule, remain inside. A very small percentage of them are able to leave the liquid. And these are the fastest molecules, whose high kinetic energy allows them to overcome the energy barrier at the border. The oil film only increases the height of the barrier because the water molecule, leaving the liquid, finds itself in the oil layer, and water and oil do not like to get inside each other.[j]

After the beginning of a violent boil, we can safely reduce the heating, but not before it. It's time to reduce the heating power. You

---

[j]This leads to a kind of "effective repulsion" of polar water and non-polar oil (for polar and non-polar dielectrics, see Section 4.2). Even if mixed vigorously, oil forms droplets inside water. This is called an emulsion. Drops of fat are easy to observe on the surface of non-fat chicken broth.

cannot miss the moment when the water boils, because the maximum fire under the pan and a tightly closed pan can provide a vigorous boil with the burner flooding. We reduce the heating power so that the water continues its boiling, but does not flood the fire.

You can relax in anticipation of victory. The main thing is not to forget to give a signal to the judge that the potatoes have become soft. And even though not the most honest methods were used and eating the winning potatoes is not recommended, the rules of the competition were not grossly violated; rather, they were slightly expanded.

5   ATOMS. SCALE: ATOMIC AND SMALLER

It's time to go to the microcosm — the kingdom of atoms, elementary particles, and quanta. It is very different from the macrocosm, which I talked about before. And its lower limit of scale moves along with progress in science, all the time being at its forefront.

In the microcosm it is impossible to neglect the quantum properties of atoms and particles. The microcosm is ruled by quantum mechanics. What does this mean? This cannot be explained in a nutshell, and I do not want to reduce everything to stupid anecdotes. So make yourself comfortable, you will listen to a long story about the quantum world and its weirdness.

5.1. How Did the Idea of Quanta Come About?

5.1.1. *Revolutions in science*

Physicists opened a window to the quantum world at the very beginning of the 20th century. To do this, they had to arrange a small revolution that affected almost all branches of physics and clearly divided this science into classical and quantum physics. On the one hand, scientists are no strangers to scientific revolutions. On the other hand, it was only this one that manifested itself not only in some section of physics but also in its very foundation. It influenced the further development of various directions of physics, from the physics of atoms, nuclei, and elementary particles to solid-state physics and astrophysics.

However, scientists do not like to use pretentious terms and modestly call it the process of changing paradigms (from the Greek παράδειγμα, "example", "model", or "sample"). What is hidden behind this sophisticated word? Surprisingly, you could find a lot of very different concepts and meanings. In linguistics, these are the rules for declension or conjugation of words; in programming, these are the principles or style of writing a program for a computer. But we are interested in science.

In science, this word means what is on the scale of authority and importance above theory and even more so hypothesis. This is a set of fundamental scientific attitudes, concepts, and terms, accepted and shared by the scientific community and uniting the majority of its members. This meaning of the term *paradigm* was proposed in 1962 by the American historian of science Thomas Kuhn in his book *The Structure of Scientific Revolutions*.

An example of a paradigm is the statement "All bodies are made of molecules, and those are made of atoms." Paradigms underlie each of the branches of physics and other sciences. Sometimes they change and these changes are called scientific revolutions. This happened with the emergence of paradigms on which general relativity and quantum theory are based. But even in one science, paradigms often change one another.

5.1.2. *Paradigm shifts in optics*

Consider optics as an example. Its paradigm is the answer to the question, what is light? For a long time, ancient thinkers answered, "We do not know," which did not prevent them from discovering the laws of reflection and refraction of light. It suffices to recall that one of the great ancient Greek philosophers argued that we see because peculiar rays come out of our eyes, as if feeling what surrounds us. To this, another, no less great ancient Greek philosopher objected that in that case we would have seen at night.

After the publication of the book *Opticks* by Isaac Newton, the corpuscular theory dominated, according to which light is a stream of the smallest particles or corpuscles. Newton was able to explain within its framework the laws of reflection and refraction of light, and additional assumptions seemed to somehow explain the interference of light. Robert Hooke's and Christiaan Huygens's alternative wave theory of light was able to explain the same processes, but there were phenomena that were not explained by any of these theories.

By the beginning of the 19th century, it became clear to physicists that the theory of the great Newton was outdated and it was necessary to somehow explain the phenomena of diffraction, birefringence of light in crystals, and many others. The Frenchman Augustin-Jean Fresnel succeeded in that brilliantly. After his works, the answer "Transverse waves in the *aether*" began to be considered the correct answer to the question about the nature of light. It was the word *transverse* that essentially distinguished the Fresnel wave theory from the previous wave theory.

Transverse waves are waves in which something moves or vibrates in a direction perpendicular to the propagation of the wave. I talked about longitudinal and transverse waves in Section 2.4. The *aether* or the *luminiferous aether* was considered as an unknown elastic medium in which these waves propagated. The word *luminiferous* means light-bearing, and *aether* comes from the Old Greek αἰθήρ, meaning upper air layer.

The last nail was driven into the coffin of corpuscular theory at the time of this rapid progress in optics. The fact is that the law of refraction of light can be obtained within the framework of both corpuscular and wave theories, but their conclusions differ greatly in one important detail: the speed of light in different media. Consider how a ray of light transfers from air to water. If this is a stream of particles, then its speed increases; if it is a wave, then its speed decreases during this transition.

Comparing the speed of propagation of light in air and in water, it remains to find out which theory is wrong. However, these speeds are so high that it was possible to do this only in the middle of the 19th century. In 1849, the Frenchman Armand Hippolyte Louis Fizeau was able to measure the speed of light in air for the first time using a rapidly rotating gear, periodically blocking the path of the light beam with teeth.

The beam passing through the cut-out reached a mirror mounted on a mountain 8.66 km away from the source and was reflected and passed back through the same gear. If during the travel of light to the mirror and back, the gear turned one tooth, then the beam freely passed through it, but if it turned only half a tooth, then the beam was completely blocked.

The resulting value of the speed of 313,000 km/s was in good agreement with both the value obtained back in 1676 by the Danish Olaf Christensen Rømer from astronomical observations and the modern accurate value of the speed of light. But it was clearly impossible to repeat this experience under water.

This difficulty was overcome by another Frenchman, Jean Bernard Léon Foucault, who in 1862 used a rotating mirror scheme that did not require light to travel such a large distance. Indeed, the distance from the light source to the mirror was only 4 m, but the mirror made 800 revolutions per second.

Incidentally, this explains why the experiments of Fizeau and Foucault could not have been carried out a century or two earlier. Without the success achieved by the then metallurgists in creating extra-strong alloys, the gear or mirror made earlier would have simply shattered to pieces when made spin so quickly.

As a result of experiments in air and in water, an answer was obtained to the question, in which medium does the light move faster? And answers to such direct questions in science are valued more than a lot of indirect evidence in favour of some theory.[a] It turned out that light in water lags behind light in air or emptiness, which meant that it was possible to give up on the corpuscular theory. The first revolution in optics is over.

Advances in explaining known phenomena were complemented by discoveries that paved the way for new subsections of optics. In the middle of the 19th century, the French philosopher

---

[a]It is interesting that direct evidence that the Earth revolves around the Sun, and not vice versa, was obtained by the English astronomer James Bradley only in 1727, i.e. two centuries after the emergence of the heliocentric theory of Nicolaus Copernicus (Mikołaj Kopernik).

Auguste Comte wrote that we cannot know anything about the stars except that they exist, so astronomy is a waste of time that cannot provide any interesting or useful results. Soon after this statement, optics, analysing the emission spectra of stars, found out what they consist of, in what proportion these elements are included in the composition of stars, and what is their temperature.

Spectroscopists discovered two new chemical elements in laboratories — cesium in 1860 and rubidium a year later. And in 1868, helium, the second most abundant chemical element in nature, was discovered in the solar radiation spectrum. On Earth, it was separated only 27 years later, in 1895.

Time passed, and James Clerk Maxwell wrote his system of equations describing all electromagnetic phenomena. From them, it turned out that there must have been electromagnetic waves propagating in a vacuum at the same speed as light and with the same properties. These waves were demonstrated in the experiments of Heinrich Hertz. It became clear that visible light is electromagnetic waves with a frequency that lies within certain limits. Soon, other electromagnetic waves were discovered that the eye cannot see, such as infrared, ultraviolet rays, or X-rays. There was a new revolution. After this, the answer "Light are electromagnetic waves in the ether" was considered correct.

At the beginning of the 20th century, two important events took place in science. The special theory of relativity (SRT) did away with the idea of *aether* and it turned out that light does not need a special medium for propagation. And after the papers of Max Planck and Albert Einstein, there appeared the idea of quanta of light — photons. After this third and, it seems, the last revolution, the correct answer to the question about light is "Light is a stream of photons." But a detailed account of this revolution is a little further ahead.

What happens to rejected theories and paradigms? Some go straight to the landfill, like the idea of *aether* or phlogiston. But in optics, the old paradigms gave rise to two sections of optics: geometric and wave. No one will calculate the optical design of binoculars or microscopes based on the idea of a photon flux. Scientists used Newtonian ideas for these purposes, but with some subsequent improvements. And the Fresnel approach is quite sufficient to explain experiments on light diffraction. Thus, old paradigms transformed into approximate theories. The same thing happened with classical mechanics, which got this status by special theory of relativity, general theory of relativity, and quantum theory.

5.1.3. *Problems that led to the idea of quantizing radiation*

Why did the idea of quantization arise? Physics was the first of the natural sciences to become a real science, and not a natural philosophy full of prejudices. This happened at the turn of the 16th and 17th centuries, when physicists developed the foundations of what is now called the scientific method. That is, the foundations of what we mean by science. These include experimental testing of hypotheses, application of mathematics, and drawing conclusions and predictions based on theoretical research. By the end of the 19th century, physics had already developed and strengthened with might and main. Scientists were able to explain many phenomena and create a lot of devices on the basis of science. Discoveries poured in as if from a cornucopia.

The cause of the scientific revolution was not the unsolved problems of science that always exist in it, but the contradictions between the theories that existed then, or rather their consequences, and the results of numerous experiments. They accumulated in different branches of physics and it was already impossible to ignore them.

A particularly tense situation had developed in two branches of physics: optics and atomic theory. It was this that led to the need to introduce the ideas of quantizing radiation and matter. The relationship between these problems was realized only later. After the ideas of quantization entered physics, they spread to other branches of physics, such as to solid-state physics, where they successfully solved a number of problems that could not be dealt with without the quantum approach.

However, the breakthrough was provided precisely by these two branches of physics, where contradictions between theory and experiment were especially glaring. Scientists trying to understand these problems have come to the conclusion that half measures will not help and the situation cannot be saved without revolutionary changes.

Let's try to understand why it was necessary to introduce photons or quanta of light instead of the wave theory of light. The latter was killed by two problems that at first seemed trivial.

Let's start with the first, more theoretical one. As you know, strongly heated bodies glow like our Sun or a spiral in an incandescent light bulb. Less heated ones, such as our skin, emit infrared waves. Our eyes cannot see IR radiation, but it can be observed with a thermal imager. It is clear that a spectrum of radiation, which is the distribution of the emitted energy by frequencies or wavelengths, depends on the temperature of the heated body.

It also depends on what material is on the surface of the body and emits electromagnetic waves. Physicists speculate about a certain ideal material that does this as much as possible and call it an absolutely black body. Any other body emits less light or other electromagnetic radiation at any of the frequency ranges.

Of the materials at hand, black velvet is suitable for the role of an absolutely black body, but for obvious reasons it is not recommended that we heat it to a temperature of several thousand degrees. For this, platinum sponge (metallic platinum in a high-purity greyish-black porous spongy form) is better suited, if you forget about its price.

The spectrum of the thermal emission of an absolutely black body is called the spectrum of blackbody radiation. Its distribution of radiation depends on temperature and does not depend on other parameters.[b] At the end of the 19th century, physicists made a lot of measurements and experiments and thoroughly studied the spectrum of blackbody radiation at different temperatures. There

---

[b]The obtained later spectra of solar emission and CMB were in some ranges very similar to the spectrum of blackbody radiation, naturally for different temperatures.

remained to obtain a formula describing the blackbody spectrum from theoretical considerations.

But some unexpected difficulties arose. The so-called Rayleigh–Jeans formula for the blackbody spectrum, based on classical physics, worked well for low radiation frequencies, satisfactorily for medium frequencies, and very poorly for especially high frequencies. At the same time, it was derived simply enough to explain the discrepancy by an error in the calculations.

But that was only half the problem. If we make a cavity filled with vacuum inside any solid body with constant temperature, then something like a gas from electromagnetic radiation exists inside this cavity. Its spectrum coincides with the spectrum of blackbody radiation with a temperature equal to the body temperature, regardless of what colour and from what the walls of this vacuum cavity are made.

Light and other types of electromagnetic waves carry energy — for example, the energy of the Sun that is transferred to our Earth. Accordingly, the radiation inside the cavity has a certain energy density depending only on the temperature of cavity walls. It is not difficult to obtain it knowing the formula of the blackbody radiation spectrum.

But if we use the Rayleigh–Jeans formula obtained in the framework of classical physics, it turns out that the energy density is infinitely high. The radiation power of heated bodies also becomes infinitely high. This happens due to radiation with huge frequencies, which is why this problem is called the ultraviolet catastrophe or the Rayleigh–Jeans catastrophe.

Why exactly a catastrophe? From these calculations, it turned out that any heated body would instantly cool down, transferring its energy to electromagnetic radiation, but would not heat it even by a millionth of a degree, because this would require infinite energy, which a heated body does not have. As a result, the temperature of everything in the Universe would drop to absolute zero. However, we see a completely different picture around us. So even without special experiments, you can be sure that no ultraviolet catastrophe exists.

The second problem was less dramatic. It is associated with experiments on the study of a phenomenon new for those times, called the photoelectric effect. What is it? Imagine two metal plates with a gap between them. This structure is placed in a vessel, for example a glass vessel, from which air is evacuated. If a voltage is applied to the resulting capacitor, then the electric current would not flow in the vacuum.

Why? Current is the ordered movement of charges, in this case conduction electrons. These electrons can move anywhere inside a metal conductor without the need for energy. We can say that electrons, like a liquid, are poured into a conductor or, like a gas, fill its entire volume.

But they do not go beyond the limits of the metal. Maybe they would like to go outside, but this requires the expenditure of a certain amount of energy for each electron emitted outside. It is called the work function. So, electrons don't get out of the metal for the same reason that swimming pool water doesn't crawl onto the surrounding walls, which are higher than the water level.

However, it is not difficult to get the water to find itself on the swimming pool walls, simply by giving it an additional portion of energy — let's say by starting to splash in the pool or by bringing down a concert grand piano from the roof into it. Splashes are guaranteed!

In the case of our plates in a vacuum, there are many ways to provide the electrons with the energy they need to escape. The plate can be heated, which was done in ancient vacuum radio tubes — diodes, triodes, and more complex tubes. This phenomenon is called thermionic emission, and the heated negatively charged plate is called the cathode. A positively charged plate is called an anode. A plate can also be irradiated with radioactive particles. This idea was implemented in 1908 by the German Hans Geiger and improved in 1928 by him and Walter Müller; therefore, the corresponding device is often called the Geiger–Müller counter. It is slightly different from the described example, in that it does not contain a vacuum, but a gas in which a discharge occurs when a particle passes through it.

You can also illuminate the metal with light. The photoelectric effect is a phenomenon in which light transfers its energy to electrons, allowing them to escape out of a metal or other substance, such as a semiconductor. Naturally, the current flowing between the plates is proportional to the light intensity. But we can put a colour filter in the path of the light and select the light of a certain length. In this case, it all depends on the frequency of the light.

Red light has the longest wavelength and the smallest frequency in the visible range. Infrared light has an even longer wavelength. The frequency of violet light is higher. It is approximately twice the frequency of the red border of visible light. UV radiation has an even higher frequency.

If you shine light of a certain colour on a plate surface, then sometimes you can achieve the escape of electrons, and sometimes not. There is a certain critical frequency above which light causes a photoelectric effect. But longer-wavelength radiation is not capable of this.

If weak violet radiation knocks electrons out of some substance, then red radiation may not knock them out at all, even if we shine a powerful searchlight through a red filter. However, in this case, red light carries much more power than violet light.

5.1.4. *Planck's idea: Quantizing the radiation process*

So, we got acquainted with two problems that could not be explained by classical wave optics. How did the idea of quantizing radiation come from these problems? To put it very roughly, the German Max Karl Ernst Ludwig Planck in 1900 proposed a formula that perfectly describes the blackbody spectrum. Since then, the spectrum of blackbody radiation is also called the Planck spectrum. Using one additional assumption, he also derived this formula within the framework of classical physics. He supposed that when light is emitted or absorbed, its energy changes in discrete portions, called quanta by Planck.

The value of the quantum of energy of light, E, is proportional to the frequency of this light v. Let us write this in the form of the formula $E = h\nu$. The coefficient of proportionality, denoted in it by the letter h, is called Planck's constant $h = 6.626 \cdot 10^{-34}\,\mathrm{J} \cdot \mathrm{s}$. Together with the speed of light in a vacuum, the charge of an electron, and the gravitational constant, it belongs to the most important, so-called fundamental constants that determine the laws of our world.

And it doesn't matter at all whether Planck chose a formula to describe the spectrum, and then figured out how to get it, or first suggested a formula for the energy of a quantum $E = h\nu$ and, on its basis, derived the formula for the Planck spectrum. It is important that an experimentally confirmed formula for the frequency distribution of thermal radiation was obtained. Naturally, when using it, no ultraviolet catastrophe occurs.

So, the energy of monochromatic radiation with one frequency, that is, light of the same colour, can be transferred to matter only in equal portions. The energy may be constant, or the energy of one, two, three, $\ldots$, 9183, $\ldots$, one hundred million quanta could be transferred. In general, any integer number of them. Energy can change discretely, in contrast to wave optics, where it can change continuously, by any arbitrary amount.

This is the main thing that Planck added to classical optics. And in 1918, he was awarded the Nobel Prize in Physics "for the services he rendered to the advancement of physics by his discovery of energy quanta".[c] This is not surprising, because the importance of his work by this time was clear to all physicists.

Planck's constant h has become a kind of emblem or symbol of the quantum world. In addition to her, physicists also use her sister, the so-called reduced Planck's constant, which is also referred to as

---

[c]Why then? According to the will of Alfred Nobel, the annual prize had to be awarded to the best works of the current year in physics, chemistry, physiology, and medicine, as well as in literature and the preservation of peace. It has been awarded since 1901, so Planck's work was ahead of it by one year. But the First World War led to a failure in the system for determining the laureates and it was decided to award prizes for the works of previous years. One of the first to receive the award under the new rules was Planck.

Dirac's constant, which is 2π times less than h. We use the special character "h-bar" for it: $\hbar = h/2\pi$.

Planck emission and absorption of light is similar to the process of trading. The buyer pays money for the goods and may receive change, but he always pays an integer number of denominations of the smallest coin. Let it be a penny, but a half-coin does not exist. The buyer cannot give, and the seller cannot receive, say, a penny and a half.

This detail does not prevent the product from costing less than a penny, say "a couple for a penny" or "a dozen for a penny". Even "8051 for a penny" is quite acceptable. Moreover, what we pay for can be continuous. For example, if we pay per minute or per second for a telephone conversation, we can talk for any period of time, but we will pay for a whole number of seconds or minutes.

If we are dealing with radiation of different frequencies, then the analogy can be continued. Each frequency has its own currency and its own "coin" of minimum denomination, that is, its own quantum, and the conversion rate between different currencies obeys Planck's formula $E = h\nu$. Well, the universal world currency is a unit of energy, equal in the SI system to one joule. It is into joules that we can convert the energy of a gift set of five red quanta with a wavelength of 656.3 nm, four yellow quanta with a wavelength of 589.6 nm, and one violet quantum with a wavelength of 404.7 nm. Naturally, the quanta themselves have no colour. We are talking about the colours of light with specific wavelengths.

The idea of a quantum also prompts an explanation of the mystery of the photoelectric effect: if light has a too low frequency and the energy of one quantum is less than the work function of the electron, then one quantum cannot knock the electrons out of the substance. And if one quantum has enough energy for this, then the number of electrons emitted per second, which determines the current in the circuit, is proportional to the number of light quanta falling per second, that is, the radiation intensity.

Note that Planck's idea consists of two parts: the existence of quanta and a specific formula, $E = h\nu$, for their energy. This formula is confirmed not only by the coincidence of the measured blackbody spectrum with the Planck spectrum calculated on the basis of this formula, but also by direct experiments on the photoelectric effect. In 1902, the German Philip Eduard Anton von Lenard, winner of the Nobel Prize in Physics for 1905, conducted these experiments, and Albert Einstein drew conclusions from them.

5.1.5. *Einstein's idea: Photons are quanta of light*

It was this great physicist who made the second step in the quantization of electromagnetic radiation in a series of papers published in 1905–1917. And although his name is primarily associated with the creation of general relativity and special relativity, Einstein received the Nobel Prize in Physics in 1921 "for his services to theoretical physics, and especially for his discovery of the law of the photoelectric effect". Einstein developed the quantum theory of light, which physicists still use today. Lasers work on its basis, for the discovery of which the Soviet physicists Nikolai Basov and Alexander Prokhorov and the American Charles Hard Townes received the Nobel Prize in 1964.

According to Einstein, light exists in the form of light quanta. We now call them photons. This is the genitive case of the ancient Greek word for light. It was used in 1926 by the chemist Gilbert Lewis in his theory, which conflicted with experimental data. The theory was forgotten, but the term remained. Einstein's theory described in detail the properties of photons. In this case, the formula $E = h\nu$ remained valid.

What is the main difference between the approaches of Planck and Einstein? Planck believed that light is emitted and absorbed by quanta; Einstein stated that it exists in the form of quanta, i.e. photons. You may not have grasped the difference. Let me explain it with the example of beer and lemonade. These drinks are available in cans and bottles. More precisely, you can only buy an integer number of cans or bottles. This is a property of trade, not a beer.

The analogue of a purchase is the process of emission or absorption of a quantum. You can't buy half a can of beer, but you can drink half a can and leave half for later. You can't buy a piece of sandwich, but you can take a bite of any size from it. This is Planck's approach. Before him in physics, it was believed that the emission of light is like pouring draft beer into a mug, where you can pour any amount.

But the analogue of Einstein's approach would be a world where you cannot drink half a can of soda: you can drink either nothing or one or several cans. Sandwiches must also be swallowed whole. It's just that cans or sandwiches in this quantum world cannot be divided into parts; these are quanta of soda and sandwiches. Rain is made of water droplets, and light is made of photons.

In experiments on the photoelectric effect, one photon can either knock an electron out of the material, transferring to it an energy equal to or greater than the work function, or not. In the first case, the current flows, and in the second, it doesn't. An electron cannot receive the necessary energy from two photons at once, only from one. Remember the splash from the piano that fell into the pool? You won't get the same hard splashing if you drop a dump truck of matches into it, even if they weigh as much as a grand piano.

The absorption of a flux of photons by a substance can be compared to the collection of rainwater in buckets exposed to rain. To fill a standard bucket, a lot of droplets must fall into it, so the water level in adjacent buckets will rise evenly and at approximately the same rate. But this is normal rain, in which the size of the drops cannot be very large.

The energy of a photon is proportional to its frequency. It can differ for different photons by hundreds and thousands of times; more precisely, it can be arbitrarily different. If instead of photons of visible light we absorb photons with high energy corresponding to X-rays or gamma rays, then their energy will be equal to the energy of many photons of visible light.

Using our rain analogy, we expose the buckets to rain, the drops of which are the size of a bucket themselves. Some end up in the bucket that has been placed. Plop! And the bucket is full. And the adjacent buckets will remain completely empty.

Physicists have long observed an analogue of this in their experiments. An X-ray tube at a distance of 1 m provides an X-ray flux carrying energy of about $0.01\,\mathrm{W/m^2}$. The cross-sectional area of one molecule is of the order of $10^{-19}\,\mathrm{m^2}$; therefore, per second, a molecule receives on average about $10^{-21}\,\mathrm{J}$ of energy, which corresponds to $6 \cdot 10^{-3}\,\mathrm{eV}$.

If you put in this place a Wilson cloud chamber, the elementary particle detector described in Section 3.2, it would immediately detect electrons with energies of 10^4–$10^5\,\mathrm{eV}$, which received this energy from X-ray quanta. But with such a rate of accumulation of radiation energy by molecules, this energy should have been collected for weeks and months. But here we are dealing with bucket-sized drops. Plop! And electrons are ready.

In a nutshell, this is the history of quantizing electromagnetic radiation. As a result, we obtain the most recent paradigm change in optics. The waves of light were replaced by photons. They are sometimes called wave–particles because of their bizarre mixture of wave and corpuscular properties, such as the ability to cause interference and diffraction and at the same time discreteness of energy. Wave–particle duality is deeply embedded into the foundations of quantum mechanics.

The new paradigm was tested in subsequent experiments, such as the scattering of photons by electrons, for which the American Arthur Holly Compton received half of the 1927 Nobel Prize in Physics.

What do we know about photons? They have the properties of both waves and particles. The photon energy is equal to $h\nu$, and its momentum is equal to the energy divided by the speed of light in vacuum. Therefore, it is equal to h/λ, where λ is its wavelength. The photon also has angular momentum, but we will talk about it separately.

Let me emphasize that the cause of the revolution was not the articles of the theorists Planck and Einstein, but insoluble problems within the framework of the paradigm that existed at that time. They are associated with the inability to explain the results of experiments with the radiation of heated bodies and the photoelectric effect. New theories explain the physics of these phenomena and their predictions are consistent with experimental results.

5.2. Physicists Build Models of the Atom

The next stage of understanding the quantum world is associated with the quantization of matter, primarily elementary particles. It demanded a scientific revolution with a paradigm shift, too. As a result, physicists consider particle–waves instead of particles. And like every revolution, there were serious reasons for this.

In 1897, the English physicist Joseph John (or J.J. for short) Thomson discovered the electron. The German physicist Johann Emil Wiechert did it independently.

This first of the elementary particles, the existence of which humankind discovered, got its name from the Greek word ἤλεκτρον — amber, like electricity in general. Electron flows arise in various processes, whose nature physicists did not initially understand. Therefore, they received several names, for example, the stream of electrons formed during radioactive decay was called beta rays.

J.J. Thomson worked with cathode rays. This term was used for the flow of electrons emitted by the cathode of a vacuum tube. Those who remember a TV set or display with a cathode-ray tube saw the picture created by these rays on the surface of a substance that glows when an electron beam hits it.

Thomson investigated the deflection of cathode rays in electric and magnetic fields, simultaneously laying the foundation for the use of picture tubes. He proved that these rays are a stream of identical negatively charged particles called electrons that move in a straight line but can be deflected by a magnetic field. They have a mass significantly less than that of atoms. More precisely, their mass is of the order of one-thousandth of the mass of a hydrogen atom — the lightest of the elements — and a charge equal to the charge of the hydrogen ion, which we now call the proton.

J.J. Thomson received the Nobel Prize in Physics in 1906. It could not be awarded directly for the discovery of the electron, which occurred before the appearance of the prize, and it was given only for the discoveries made in the current year. But the wording "for his theoretical and experimental investigations on the conduction of electricity by gases" allowed Thomson to receive the award.

Everyone understood that it was the discovery of the electron that made a decisive contribution to the noted research.

It became clear to physicists that electrons must exist inside a substance (otherwise where would they come from in the cathode?), but they cannot consist of heavier atoms. This means that they must somehow be part of these atoms. Since atoms are electrically neutral and electrons are negatively charged, some parts of the atoms must be positively charged in order to compensate for the negative charges of the electrons.

5.2.1. *Pudding model*

J.J. Thomson proposed his model of the atom, nicknamed the plum pudding model, in 1904. In it, the atom consists of electrons located inside a positively charged medium, like negatively charged "plums" in a positively charged "pudding". Electrons in different versions of the model could be at rest or somehow move, but these were just speculative assumptions. The pudding model only survived until 1911.

I note that during this period, it competed with other models, for example, the model of the atom of the Japanese physicist Hantaro Nagaoka, in which electrons formed something like rings around a massive central positively charged body. Two assumptions present in this model have been preserved in later descriptions of the structure of atoms.

This is the existence of a massive positively charged nucleus and the rotation of electrons around it. Now physicists agree with the first part and to some extent with the second, but the name of the Japanese is not often remembered when talking about the models of atoms. This is because his model was the fruit of pure guesswork, inspired by Maxwell's work on the stability of Saturn's rings.

5.2.2. *Planetary model*

The main argument against Thomson's model, which sent it into retirement, was naturally related to new experiments. In 1911, the New Zealand-born British physicist Ernest Rutherford passed a stream of alpha particles through gold foil and measured the distribution of their deflection angles. Most alpha particles were deflected at very small angles. However, a small part was reflected in a direction almost opposite to the initial. And it was very strange.

Imagine that you are firing a large-calibre machine gun at a bunch of children's balloons. It is natural to expect that ordinary balls hardly change the direction of the bullets' flight. If the balls are filled with metal, then the bullets would ricochet from them in all directions. And a scattering pattern similar to that observed by Rutherford is only possible if the balls are mostly filled with air but contain small metal balls heavier than bullets.

He suggested that an atom has a very small nucleus containing a large positive charge. Positively charged alpha particles deflect from it due to electrical repulsion. It becomes very strong near the nucleus, and alpha particles may even fly back. The positive charge of the nucleus is compensated by the negative charge of the electrons. Rutherford suggested that electrons revolve around the nucleus in orbits similar to how planets orbit the Sun. This model is called the planetary model of the atom.

The Dutchman Antonius Johannes van den Broek immediately hypothesized that the number of electrons that an atom has, which is equal to the charge of the nucleus, expressed in units of electron charge, matches the atomic number of a chemical element in the periodic table. In 1913, Henry Moseley confirmed it experimentally.

Rutherford was awarded the Nobel Prize in Chemistry in 1908 "for his investigations into the disintegration of the elements, and the chemistry of radioactive substances." The person who wrote "All science is either physics or stamp collecting" clearly regarded this as an irony of fate. We have been using graphic illustrations of the planetary model for a long time.

The planetary model of the atom immediately alerted theorists because it had an Achilles heel. Electrodynamics asserts that electrons orbiting around the nucleus should emit electromagnetic waves, like any charged body moving with acceleration.

In atoms, the periods of revolution of electrons had to be very small, and their accelerations huge. Accordingly, energy losses due to electromagnetic waves emission must also be huge. Theorists calculated that electrons would emit all their energy and fall onto the nucleus very quickly, in a time significantly less than a second. But atoms do exist! And after Rutherford's experiments, it was no longer possible to abandon the concept of the atomic nucleus and return to the pudding model, in which there is no radiation.

This contradiction was not resolved in any way. It was this that led to a new revolution in physics, which revealed the quantum properties of matter. Before it, all models of atoms were developed and described in the framework of classical physics. But at the stage of the planetary model, contradictions arose between the theory, specifically classical electrodynamics, and the experiment, which required both a compact atomic nucleus and electrons that do not fall on it.

So in this case, physicists were faced with a sequence that is quite characteristic for the situation before the paradigm shift in science. New experiments (such as Rutherford's) led to the emergence of new concepts (such as the compact atomic nucleus), which were incompatible with the traditional paradigm (with the loss of energy by the emission of electromagnetic waves when charges move with acceleration). Sometimes, in similar cases, the problem is resolved in some other way. It turns out that the experiments were not carried out entirely correctly, or they can be explained alternatively. But in the case of the atomic nucleus, scientists understood that a scientific revolution was indispensable.

5.2.3. *Bohr's model*

The first step in a new direction was taken by a Danish physicist, Niels Henrik David Bohr, a young theorist working for Rutherford. He decided that if the problem is that atoms do not emit electromagnetic waves, despite the fact that electrons move with acceleration, then it makes sense to find out exactly how they emit electromagnetic waves, that is, light, in the cases where they do. Let the atom not radiate all the time, but sometimes. This seemingly simple and obvious idea has unexpectedly been very useful.

I have already mentioned that each element has its own spectrum of radiation. However, some of the nuances of spectroscopy go beyond what every physicist was taught. Bohr was sure that the radiation of atoms of some specific chemical element forms a rather random picture of spectral lines. After speaking with experts in these matters, Bohr unexpectedly found out that there is an empirical formula (i.e. a formula that works, but it is not clear why) that describes the frequencies of all hydrogen emission lines. It was found by the Swedish scientist Johannes Robert Rydberg back in 1888. And similar formulas also describe the emission lines of some other elements, such as sodium or mercury.

Bohr suggested that electrons can revolve around the nucleus not in any orbits, but only in some specific, somehow selected ones. This detail makes the situation fundamentally different from the motion of planets around the Sun or satellites around the Earth, the orbits of which can be very different, almost arbitrary, as long as they satisfy the laws of celestial mechanics. These admissible specific orbits can be numbered and the energy of an electron in a given orbit can be determined for them.

When electrons jump from a higher-energy orbit to a lower-energy orbit (sometimes scientists talk about a transition from a higher orbit to a lower one, but usually about a transition from a higher-energy state to a lower-energy one), then the released energy is radiated as a photon. Its energy, and hence its frequency, related by Planck's formula, is determined by the difference in the energy of the electrons in these orbits. This value is the same for all hydrogen atoms in the world if we consider the transition of an electron from one specific orbit to another.

Therefore, radiation has a certain set of frequencies, which determines the radiation spectrum. When a quantum of light is absorbed, its energy should be equal to the difference between the energies of two states, new and old. Therefore, the absorption spectrum of light should be exactly the same as the emission spectrum — a fact long established by opticians.

But why don't electrons emit synchrotron radiation as they move? Bohr reasoned that if they radiated, then the parameters of their orbit would change continuously. He himself considered motion only along circular orbits, which have one parameter — its radius. It would continuously decrease when the electron approaches the nucleus. But what if, for some reason, it can only change discretely? What if the orbits are "quantized"? This could prevent an electron from falling onto an atomic nucleus, as predicted by classical electrodynamics.

It is essential to use the combination of Planck's formula and Rydberg's law in order to find out the energy differences of some levels and then the energies of these levels, and then to understand what exactly distinguishes the orbits on which electrons can move among all possible orbits. This question can be phrased even more succinctly: what exactly is quantized?

The answer was surprisingly simple. The moment of momentum of an electron in a hydrogen atom, which in a circular motion is equal to the product of its velocity and mass and the radius of the orbit, must be equal to an integer number of the reduced Planck constant $\hbar$.

If it is equal to $\hbar$, then the orbit will correspond to the state with the lowest possible energy, and the electron on it cannot emit light or fall on the nucleus. It is called the ground state of the electron. Orbits with angular momentum $2\hbar$, $3\hbar$, and so on are allowed by the laws of nature, but an electron on them can go to a lower orbit while emitting a photon of a certain frequency. Orbits with an angular momentum not equal to an integer multiplied by $\hbar$ are forbidden; an electron cannot move along them.

So, Bohr suggested that the angular momentum of an electron in a hydrogen atom is quantized and found what the magnitude of the quantum is equal to. From this one assumption, conclusions were drawn about the emission and absorption spectrum of hydrogen and the Rydberg formula, which was repeatedly verified experimentally. Therefore, the hypothesis was greeted with enthusiasm. Physicists concluded that Bohr did indeed establish rules that described the permissible orbits of electrons in atoms. In 1922, Niels Bohr received the Nobel Prize in Physics "for his services in the investigation of the structure of atoms and of the radiation emanating from them".

Bohr's idea was developed by the German physicist Arnold Johannes Wilhelm Sommerfeld. He considered not only circular but also elliptical orbits, took into account the effects of STR, and predicted a number of effects, confirmed experimentally. Sommerfeld achieved many other important results in theoretical physics, but was never awarded the Nobel Prize, although he was nominated for it a record number of times.

The Bohr–Sommerfeld theory perfectly described the radiation of a hydrogen atom and other atoms with a single electron, which are called hydrogen-like. Strictly speaking, these are not atoms, but ions: a singly ionized helium ion, a doubly ionized lithium ion, up to sevenfold ionized oxygen ions. But if there are at least two electrons in an atom, this theory was no longer working so confidently. However, physicists understood that it was not complete, despite all its successes. After all, the quantization of the angular momentum in this model was introduced artificially to fit the conclusions to the empirical Rydberg formula, reflecting the results of the experiments.

Now we call this theory a quasi-classical or semiclassical one. The beginning of the word, i.e. "quasi", means "similar", "nominal", "almost", or "partially". It is almost classical, but the quantization rule was added to classical mechanics and electrodynamics. However, Bohr's theory clearly indicated in which direction to move on: quantumwards!

5.2.4. *The wave properties of the particles*

The next step towards understanding the truly quantum properties of particles was made in 1924 by the French physicist Louis de Broglie, or more formally Louis Victor Pierre Raymond, 7th Duc de Broglie.[d] He received the Nobel Prize in Physics just five years later, in 1929 "for his discovery of the wave nature of electrons".

---

[d]Strictly speaking, de Broglie became Duke only in 1960 after the death of his elder brother Maurice de Broglie, also a physicist, who was the 6th duke in this family.

De Broglie's idea was that particles should have some properties of waves. More precisely, the properties of particles and waves should be combined into a kind of quantum object, often called a particle–wave. He assumed that a particle with momentum p behaves like a wave with wavelength λ, related to the momentum as $\lambda p = h$. Here h is the Planck constant.

From de Broglie's theory, one can understand where the quantization rules in Bohr's theory come from. If the circular trajectory of an electron in Bohr's orbit is likened to a ring, then the wave describing the electron will correspond to the bends of this ring. This standing wave with length λ must fit an integer number n of times along the length of the ring in order to match itself, as shown on the left side in Fig. 5.1.

Waves with different numbers of nodes, n, have different orbital radii, denoted by r_n, and momenta, denoted by p_n. So the orbital circumference $2\pi r_n = n\lambda_n$, where λ_n is the wavelength of the electron on the n-th orbit. If this condition is not met, then, as in the right figure, instead of a standing wave we get something unintelligible, or rather impossible. Substituting the expression for the de Broglie wavelength for the n-th orbit, $\lambda_n = h/p_n = 2\pi\hbar/p_n$, we get $p_n r_n = n\hbar$, i.e. Bohr's quantization condition, because on the left is the angular momentum of an electron in the n-th orbit.

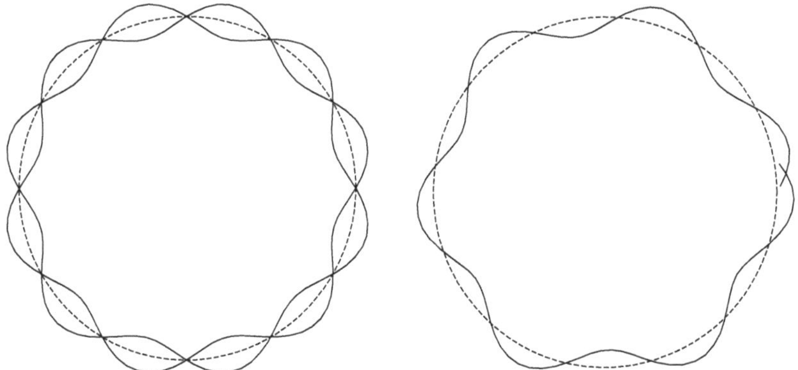

Figure 5.1.   The wave describing an electron in de Broglie's theory should match itself when going around the atomic nucleus in the middle, as shown on the left.

The arguments in favour of de Broglie's idea were theoretical or indirect, so interest in them increased sharply after the appearance of its direct experimental confirmation. George Paget Thomson, a son of J.J. Thomson, passed an electron beam through a thin metal foil in a laboratory at the University of Aberdeen in the UK and observed the pattern of electron interference predicted by Louis de Broglie.

Interference is a purely wave phenomenon, so Thompson demonstrated the wave properties of the electron, a particle discovered by his father. At the same time, employees of Bell's laboratory, the Americans Clinton Joseph Davisson and Lester Halbert Germer, conducted similar experiments with crystals. In 1937, Thompson and Davisson shared the Nobel Prize in Physics for this discovery.

By that time, theoretical physicists formalized de Broglie's idea of a particle–wave mathematically and obtained the equations that formed the basis of quantum theory. Of the many physicists who participated in this breakthrough, one can single out the Austrian Erwin Rudolf Josef Alexander Schrödinger and the German Werner Karl Heisenberg, the authors of two approaches to the quantum description of the world, which ultimately turned out to be equivalent to each other. Heisenberg was awarded the Nobel Prize in Physics in 1932 "for the creation of quantum mechanics, the application of which has, inter alia, led to the discovery of the allotropic forms of hydrogen". A year later, the Nobel Prize in Physics was split up between Schrödinger and the Englishman Paul Adrien Maurice Dirac, another of the fathers of quantum mechanics "for the discovery of new productive forms of atomic theory".

Note that in its early years, quantum mechanics was also called wave mechanics, and it intensively borrowed the methods of describing waves from optics, applying them to particle–waves. The problem of the motion of an electron in a hydrogen atom and hydrogen-like ions was solved within the framework of quantum mechanics. The energy levels, and hence the radiation spectrum, completely coincided with those predicted by Bohr's semiclassical theory.

One could talk for a long time about how quantum mechanics developed further, but this book is not about the history of science. A long historical excursion was needed in order to show that physicists wandered into the quantum world not by chance, and not because someone so much wanted to. They were driven there by the results of experiments, or rather the laws of nature that had not yet been discovered at the time.

And they entered the mysterious quantum world twice, each time to a different part of it. First, physics quantized radiation, and then matter. But the world still turned out to be one. The same formula that is applicable to de Broglie's wavelength is perfectly applicable to photons; moreover, it was borrowed from there.

I would like to point out one non-trivial circumstance. Everything that I wrote about, except for the very fact of the existence of atomic nuclei, has nothing to do with either strong or weak interactions. The electron shell of an atom is determined by electromagnetic forces and unusual properties of the quantum world, which are not considered as a separate interaction.

So let's face the fact that our world requires a quantum description and that we get acquainted with what follows from this — more precisely, with those details that seem unusual for classical physics.

5.3. The Weirdness of the Quantum World

5.3.1. *The realm of probability*

Let's try to talk about how the quantum world differs from the world described in classical physics. In this we will be helped by two inhabitants of the microworld, namely a photon and an electron. Physicists conducted experiments that made them recognize the quantum properties of these elementary particles. For symmetry, consider two objects of the macrocosm: a billiard ball and a cat.

In the macrocosm, a billiard ball moves along a certain trajectory, which is used by champions in snooker or billiard tricks to amaze us with their skill. A cat that has run through freshly fallen snow will leave clear traces on it, thus capturing its trajectory.

And in the quantum world, nothing has a position, trajectory, and so forth, on which classical mechanics is based. Quantum objects cannot be considered a "material point". They cannot be in a certain place, but are somewhat blurred in space.

The electron in the hydrogen atom is somewhere around the nucleus, but we cannot say exactly where. Rather, it should be represented as a cloud, which graphically depicts the probability density distribution of finding this electron in a given location. Physicists, based on the equation deduced by Schrödinger, can calculate how the quantity describing this probability density changes in space. And since the electron at each moment of time does not have a definite position, it is strange to talk about its trajectory.

We can say that in the process of observation, an electron appears in some place with the probability described by this distribution. The probability cloud suddenly "collapses" at some random point inside it. And the location of this point could vary for different observations.

Physicists have learned how to irradiate hydrogen atoms with high-energy particles at an accelerator and determine where exactly they hit an electron. The places in which the electron "manifests itself" are different for absolutely identical hydrogen atoms in the same ground state. This fact has been proven experimentally.

So, an electron is described by a cloud of probability. However, in the process of interaction, it "appears" in some place in accordance with the distribution of probability, and it is impossible to predict exactly where this will happen. The characteristic dimensions of the area occupied in space by this cloud are about the wavelength of an electron. For an electron cloud in an atom, they determine the size of this atom.

Since both the cat and the billiard ball have a much greater mass than an electron, the size of their clouds is much smaller than the size of atomic nuclei or elementary particles. Therefore, these dimensions can be disregarded. Although formally they are not zero, they are so small that they do not manifest themselves in experiments.

5.3.2. *Quantum tunnelling*

The blurring of a quantum object leads to the possibility of a quantum tunnelling transition. What is it? In the macrocosm, this would be called passing through a wall, more precisely, instantaneous relocation behind it. But a billiard ball, a cat, or you yourself are not capable of this, unless you break through a not very strong wall. It is easy in the microcosm. The atomic nucleus emits an alpha particle precisely through quantum tunnelling.

This particle, being a quantum object, is described by a probability cloud. Its edges are capable of going beyond the atomic nucleus. This means that it can be localized, i.e. it appears outside the nucleus with a small but nonzero probability. And then it would be carried away by the electrical repulsion of two positively charged bodies: an alpha particle and an atomic nucleus.

This is an example of quantum tunnelling. It is not so difficult to be on the other side of the wall if it is not entirely clear which side you were originally on.

5.3.3. *Influence of the measurement process*

Let's continue our acquaintance with the features of the quantum world and discuss one more of its nuances. This is the need to take into account the measurement process. In the macrocosm, this principle is not so important. By measuring the sample with a ruler, we can at most scratch it. If you measured it accurately and did not scratch it, then it seems as if no influence was exerted. In fact, it exists, but so little that it can be neglected.

And in the microcosm, where everything is much less, even a small influence can be significant. If we are trying, after reading Confucius, to catch a black cat in a dark room, then it will be difficult, especially if it really is not there. To facilitate this process, we turn on the light to see this cat. It's over! We have already had an impact on the cat (if there is any).

The light has a small, but nonzero pressure, so by turning it on we provided a small force pushing the cat away from the light bulb. It is possible that the light acts on the cat in a thousand other ways — for example, it just woke it up or frightened it, but this light pressure alone is enough to dismiss the opportunity to see the cat without affecting it. If instead of a cat we are trying to see an electron, then even one interaction with a high-energy photon is enough to affect the motion of an electron.

The influence of measurement can be masked, for example, when discussing a rather complex problem called quantum entanglement. Suppose that some particle with zero angular momentum has decayed into two spinning particles. It is clear that the directions of their rotation must be opposite and the total angular momentum must be equal to zero, because it is conserved.

At the same time, it cannot be predicted how exactly each of the particles rotates due to quantum uncertainty. But if you know the angular momentum of one of the particles after its measurement, i.e. interaction with some device, then it will instantly become clear how the second particle spins, although the distance between them can be arbitrarily large.

Some physicists do not see this as a problem, citing the example of a pair of socks. If one of the socks is put on the left leg, then it will immediately become left, and the second, respectively, right, although the distance between them can be as large as you like. Others, on the contrary, consider this an important thing and propose to arrange on its basis a channel for transmitting secret information that can be intercepted, but cannot be deciphered.

However, things are not as simple as they seem. Suppose that the decay occurred in the middle between the Earth and the Andromeda Nebula, that one of the particles flew to us and we learned the direction of its rotation, and that the second is somewhere in the Andromeda Nebula. Does this mean that we have learned how this distant particle rotates?

Yes, based on the simplified formulation of the *gedanken* experiment. But if you apply it to the real Universe, for example, why are we sure that both particles on their way did not interact with any other particles? These other particles are not mentioned in the description of the experiment, but they exist in nature. Each cubic centimetre of our world contains about 500 relic radiation photons alone. What if one of them somehow influenced the rotation of one of the particles?

If along the way the particles enter into interactions not foreseen by the thought experiment, then their angular momentum could change. And to be sure that there were no illegal interactions, the particles and their neighbours must be observed all the time, for example, by illuminating them with a stream of photons.

But then, their angular momentum can change when interacting with a photon from the stream. That is, trying to make it impossible to change the angular momentum of particles, we are guaranteed to change it. In any case, the measurement process itself affects the properties of what is being measured in the quantum world.

5.3.4. *Heisenberg's uncertainty principle*

If we are interested in the trajectory of a point object in the framework of classical mechanics, then we need to act in the usual way for it: enter a coordinate system, for example, one whose axes look north, east, and up, and find out how each of these coordinates of the object changes over time. From here, you can calculate the three components of the object's velocity along the axes we have chosen and report on the work performed.

In the quantum world, this will not work. There is the uncertainty principle formulated by Werner Heisenberg in 1927. It states that if we can measure the x-coordinate of an object with an error of Δx and the component of its momentum along this axis p with an error of Δp, then these errors must satisfy a fundamental inequality, $\Delta x \Delta p \geq \hbar/2$. We see that it is impossible to accurately determine the coordinate of an object without having complete uncertainty about its momentum, or to accurately determine the momentum of an object without having no idea of its coordinate. A similar relationship exists between the uncertainty in measuring the energy of a particle and the time interval of this determination. It underlies the existence of virtual pairs and particles, which I will talk about a little later.

The uncertainty relation is proved in quantum mechanics textbooks, and I am not going to repeat it in this book. I'll just illustrate it with two examples. The first is the cat catching described above. Let us have such a keen eye that we can tell where in the dark room the cat is located by seeing a single photon reflected from it. Let it be a photon of light with a frequency v; then its wavelength is $\lambda = c/v$, where c is the speed of light.

This photon, hitting the cat, imparts to it an impulse not less than the photon's own impulse, equal to $p = hv/c$, and provides an error Δp in the cat's impulse close to this value. And the error in the cat's coordinate, Δx, is close to the photon's wavelength, λ. By multiplying these errors, we get an estimate that is very close to that mentioned above.

The second example illustrates the case when a photon passes through a narrow slit of width a in an opaque wall. Initially, it moves perpendicular to the wall, and the momentum component in the direction of the x axis, perpendicular to both the slot and the direction of the photon's motion, is exactly zero until the photon reaches the slit and is diffracted by it.

After that, it can change the direction of its movement by deviating by some angle from the original direction. This angle does not exceed $\lambda/2a$ radians for the main diffraction maximum. So, after diffraction, the photon will have some component of the momentum along the x axis that does not exceed $p \cdot (\lambda/2a) = hv/c \cdot \lambda/2a = h/2a$.

We can consider the passage of a photon as the process of measuring its x-coordinate, which is in the range $-a/2$ to $a/2$ since the photon has passed through the slit, so $\Delta x < a$. For the main diffraction maximum, the component of its momentum along this axis lies in the range $-h/2a$ to $h/2a$, so $\Delta p < h/a$. It is easy to verify that the relation between the errors is close to the one given above.

To obtain an exact inequality, including a numerical coefficient, it is necessary to strictly define what is considered an uncertainty or an error of coordinates and momentum. Those wishing to learn all these details can refer to any good textbook on quantum mechanics.

The uncertainty principle was initially rejected by Einstein. In a letter to Max Born, he wrote: "*Gott würfelt nicht* (God does not play dice)." Einstein came up with a thought experiment that, in his opinion, made it possible to accurately measure the coordinate and momentum of a particle. But Bohr had found a mistake in it. Then more than one attempt followed, but each time Bohr found a hole in it, an error, a nuance that prevented accurate measurements. The correspondence between Einstein and Bohr on this matter has been published and allows amateurs to enjoy the intricacies of the foundations of quantum theory.

5.3.5. *Zero-point energy*

Any macroscopic body strives to reduce its potential energy. The metal ball released from the hand falls down, and water also flows down, turning the turbines of hydroelectric power plants on occasion. But if the same ball lies in a bowl, and water is poured into a cup, then they will remain there, although it is energetically more advantageous for them to be a couple of floors below.[e] The ball will stop at the lowest point of the bowl bottom if you do not take into account its quantum properties. In this case, at the lowest point of the bottom, the coordinate and speed of the ball would be known exactly, which is prohibited by the Heisenberg uncertainty principle.

Quantum mechanics claims that the ball actually makes so-called zero-point oscillations with very small amplitude. At the same time, its average kinetic energy is positive, and the potential energy is slightly greater than the energy of the classic (non-quantum) ball lying at the lowest point. The sum of the potential energy and the kinetic energy is equal to $h\nu/2$, where ν is the frequency of zero-point oscillations. This is the energy of the ground state of the ball, or its zero-point energy. It is very small and can be attributed to quantum fluctuations.

This results in motion even at absolute zero. It is precisely the zero-point oscillations of atoms that prevent liquid helium from

---

[e]The only liquid that would flow out of the cup is the superfluid helium, but it could do so due to its quantum mechanical properties. To some extent, such helium is not quite an object of the macrocosm.

becoming solid at ordinary pressure, even at a temperature of absolute zero. The existence of solid helium is possible only at a pressure of more than 25 atmospheres. So, near the absolute zero temperature, helium behaves like a quantum liquid. It not only does not freeze, but also demonstrates superfluidity and some other properties that are impossible for ordinary liquids.

5.3.6. *Virtual particles and pairs*

Another purely quantum feature is associated with the existence of virtual particles and virtual particle–antiparticle pairs. What is it? A quantum particle is capable of behaving as if it has more energy than it actually has, but for a very short time. The reason for this lies in the Heisenberg uncertainty principle, which does not allow determining the energy with an error less than $\Delta E = h/\Delta t$, where Δt is the time interval during which this energy was measured.

The particle, as it were, makes possible (physicists prefer the sophisticated word *virtual*, from Latin *virtualis* — possible) jumps in place, with energy occasionally exceeding its average value. If the energy of a particle temporarily becomes so large that in this excited state it can give birth to a particle (virtual) or a particle–antiparticle pair (also virtual), then they will exist for the smallest fraction of an instant and then disappear. The law of conservation of energy does not allow them to appear constantly.

This is a very non-trivial process, and it is difficult to explain it in simple terms. Let me try to use an analogy in which money plays the role of energy. A person or a group of people are dreaming of a computer. They can buy it for 120 gold, but have only 100 gold, so they cannot become the proud owner of a computer. It would seem that everything that requires the purchase of a computer is inaccessible to them: games, the Internet, e-mail, and hundreds of other ways to use it.

However, the company is based in a country which has a law stating that every buyer can return any purchase back within a week of having received its value back. And this law is strictly observed. If they scrape up an additional 20 gold in debt, then they can have fun with the computer for almost a week, and then hand it back, get their 120 coins, pay off the debt, and break even. At the same time, instead of sad computerless sufferers, they are happy users, that is, creatures with a slightly different mood and properties.

But where to borrow the money from? Let's imagine that the country is in the quantum world, where the principle of uncertainty reigns. It states that if the energy E of any system is measured during the time interval Δt, the measurement accuracy cannot exceed the minimum value of $\Delta E = h/\Delta t$.

Recall that in our analogy, money plays the role of energy. This means that if you count the money in your pocket every minute, then the number of gold coins counted changes. Sometimes there will be less than a hundred, sometimes more. If the monetary analogue of the value ΔE at Δt equal to a minute exceeds 20 gold coins, then sometimes more than 120 coins can be found in the pocket. This quantum excess will not last long, only a minute, and then the money must be counted again. If a minute is a lot and the analogue of ΔE is not large enough, we could recalculate the money every second.

Now, you need to quickly buy a computer, use it, and hand it over to the store — and do all this in a minute or even in a second. Fortunately, in the quantum world, sellers and cashiers work very quickly, and a second is enough for them. There will be half a second left for the Internet and games. Of course, this is not an eternity or even a week, but it will cheer you up.

The money received is put into your pocket and its amount again begins to fluctuate around the original 100 gold, in accordance with the law of conservation of energy–money. Physicists call these vibrations quantum fluctuations. Now you can start counting the money in your pocket again, catching the moment of the next arrival of wealth.

I hope this analogy has helped you understand the principle of the brief appearance of virtual pairs. They seem to be absent in nature, but the possibility of their real existence somehow affects the properties of existing particles. It happens like the ability to buy a computer with money that is in short supply affect the user's mood in our strange example.

5.3.7. *Physical vacuum*

For all their short duration and illusory nature, quantum fluctuations are capable of interacting with objects of the microworld, influencing their properties. The most famous object of such influence is the physical vacuum. The adjective is added in order not to confuse it with emptiness or vacuum, in which there is nothing. And in a real vacuum, due to its quantum properties, there are virtual particles. Moreover, it can contain only such virtual particles in which their real counterparts can actually exist.

In one popular science book, this is illustrated with a story, which I reproduce briefly. In the old days, a beautiful young lady volunteered to sell water and syrup at a charity ball. According to the instructions, she had to take a donation for charity and ask the donor which syrup they preferred with water: strawberry, raspberry, or currant. Once, when a donor told her to provide water without syrup, she asked what kind of syrup he needs water without: strawberry, raspberry, or currant.

Surprisingly, there are no real particles in the physical vacuum,[f] but its properties depend on what particles are not in it, but they are possible in nature. The situation is similar to the one in which the taste of soda would depend on what kind of syrup is not in it. Because of this, physicists really want to know all the existing types of particles, among other reasons.

If it seems to you that all this reasoning about the physical vacuum is just feverish delirium, I have something to say in its favour. Willis Eugene Lamb was awarded the Nobel Prize in Physics in 1955 "for his discoveries concerning the fine structure of

---

[f]If you place a vacuum in a very strong electric, gravitational, or other field, virtual pairs can sometimes turn into real particle–antiparticle pairs. This is called quantum pair production in an external field.

the hydrogen spectrum" — more precisely, for the discovery of the smallest shift of the electron energy level in the hydrogen atom. This phenomenon, called the Lamb shift, is caused by the interaction of the electron with the quantum oscillations of the electromagnetic field, that is, with the physical vacuum.

In the Philips laboratory, another similar effect was confirmed, called the Casimir effect. Two parallel, uncharged metal plates begin to attract each other with a very, very weak force due to the fact that these plates affect the properties of the vacuum.

Be that as it may, the physical vacuum turns out to be a very complex object. It can have a certain energy density, which we will simply call energy for the sake of brevity. If we are talking about the energy of a vacuum with some kind of additives like fields and particles, it can change with a change in external fields or some global quantities.

In scientific articles and books on physics, you can find reasoning about false and true vacuum, vacuum decay, and other strange things. These terms become clearer if we explain that the true vacuum is an analogue of the ground state, and the false one is an excited one. The transition from the state of a false vacuum to a true state is called the decay of a false vacuum and occurs with the release of energy. I used these concepts in Section 1.2 when discussing the possible cause of cosmological inflation.

5.3.8. *Particle spin*

When we discussed the properties of a photon, I talked about its energy and momentum, but I did not say anything about its angular momentum. The point is that this value deserves a special discussion. Each elementary particle has a unique angular momentum, measured in the frame of reference in which it is stationary. It is called the spin of the particle. The concept was coined by two Dutch Americans, Samuel Abraham Goudsmit and George Eugene Uhlenbeck, in 1925.

The frame of reference is chosen so that the spin characterizes only the internal angular momentum of the particle, but not the angular momentum due to its motion.[8] Note that the spin of the particle may be simplistically considered as its rotation around its axis, but this is not necessary. Spin is a kind of property inherent in elementary particles along with charge or rest mass. The spin of an antiparticle coincides with the spin of the corresponding particle.

Physicists have proven that the spin of a particle must be proportional to $\hbar/2$, i.e. half of the reduced Planck constant. Therefore, it can be written in the form $s\hbar$, where s is some number that can be an integer, including zero, i.e. $s = 0, 1, 2, 3, \ldots$, or a half-integer, i.e. $s = 1/2, 3/2, 5/2, \ldots$. It is called the spin quantum number.

Physicists talk about particles with integer and half-integer spin, meaning exactly s. Naturally, we can introduce the spin quantum number for macroscopic objects, such as a spinner or the Earth, but this makes no sense, because their quantum properties are practically not manifested.

5.3.9. *Bosons and fermions*

The properties of particles with integer and half-integer spins are fundamentally different. Therefore, I will tell you about them separately. I'll start with particles with integer spin. These include quanta of fundamental interactions: photon, quanta of weak and strong interactions (they are characterized by $s = 1$), graviton — a hypothetical quantum of the gravitational field with $s = 2$, and Higgs boson. The last particle, the story of which is told ahead, has zero spin. Let's add some compound particles like alpha particles to the boson group.

All particles with integer spin obey the so-called Bose–Einstein statistics. It is named after Albert Einstein and the Indian physicist

---

[8]For a photon, which always moves with the speed of light, such a frame of reference is impossible, but instead of the photon's spin, its projection along the direction of motion is introduced. It's called helicity, and it replaces the spin for the photon. If a particle's spin vector points in the same direction as the momentum vector, the helicity is positive, and if they point in opposite directions, the helicity is negative. In what follows, I will still refer to this value as spin to avoid unnecessary words.

Satyendra Nath Bose, who discovered it. Such particles are called bosons in honour of Bose.

What do all bosons have in common? Simply put, they all love to hang out. In one famous quantum mechanics book,[h] each chapter has an epigraph. In the chapter on bosons, it is: "The more we are together, the merrier we'll be." Bosons not only can get together and be in the same state, but also willingly join the crowd.

The emission of photons into a state in which there are already many of them is called stimulated or induced emission. Its prediction, followed by experimental confirmation, played a decisive role in awarding Einstein the Nobel Prize. It is stimulated radiation that underlies the operation of lasers. The very word *laser* is an abbreviation of the expression "light amplification by stimulated emission of radiation".

But what about particles with half-integer spin? They obey other laws known as Fermi–Dirac statistics after the Italian Enrico Fermi and the Englishman Paul Dirac (Nobel Prizes in Physics in 1938 and 1933). Such particles are usually called fermions. What is hidden under this word, and how do they differ from bosons? Fermions are misanthropic by nature and avoid each other. Two fermions can never be in the same state. This ban was established by the Swiss Nobel laureate Wolfgang Ernst Pauli in 1945. It is called the Pauli principle.

Why can't there be more fermions? Two ladies at a party, finding themselves equally dressed, may be upset, but nothing more. But two fermions simply cannot find themselves in one state. This is how nature works for some reason. An analogue of Pauli's principle in the example with the party would be a situation in which the doorman at the entrance would not let the second of the equally dressed ladies into the reception until either she or the lady arriving first changed.

Most of the known elementary particles which are true fermions, such as electrons, protons, and neutrons, have spin $1/2$.

---

[h]John Ziman, *Elements of Advanced Quantum Theory* (Cambridge University Press, 1969).

All of them have two possible directions, conventionally "spin-up" and "spin-down"; therefore, no more than two particles with opposite spins can be in a quantum state identical in other parameters. For example, in one orbit in an atom, there can be no more than two electrons. Some rapidly decaying particles have a spin of 3/2. Atomic nuclei, consisting of many protons and neutrons, can have a spin of 5/2 and more.

As you have seen, the properties of bosons and fermions are quite different. The nuclei of isotopes of one element can belong to different groups. The proton is a fermion and the deuteron with $s = 1$ is a boson. An alpha particle with $s = 0$ is a boson, and a ^{3}He nucleus with $s = 1/2$ is a fermion.

5.3.10. *Schrödinger's cat*

It is possible that the reader, having found the mention of a cat in the section on quantum mechanics, is looking forward to the appearance of the most famous quantum mechanical cat. It entered popular culture as Schrödinger's cat, becoming a meme and the object of attention of cartoonists. However, in the original German text, she is *die Katze* (a she-cat).

Here is the beginning of this meme from Erwin Schrödinger himself:

> One can even set up quite ridiculous cases. A cat is penned up in a steel chamber, along with the following device (which must be secured against direct interference by the cat): in a Geiger counter, there is a tiny bit of radioactive substance, so small, that perhaps in the course of the hour one of the atoms decays, but also, with equal probability, perhaps none; if it happens, the counter tube discharges and through a relay releases a hammer that shatters a small flask of hydrocyanic acid. If one has left this entire system to itself for an hour, one would say that the cat still lives if, meanwhile no atom has decayed. The first atomic decay would have poisoned it.

Many interpreters usually say that the cat should be considered half alive and half dead after this hour. Note that in the above example there is nothing quantum other than a radioactive substance, but it is easy to replace it with, for example, a coin toss.

Uncertainty arises due to the fact that we simply do not know whether the decay of the atom has occurred or whether the coin fell heads or tails. We are not saying that the coin, hidden after the throw with the palm, fell half with the heads and half with the tails. Likewise, we should not seriously argue that the cat is half alive and half dead — just that it is in one of these two states with the same probability. We simply do not know the outcome of the experiment; for this, we must raise the palm hiding the coin, look into the camera with the cat, and at least hear meowing or angry hissing from there.

5.3.11. *Features of the microcosm: No diversity, no evolution*

I will use this example to remind once again the differences between the micro- and macrocosm using the example of cats and electrons. All cats are different. They have different breeds. The fluffy Persian cat cannot be confused with the hairless Sphynx. Cats of the same breed have different colours, weight, proportions, habits, voice, and sex. There are kittens and old cats. The owner recognizes his cat.

And all electrons are the same. It is impossible to mark them somehow in order to later identify the selected copy of the electron, among others. They can differ only in a small number of physical parameters. For a freely flying electron, these are its position, speed, and direction of motion. And even then they cannot be measured simultaneously due to the Heisenberg uncertainty principle. For an electron in an atom, there are parameters of the orbital in the electron shell. And for any variation, you also need to add the spin orientation: the notorious "spin-up" and "spin-down". An electron or a photon can do nothing but interact with the fields surrounding it. This greatly simplifies the construction of the theory of particles, but sometimes interferes in experiments, because the object under study cannot be marked.

I formulate a few more obvious differences between the microcosm and the macrocosm we are accustomed to. The first is the absence of history, or immutability in time. A conclusion from the identity of particles is the independence of their lifetime from the time of their previous existence, which I wrote about in Section 1.4. At the qualitative level of understanding, it follows from this that there are no "old" and "young" particles. And the quantitative law of decay of particles can be unambiguously obtained, which describes the decrease over time in the percentage of non-decayed particles or atoms of radioactive isotopes.

This feature seems obvious and is rarely emphasized. However, in the sections describing the Universe, the Sun, the Solar System, and the Earth and its surface, we have been discussing their evolution all the time. But there is no evolution in the microcosm. Note that all attempts have failed to detect a change in time of either the laws of physics or at least fundamental constants, such as the speed of light, charge, or mass of an electron, etc. This does not allow anything to change in the microcosm, this kingdom of true conservatives.

The situation is similar to a change in position in space. We, the inhabitants of the macrocosm, sometimes go somewhere on weekends or on vacation to enjoy new views, monuments, and spectacles, or just to change the scenery. At our service are forests and mountains and the sea, the cold north and warm south, museum halls, stands of stadiums, and dance floors of discos.

In the microcosm, everything is the same everywhere, and it will not be possible to change the situation, unless one throws an atom into the centre of a neutron star. The laws of the microworld are the same everywhere, including the centre of a neutron star; just an atom will not exist there for a long time.

The invariability of the laws of nature was checked more than once by the radiation that came from especially distant objects. If at the moment of light emission there, far from us and at the same time in a rather distant past, different laws or the same laws, but with different constants, were in effect, this would manifest itself in the physical characteristics of the radiation that astronomers observe here and now.

5.4. An Electron Shell of the Atom

Only electromagnetic forces act on the electrons that form the shell of atoms. For electrons, a nucleus is simply a very small, positively charged object to which they are attracted. Strong and weak interactions are concentrated in this nucleus, without affecting the electrons in any way.

If it were not for the laws operating in the microcosm, then the electrons in a split second would fall on the nucleus of the atom, emitting their energy in the form of electromagnetic waves. But quantum mechanics forbids such a fall and the shell not only exists, but also determines all the chemical properties of the atom. Yes, exactly chemical, because chemistry is based on the properties of the electron shell of an individual atom and the interaction of two or more shells of different atoms, known as a chemical bond.

Physics explains the properties of the electron shell, and chemistry uses them. The question of why an explosion would be heard and a lot of smelly gas would be released if you merge green and red liquids from test tubes together belongs exclusively to the competence of chemistry. But only the person who had sniffed a lot of this gas could begin to tell the story about the causes of the explosion from the properties of the electron shells of atoms.

5.4.1. *Atom size*

By the way, what are the dimensions of the electron shell of an atom? They vary for different chemical elements and lie in the range 30–300 pm (1 pm = 10^{-12} m). This is a thousand times shorter than the wavelength of visible light and more than 10,000 times the size of an atomic nucleus. If the apple could be enlarged to the size of the Earth, then the atoms would have reached the original size of the apple.

In the periodic table of elements, the size of an atom usually decreases as it moves along a line (period) from left to right, from alkali metals to noble gases. This is due to an increase in the charge of the nucleus and the force of electric attraction of electrons to it. It increases sharply when moving from one line to another, from noble gases to alkali metals. This is due to the appearance of electrons at higher energy levels. Details are explained below. Within one column (group), the size increases as you move from top to bottom.

Accordingly, the smallest atom is a helium atom, with a radius of 31 pm. The largest atom is possibly the francium atom, and its radius is equal to 290 pm. The second place is occupied by the cesium atom, with a radius of 267 pm.

Can an atom be seen or photographed? The answer is ambiguous. Because of their smallness, atoms cannot be seen through an optical microscope. However, the shape of individual atoms can be reproduced by using a scanning tunnelling microscope. Scientists from the Kharkiv Institute of Physics and Technology have learned how to get something like a photograph of an electron cloud of individual atoms using an electron microscope.

This method has become popular and is now used in laboratories in many countries. This is how the "photograph" was obtained. Keep in mind that the colour in it was added just for clarity; it illustrates how the probability density of finding an electron at a given location changes in space. The electronic shell has no colour, as well as no smell; these concepts simply do not apply to it.

5.4.2. *Orbitals and their quantum numbers*

The properties of the electrons in a shell cannot be explained without involving quantum mechanics. Everything becomes explicable this theory, which gives results that, both qualitatively and quantitatively, are in excellent agreement with the experimental data. I will describe these results, at least qualitatively.

Let's start with the concept of the energy levels of electrons in a shell. All of them are below the minimum energy of a free electron far from the nucleus. If we take this energy as zero, then all levels correspond to negative values of energy. The energy difference prevents electrons from leaving the shell, moving away from the nucleus at distances significantly larger than the size of an atom.

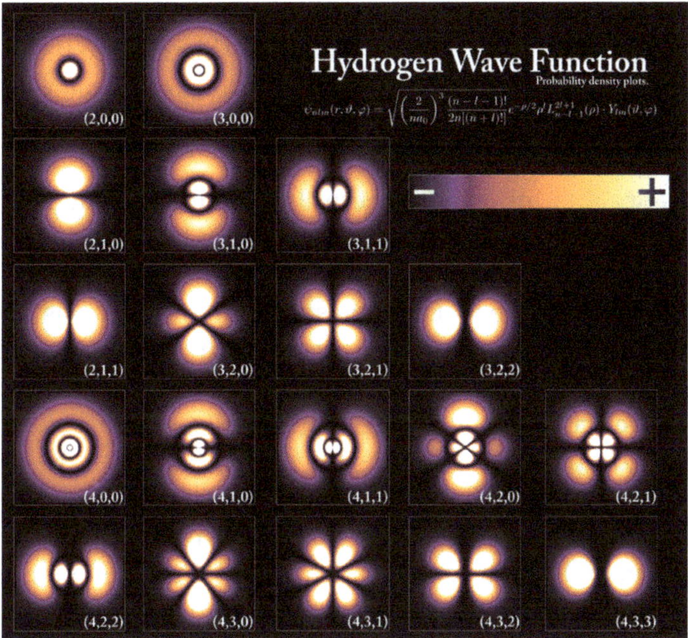

Figure 5.2. Distribution of the probability density of finding an electron for several lower excited states of the hydrogen atom (orbital). Different orbitals are depicted at different scales. For each orbital, there are three numbers below, the so-called quantum numbers, by which a physicist can find its shape. For true connoisseurs, in the upper right corner, there is a complex formula that describes any orbital. Extracted from Wikipedia, public domain. Author PoorLeno.

An electron can be in states with different energies, that is, at different energy levels. Figure 5.2 shows some of the probability distributions, called orbitals, for the lowest excited energy levels in the hydrogen atom. These are the very clouds of probability that I wrote about above. It does not depict the spherically symmetric cloud for the ground state. The brightness at each point in the figure is proportional to the likelihood of finding an electron in that exact spot.

All these distributions are obtained theoretically, based on quantum mechanics. As you can see, physicists do not confine themselves to general arguments about quantum blurring of particles, but move to the probability distribution for a specific orbit.

Three integers are indicated near each cloud, shown in Fig. 5.2, that fully describe it. In the upper right corner there is also the formula for probability distribution. It is written for the true fans of quantum mechanics, but other readers are not required to even look at it.

Let's speak a little about these three numbers. The first is the principal quantum number n or the level number. This is an integer value: $n = 1, 2, 3, \ldots$ In hydrogen-like atoms, energies of different levels depend only on it. The corresponding formula was obtained by Bohr. The quantum number l was introduced by Sommerfeld. It is also an integer and varies from 0 to $n - 1$. The moment of momentum of the electron in the orbit is equal to $l\hbar$.

At the ground state, i.e. at the lower energy level of the electron, we have $n = 1$; therefore, $l = 0$ and the electron does not orbit the nucleus at all. It can have a nonzero orbital angular momentum only in excited states. The talk about electrons moving around the nucleus describes not the atom, but the old semiclassical Bohr model, from which atomic physics has long grown.

Spectroscopists grouped the emission lines of different chemical elements into several series, which are called sharp, principal, diffuse, fundamental, etc. There are some empirical formulas describing their spectra, which include several constants denoted by the first letters of the series names.

They gave names to orbitals for which the Latin letters s, p, d, f, etc., are used, corresponding to the values $l = 0, 1, 2, 3, \ldots$ These designations are used by chemists, but not all of them know their origin. So the 3p orbital of an electron is a state with quantum numbers $n = 3$ and $l = 1$ (because the letter p is indicated).

The third number, called magnetic, is denoted by the letter m. It defines the projection of the angular momentum of the electron onto the direction of the selected axis equal to $m\hbar$. This integer ranges from $-l$ to l and can take $2l + 1$ values. Figure 5.2 does not show the ground state with the set $(1, 0, 0)$, but shows the states with zero and positive magnetic quantum numbers and $n = 2$, $n = 3$, or $n = 4$. States with negative m look like the states with a set of quantum numbers $(n, l, |m|)$.

Note that some orbitals look different, but have the same or nearly the same energy. For hydrogen-like atoms, these are states that have the same principal quantum numbers, but a different set of the other two. Due to the Pauli principle, each orbital can contain no more than two electrons, and they must have the opposite directions of their spin.

5.4.3. *Quantum mechanics explains the periodic table*

One electron tries to be on the ground level with the minimum energy in the hydrogen atom or the helium ion. The second electron in the helium atom or the lithium ion does the same, and certainly, they have opposite spins. The third electron in the lithium atom would also like to be in the orbital with the minimum energy, but there are no empty low-energy orbitals for it. Therefore, it has to occupy a free orbital with the minimum energy, but among the remaining ones ("the best among the worst").

If something in these arguments is not clear to you, then I will use an analogy. The electron strives to minimize its energy by being closer to the nucleus. The viewer at an entertainment event wants to sit closer to the stage, arena, or ring. But the number of such places is limited. There is a bench that is closest (the ground hydrogen state 1s), but no more than two people can sit on it.

The other benches are farther away and there are also no more than two people sitting on them. Everyone must have a baseball cap on their head, either red or blue; a person cannot exist without a baseball cap (or they are simply not allowed into the hall). This is an analogue of the two possible directions of the electron spin. On each bench, there cannot be two spectators in the same baseball caps; the manager Wolfgang with a big cudgel is constantly monitoring this.

So, the first electron appears, that is, the viewer. He or she wants to sit on the nearest bench. His baseball cap can be any colour. This is an analogue of the ground state of a one-electron atom, that is, hydrogen or the He^+ ion.

Of course, the first viewer could take a seat on another bench, say, after mistaking them with excitement. But very soon he would realize his error and move closer to the stage. In this case we get an analogue of the excited state of the hydrogen atom. But for now, let us forget about the excited states of atoms, limiting ourselves to only ones having the lowest possible energy.

The second visitor has to sit on the nearest bench near the first, wearing a baseball cap of a different colour than the first one. The third would like to sit down next to them, but, looking at the club in the hands of Wolfgang, sadly trudges to the bench in the second row. This is an analogue of the lithium atom, the size of which is clearly larger than that of a helium one.

In the second row, there are not one, but four benches (one of them has a "2s" label, three others[i] are labelled "2p") at approximately the same distance from the stage. No more than two spectators can sit in each of them. But these are different benches, so the Pauli principle allows up to eight spectators in the second row, both in our example and in the periodic table of elements.

In the second row are atoms from lithium to neon. The eleventh visitor, muttering something unintelligible, but clearly expressive, sits down in the third row, in which there are already nine benches. We got an analogue of the sodium atom. Further, the principle of filling the hall becomes somewhat more complicated, but the essence, I hope, is already clear. In the end, we can arrange in order all the chemical elements we know in the periodic table.

Having dealt with the basic states of atoms, let us also recall the excited ones. An electron can go to a level with higher energy when this atom collides with another atom, ion, electron, or other particle, or by absorbing a photon of the corresponding energy. It is not at this excited level for long[j] but quickly goes to a lower level with the emission of a photon. Its energy E is equal to the difference between the energies of the electron at these levels, and the frequency of the emitted radiation, ν, is obtained from Planck's relation $E = h\nu$.

---

[i]This corresponds to three possible values of the magnetic quantum number m.
[j]In lasers, electrons are constantly in an excited state, where they are transferred by the so-called pumping. Therefore, the excited level is not empty there for a long time.

If we know the energies of all atomic levels, then we can calculate the energies and frequencies of photons emitted during the transition of an electron between them, that is, the emission spectrum of an atom. We can use the zero energy level as the minimum possible for the final state of the electron when it escapes after receiving some energy. It leaves the atomic shell and becomes a free electron. Therefore, the energy levels of electrons, taken with the opposite sign, are equal to the minimum ionization energy required to transfer an electron from this level into a free state.

5.4.4. *Exchange interaction*

We have considered a single atom. But it is not alone in our world. Somewhere, maybe very close, there are other atoms and ions. What happens when they meet? Let's use the analogy again and compare a single atom to a table in a cafe or restaurant. The table itself is the nucleus and the visitors, i.e. electrons, are located around it.

If there is another table nearby with other clients, then they begin to interact. Let's say that some wine is passed from one table to another, and they send something else in return. Or they just exchange glances and smiles. The two companies are clearly drawn to each other. This is an analogue of what is called exchange interaction in physics.

Sometimes one or more people sit down with the company at the next table and instead of two atoms we see two ions, positive and negative. Companies can slide their tables together and form an analogy of a molecule. This is how different types of chemical bonds work. It is thanks to it that the atoms unite into bodies, including ours.

5.4.5. *Science and fiction*

At the end, I want to make one trivial conclusion. Look at the periodic table. The first hundred cells in it are completely filled. This means that chemists know all chemical elements with the number of electrons $1, 2, 3, \ldots, 99, 100$. Other elements with the number of electrons no more than a hundred cannot exist in nature. So there are no new elements with unusual properties and cannot be. Let me clarify that we will talk about the atoms of transuranic elements with a large number of electrons separately.

In science fiction, you can regularly find stories about new elements, usually metals, which are mined somewhere on distant planets. Alas, if they exist, it happens only in the imagination of authors and readers. All elements must be in the periodic table, and there is no place for this exotic in it.

It's funny that one could write about a unique compound or allotropic modification of known elements. After all, diamonds are clearly more expensive than graphite, and these are all carbon in different forms. But the story about the mines on Jupiter's moon, Ganymede, where "a unique metal that reflects all types of radiation" is mined, excites inexperienced minds much more than the story of an unusual allotropic modification.

But maybe this new metal has 200 electrons or even more and has not yet been received on the Earth's accelerators? But in the mines on Ganymede ... Alas, this is very unlikely. The higher the number of transuranic chemical elements, the greater the charge of its nucleus and the faster its decay due to electrical repulsion. This is what physics says.

Let's consider the transuranic elements in more detail. The names of some new ones were officially approved by the International Union of Pure and Applied Chemistry in 2016. The 113th element was called nihonium (Nh), the 115th became moscovium (Mc), the 117th tennessine (Ts), and the name oganesson (Og) was chosen for the 118th element of the periodic table.

The half-life of the most stable isotope, ^{289}Mc, is estimated to be 156 ms. This means that $1{:}0.156 = 6.41$ half-lives will pass per second, and during this time the amount of Muscovy will decrease by $2^{6.41} \approx 85$ times — or decrease by $85^{60} \approx 5.8 \cdot 10^{115}$ times during a minute. So it is quite possible to call it disintegrating almost instantly. What mining for moscovium could there be, even on Ganymede?

But maybe, among even heavier atoms, you can find stable or at least not very quickly decaying ones? Enthusiasts sometimes talk about a hypothetical island of stability in the far part of the periodic table, but physics gives us no reason to hope for its existence.

I will say that carefully: it cannot exist unless scientists discover something fundamentally new, completely changing our ideas about the atomic nuclei — for example, new fundamental interactions that will make such nuclei stable. And until they are found, the departure of the rocket with the miners to Ganymede must be postponed indefinitely.

Well, we admit that we will not find new chemical elements in space. But maybe already known ones will show themselves there in some unusual way? Such statements can also be found in science fiction. What does physics think about this? The answer is rather negative. In Section 3.2, I have already talked about the states of aggregation and allotropic modifications of chemical compounds, including elements such as carbon or tin.

They exist at different temperatures and pressures. Therefore, in space, where these parameters can change very much, theoretically, you can find exotic modifications, such as the metallic hydrogen mentioned in Section 3.3. It is possible that hydrogen exists in the interior of Jupiter in the form of a metal. But if you deliver a sample of such exotic things to Earth, then it will return to the state we are accustomed to. One can talk about metastable states, but there is almost no hope for that.

And what will definitely not happen is that extra electrons enter the energy levels in the atomic shell. In fiction, especially very old fiction, one can find the reasoning that if lithium is strongly compressed, then the third electron can be pushed onto the first level. If you squeeze sodium, then the eleventh electron gets to the second level, writhing in pain and anger.

No, it is impossible to violate the Pauli principle without a complete restructuring of all physics, even if you really want to. But against the background of discussions about the squeezing of the atom's electron shell, the story about the love of the protagonist and the daughter of the leader of the natives with an exotic name and lovely slender tentacles will be considered science fiction.

6 ATOMIC NUCLEI, SUBATOMIC PARTICLES, AND FIELD QUANTA. SCALE: THE SMALLEST

Each atom has a nucleus consisting of nucleons (protons and neutrons), except for the nucleus of the hydrogen isotope ^{1}H, which is simply a proton itself. The lion's share of the atomic mass is concentrated in its nucleus. At the same time, the size of the nucleus is about several femtometres (i.e. 10^{-15} m), which is tens of thousands of times smaller than the size of the atom. Events inside the nuclei are in full swing. They have their own life, completely different than in the electron shell.

The atom is dominated by strong and weak forces, but the electromagnetic interactions are not to be forgotten either. Positively charged protons tend to scatter in all directions, and if it were not for nuclear forces (also called strong interactions), then atomic nuclei simply would not exist. Nuclei with a huge number of protons do decay, precisely because of their electrical repulsion.

As the number of protons in a nucleus increase, protonic forces grow faster than nuclear forces and eventually win the battle between the forces of attraction and repulsion. As a result, the atom falls apart or emits a positively charged particle, most often an alpha particle, thus shedding excess charge. But let's digress for now from the decay of nuclei. I will talk about it a little later.

What kind of interaction is called weak? To answer this question, you need to talk about radioactivity. It was discovered in 1896 by the French physicist Antoine Henri Becquerel, who found out that some uranium salts were capable of blackening a photographic plate wrapped in black paper. They were emitting something that affected the photographic emulsion, some kind of mysterious radiation called radioactive from the Latin word *radius*, meaning a ray. For this discovery, Becquerel received the Nobel Prize in Physics in 1903. Besides, physicists learned about an effect that requires urgent study and explanation.

We now distinguish between many types of radioactivity, but the first experiments found only three of the most common types of radioactive radiation. They were named alpha, beta, and gamma radiation. Their particles have positive, negative, and zero electric charges, respectively. Gamma rays turned out to be electromagnetic radiation of high frequency (more about this in Section 1.6), beta rays are a stream of electrons, and alpha rays are a stream of alpha particles consisting of two protons and two neutrons.

They all radiate from the nuclei. In principle, electrons are in the atomic shells, so there could be subtle indications of the presence of a source of gamma radiation. But there are definitely no helium nuclei or their parts in the shells. Therefore, from the very beginning, scientists understood that alpha radiation knowingly comes from atomic nuclei. The cause of radioactive decay, then still unknown, was later called weak interaction.

Strong and weak interactions are the short-range forces, because they quickly decrease with the distance from their source and are significant only at very small distances. Gravity and electromagnetism are considered long-range forces.

The gravitational interaction in the microcosm becomes the weakest one and practically does not affect the behaviour of atoms and elementary particles. But it does exist. When we weigh a watermelon on a balance, the Earth's gravitational field acts on every elementary particle that makes up this watermelon. Summing up all these negligible forces, we get a force called the weight of a watermelon.

But if we study the behaviour of nucleons in the nucleus or the collision of elementary particles in an accelerator, then we can safely forget about the gravitational interaction of particles.

Theories of strong and weak interactions are exclusively microscopic theories, originally based on quantum principles. Starting from electromagnetic interactions in the shells of atoms, quantum mechanics soon extended its focus of interest to nuclear forces.

These theories appeared and developed rapidly in many respects, due to the fact that there are no fundamental problems in performing experiments in this area. More precisely, they are limited only by our capabilities, both technical and financial.

The most sophisticated particle accelerator currently available is called the Large Hadron Collider (LHC). It is located at CERN (*Conseil européen pour la recherche nucléaire*, the European Organization for Nuclear Research) in Switzerland and is jointly owned by many countries. Even the richest country cannot afford to own such an accelerator, let alone a more powerful one. In addition to it, there are many weaker accelerators in the world. Their results are processed by an army of scientists using the most powerful supercomputers. Figure 6.1 depicts a mural inspired by the LHC.

Don't expect your guide to tell you a story in this section that is as detailed as about the more familiar forces of gravity and electromagnetism. After all, even if the reader, following Dr Watson, exclaims: "But how you guessed it, Holmes?" I can hardly explain in detail, step by step, how he came to the conclusion that it was John the butler who stole the family diamonds and poisoned the witness parrot. At the same time, the simple answer "Elementary" is unlikely to suit the reader. I'll try the middle way, which is not very superficial, but not very deep either.

6.1. Subatomic Particles

The proton, the neutron, and the electron are examples of the so-called subatomic or elementary particles, which have long been considered the smallest indivisible constituents of all matter in

Figure 6.1.   Mural at the Centre for Culture and Arts of the Kyiv Polytechnic Institute, created by the Canadian street art specialist Aaron Lee-Hill in 2016. It is noted that this is an image of two protons that collide in the Large Hadron Collider. Adapted from: https://kpi.ua/en/ckm-mural.

nature. There are many of these particles, many more than was originally thought.

There was a period when the discovery of every new elementary particle was awarded the Nobel Prize in Physics. Thus, it was received in 1935 by the Englishman James Chadwick "for the discovery of the neutron", in 1936 by the American Carl David Anderson "for his discovery of the positron", in 1959 by the Americans Emilio Gino Segre and Owen Chamberlain "for their discovery of the antiproton", and in 1995 by the Americans Martin Pearl "for the discovery of the tau lepton" and "for pioneering experimental contributions to lepton physics" and Frederick Reines "for the detection of the neutrino" and "for pioneering experimental contributions to lepton physics".

In 1949, this prize was received by the Japanese theoretical physicist Hideki Yukawa "for his prediction of the existence of mesons on the basis of theoretical work on nuclear forces". Nobel Prizes were given in 1976 to the Americans Burton Richter and Samuel Thing "for their pioneering work in the discovery of a heavy elementary particle of a new kind", in 1984 to the Italian Carlo Rubbia and the Belgian Simon van der Mer "for their decisive contributions to the large project, which led to the discovery of the field particles W and Z, communicators of weak interaction", and in

1988 to the Americans Leon Lederman, Melvin Schwartz, and Jack Steinberger "for the neutrino beam method and the demonstration of the doublet structure of the leptons through the discovery of the muon neutrino".

In 1968, the American Louis Walter Alvarez received the Nobel Prize in Physics "for his decisive contributions to elementary particle physics, in particular the discovery of a large number of resonance states, made possible through his development of the technique of using hydrogen bubble chamber and data analysis", i.e. for the discovery of a whole group of exotic elementary particles.

For every elementary particle existing in nature, there is an antiparticle with the same mass. Physicists have learned to make artificial antihydrogen atoms, in which there are positron shell clouds around antiproton nuclei. They have also learned to make antideuterium, antithritium, and antihelium atomic nuclei, which consist of antiprotons and antineutrons.

Antihydrogen is the most expensive substance known to mankind. According to 1999 estimates, one gram of antihydrogen would be worth $62.5 trillion, which was the budget of the United States for five years! Naturally, much less of it was obtained in the experiments.

But investing your savings in antihydrogen would be reckless. The fact is that the atoms of antimatter do not exist for long, but not because of some fundamental reasons, but simply because it is not possible to rid them of their proximity to ordinary atoms with which they annihilate. The maximum antihydrogen retention time in 2011 reached 17 minutes. But science fiction writers fell in love with antimatter. Let us recall at least the positronic brains of Isaac Asimov's robots, or antimatter as fuel or weapons in . . . Yes, everyone has them!

Physicists have discovered many particles called subatomic because they are lighter and smaller than atoms. Most of them are not found in nuclei, but can be obtained as a result of various reactions, such as the collision of accelerated protons, electrons, or atomic nuclei. Or they fly to us from the depths of outer space. Once, all of them were called elementary particles, but now the situation is confused. How is it shown?

When the number of discovered elementary particles and their antiparticles exceeded many tens, the Nobel Prize began to be given only for the discovery of especially important ones. Moreover, theorists had a growing thought that some of the discovered particles are composed of even more fundamental particles. Subsequently, this idea received experimental confirmation.

6.1.1. *Elementary particles and fundamental ones*

Many people get their information from Wikipedia articles. Indeed, mistakes are not very common there, so such a decision is not devoid of sense. Let's try to take the place of such a seeker of knowledge and search in Wikipedia if the proton belongs to elementary particles. We look at the article in Russian and find out that the proton is an elementary particle. The same is written in the Ukrainian, Belarusian, and other languages of the former USSR countries. But in articles in English, French, German, Polish, Portuguese, etc., the proton is referred to as a subatomic particle, but is not called an elementary one. It is not in the list of elementary particles either.

What's the matter? Is the proton really behaving badly and is deprived of the honorary title of an elementary particle, but only in Western countries? Let's find out the definition in the English-language Wikipedia, taken from modern scientific literature. According to it, an elementary or fundamental particle is a subatomic particle that has no structure, i.e. not composed of other particles. As I'll explain shortly, a proton is made up of three particles called quarks. Therefore, according to this modern definition, this particle is certainly subatomic, but not elementary.

However, half a century ago, a proton was considered an elementary particle, and in the definition from the Russian-language Wikipedia, they decided not to break with the traditions and precepts of their ancestors. According to the Russian Wikipedia, "an elementary particle is a collective term referring to micro-objects on a subnuclear scale, which *in practice* cannot be split into their component parts".

The fact is that you cannot break a proton into quarks due to the unusual properties of these particles, so you can use all sorts of casuistry in order to continue to call the proton, neutron, and many other particles elementary. Therefore, the words "in practice" were highlighted in italics in the original source in Russian.

The authors of this definition were a little sorry that the proton and the neutron do not belong to the elementary particle group. They can't be blamed for being made up of quarks. Additionally, the concept of a fundamental particle was introduced. Its definition completely coincides with the definition of an elementary particle from the English Wikipedia.

To avoid confusion, I refer to any subatomic particle that is elementary and fundamental as fundamental when using the definition from the English Wikipedia. And the particles that meet the criteria of the Russian Wikipedia, that is, elementary particles according to the definition given half a century ago, are what I refer to as elementary particles. Alas, I couldn't think of anything cleverer. At the same time, the text of this part will not sufficiently contradict old popular science books.

6.1.2. *Leptons*

Everything is OK with an electron; it is both an elementary and a fundamental particle. As is the entire group of particles to which it belongs. They are called leptons from the Greek word *leptos* ($\lambda\varepsilon\pi\tau\acute{o}\varsigma$), meaning fine, small, thin. This term was first used by the physicist Léon Rosenfeld in 1948.

The group of leptons includes an electron and two charged nonstable particles similar to it. One is called a mu-meson or just a muon, and the other is called a tau-meson or a tau lepton (it is rarely called a taon). If you are interested in numbers, the masses of the muon and tau lepton are 207 and 3477 times greater than the electron mass, and their lifetimes are, respectively, $2.2 \cdot 10^{-6}$ s and $2.9 \cdot 10^{-13}$ s. Considering that the muon and tau-meson decay rapidly, the question naturally arises as to which of their variants should be considered a particle and which antiparticle. Physicists decided that the lepton must have a negative electric charge, like an electron, and its antiparticle must be positively charged.

The lifetime of an electron is not limited by anything. An electron is stable by itself. Unless it gets into bad company, enters into some kind of reaction along with the rest and disappears without a trace, turning into something else. Or it will meet its antiparticle — the positron and they will annihilate.

In Chapter 5.2, I talked about how the electron was discovered at the end of the 19th century, and in Section 2.9, about the fact that positrons and muons were discovered in cosmic rays. But in 1977, an accelerator was needed to detect the tau-lepton. For this discovery, Martin Lewis Perl received the 1995 Nobel Prize in Physics.

The electron, the muon, and the tau-meson can be considered three brothers, one of whom is very thin (electron), the second significantly thicker (mu-meson), and the third stout to the limit (tau-meson). A tau-lepton is approximately twice as heavy as a proton, which makes the name lepton not entirely adequate for its mass.

And the name meson itself used in mu- and tau-mesons is purely historical and they do not belong to the group of mesons, the story of which is ahead. So here, too, confusion arises with the names from which the rapidly maturing physics of elementary particles has grown. Booties are often worn on the baby's feet, but when a person grows up and needs size 10 booties, they are usually called something else.

Physicists have learned how to obtain meso atoms, in which muons form the negatively charged shell around the nucleus instead of electrons. Naturally, the lifetime of meso atoms is extremely short, because muons decay in a millionth of a second. But in this short moment, physicists have time to investigate the properties of these unusual objects.

The three particles I have described and their three antiparticles are capable of weak, gravitational, and electromagnetic interaction. The latter is clear because they have an electric charge. But leptons include not only them, but also neutrinos that have no charge. The absence of an electric charge explains the colossal penetrating ability of these particles, which are indifferent to both strong and electromagnetic interaction. To stop a proton, you need an iron plate several centimetres thick. And a neutrino can fly through a plate billions of kilometres thick, of course, if such is made by someone for some reason.

Neutrinos quietly pass through the entire Earth. It is difficult to register them with instruments, because almost the entire neutrino flux will pass through the instrument without interaction. But this is still possible if there are a lot of these neutrinos, especially in special underground laboratories, in which the influence of cosmic rays, described in Section 2.9, is greatly weakened.

There are three different kinds of neutrinos and three kinds of their antiparticles called antineutrinos. Each is associated with a matching charged lepton. They form three sets, called three generations of leptons. These are an electron and an electron neutrino, a muon and a muon neutrino, and a tau-lepton and a tau neutrino. When a neutrino is mentioned without a preceding adjective, it means an electron neutrino.

Electronic antineutrinos can be found near a nuclear reactor. It is the reactors that are the most powerful sources of these particles on Earth. In a reactor with a power of 1000 MW, 10^{20} electron antineutrinos are born per second.

All types of neutrinos have a small but nonzero mass and they move at a speed slightly less than the speed of light in vacuum. It was once thought that they have no rest mass and move at the speed of light. But now the opinion of scientists has changed. Here the Sun suddenly came to the aid of physicists.

Not so long ago there was a mystery of solar neutrinos. We are sure that the source of the Sun's energy is thermonuclear or fusion reactions, and quite specific reactions. We know everything about each of them, including the energy released and the number of emitted neutrinos. Knowing the power emitted by the Sun, one can find the neutrino flux from the Sun, equal to $6 \cdot 10^{10}$ neutrinos per cm^2 per second.

However, the flux of electron neutrinos recorded in an experiment, for example, on the Japanese Super-Kamiokande facility, is three times lower. After testing various hypotheses, physicists agreed on the theory for which Takaaki Kajita and Arthur MacDonald received the Nobel Prize in Physics in 2015 "for the discovery of neutrino oscillations, which shows that neutrinos have mass".

If the currently known three types of neutrinos were massless, then the interval between the moment of radiation on the Sun and hitting the detector on Earth would be equal to zero, if measured by the readings of a clock moving with neutrinos with the speed of light in a vacuum. But if there is a mass, albeit very small, then their speed is less and this time interval is nonzero.

During it, these very neutrino oscillations occur, in which neutrinos of one type transform into a stream of three different types of neutrinos. Thus, electrons, muons, and tau neutrinos are constantly transforming into one another.

As a result, the electron neutrinos emitted on the Sun arrive at the Earth in the form of neutrinos of three different types. Two types of neutrinos fly farther, and a stream of electron neutrinos spends a very small fraction of their particles interacting with detectors in neutrino laboratories. The antineutrino flux from the reactor turns into a mixture of three types of antineutrinos. If the neutrino were massless, it would not have time for transformations. So the study of solar neutrinos ultimately told us more about neutrinos than about the Sun.

But not only reactors and the Sun supply us with neutrinos and antineutrinos. There are several unique installations on Earth that catch neutrinos from distant cosmic sources. Therefore, it can be argued that neutrino astronomy has already made its first steps and helped to obtain new information about the sources of neutrino radiation.

Both neutrinos and antineutrinos of different types appear when the muon or taon decay. A muon decays into an electron, an electron antineutrino and a muon neutrino. A tau-lepton can decay into an electron, an electron antineutrino, and a tau neutrino, or into a muon, a muonic antineutrino, and a tau neutrino. This heavy lepton can also decay into other lighter particles, which will further decay according to their own laws and customs.

6.1.3. *Quarks*

In the meantime, let's get acquainted with quarks, which have the property, unique among fundamental particles, to participate in all four fundamental interactions. It is clear that this will be just a vague picture of the most basic properties of these not fully investigated objects.

To begin with, the idea of quarks was proposed more than half a century ago, in 1964 by two American physicists: Murray Gell-Mann and George Zweig. Gell-Mann was awarded the Nobel Prize in Physics for 1969 "for his contributions and discoveries concerning the classification of elementary particles and their interactions". Since then, quark theory has withstood all experimental tests and perfectly describes the properties of hadrons and strong interactions. Gell-Mann took the name "quark" from the poem *Finnegans Wake* by James Joyce:

> Three quarks for Muster Mark!
> Sure he hasn't got much of a bark
> And sure any he has it's all beside the mark.

Initially, in the original version of the theory of quarks, there were only three and this was enough to make up all the hadrons known at that time. However, over time, physicists had to add a new type of quark to the set three times. So far, the current set of six quarks has completely solved all problems, but it is not certain that it will not need to be expanded over time.

These six quarks are called up, down, strange, charm, bottom, and top, giving the designations for the species or, as physicists say, the flavours of the quarks: u, d, s, c, b, and t. They are divided into three generations, each with a pair of quarks. The first of the quarks of each generation, i.e. u, s, b, has a negative electric charge three times less than that of an electron, and the second, i.e. d, c, t, is a positive charge equal to 2/3 of the charge of a proton or positron, i.e. of the same electron charge, but with the opposite sign.

The physics of elementary particles has its own traditions. Instead of the rest mass m, its adherents prefer to use the rest energy obtained from the mass by the Einstein formula $E = mc^2$.

In addition, this energy is measured not in joules, but an off-system unit of energy measurement is used, called the electron-volt and its derived units: $1\,\text{MeV} = 1.6 \cdot 10^{-19}\,\text{J}$. The electron-volt or eV is the energy acquired by an electron accelerated by an electric field with an electrical potential difference of $1\,\text{V}$. A proverb says, "When in Rome, do as the Romans do." So, I use electron-volts in this part to describe both the energy and the mass of particles.

Quark masses are very different from each other. The lightest u-quark has a mass about five times the mass of the electron. It corresponds to an energy of $2.3\,\text{MeV}$. And the mass of the heaviest t-quark is about $173200\,\text{MeV}$. Three of the six quarks have masses that exceed the mass of a proton.

Quarks have a special characteristic called their colour, although this, of course, has nothing to do with the real colour of anything. There are three colours, conventionally called "red", "green", and "blue". They are often denoted by the letters R, G, and B, respectively.

There are also six antiquarks, antiparticles of quarks. All of their charges are opposite, including the "colour". So three "anticolours" are introduced, which can be seen in Fig. 6.2. Note that the shades of "anticolours" sometimes differ in the pictures of different authors, which can become a source of additional difficulties in their perception. Anticolours are indicated by an overline above the colour designation.

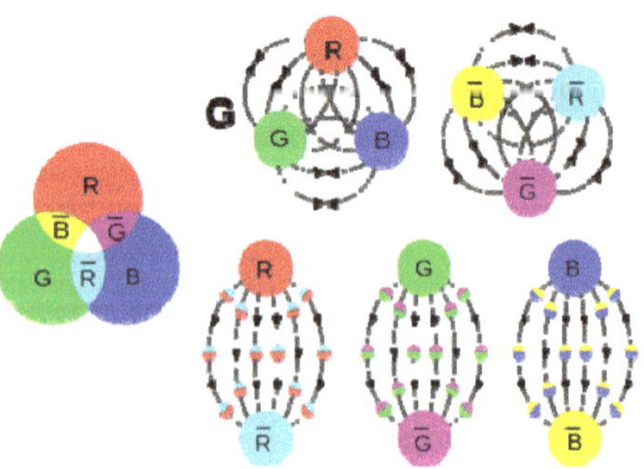

Figure 6.2.   Colour charges of R, G, B quarks, their anticolours (indicated by an overline) and how baryons, antibaryons and mesons are formed from quarks and antiquarks. Arrow lines represent the exchange of gluons. In the public domain.

Colours play the role of charge; they are also called colour charge. The theory of strong interaction is called quantum chromodynamics from the Greek word $\chi\rho\tilde{\omega}\mu\alpha$ (*chromium*) — colour. In electricity we can compensate for the positive and negative charges, getting an uncharged body in total. In quantum chromodynamics, by adding the red, green, and blue charges, we get a colourless or "white" object without colour charge. From this, we can conclude that the sum of the red and green charges gives an anti-blue charge, etc.

6.1.4. *Hadrons: Baryons and mesons*

Particles formed from quarks are called hadrons from the Greek word *hadrós* ($\dot{\alpha}\delta\rho\acute{o}\varsigma$), i.e. large or thick. They gave the name to the most famous of the particle accelerators at CERN. Quarks are fundamental particles, and hadrons are elementary ones. It is important that the particles that we can observe must be colourless. Therefore, we cannot detect one quark separately from the others; it cannot be colourless. But the hadrons can, being a combination of quarks.

There are two simplest colourless combinations. You can take three quarks of different colours and get a particle belonging to the group of baryons, so named from the Greek word *barýs* ($\beta\alpha\rho\acute{u}\varsigma$) — heavy. The name was introduced by Abraham Pais. Protons and neutrons are examples of baryons. Therefore, ordinary matter, which includes everything we see around us, is called baryonic matter.

Or you can take a quark and an antiquark so that the colour of the quark is compensated by the corresponding anti-colour of the antiquark, as shown in Fig. 6.2. This combination is called a meson. The name is associated with the Greek word $\mu\acute{\varepsilon}\sigma\sigma\varsigma$, i.e. average. Some types of mesons hold together nucleons in the nuclei of atoms. The most famous of them are called pi-mesons or pions for short.

So hadrons are divided into baryons and mesons. Let me remind you once again that the latter do not include mu- and tau-mesons, named so historically, but related to leptons in their essence, more precisely because of the inability to participate in strong interactions. Now it is customary to call them muons and tau leptons, although the former names have not disappeared.

There are also quite exotic options that have recently been discovered. They all disintegrate almost instantly. In 2014, the so-called tetraquarks, that is, mesons consisting of two quarks and two antiquarks, were discovered at the Large Hadron Collider. To some extent, they can be considered a mesonic molecule in which a strong interaction holds two mesons together.

In 2015 and 2019, two pentaquarks were discovered, consisting of four quarks and one antiquark, which can be considered a rapidly decaying combination of a baryon and a meson. Further in the text of the book, in order not to complicate the already not simple presentation, I ignore them and write only about mesons consisting of one quark and one antiquark, and baryons of three quarks. In addition, baryons are fermions, and mesons (but not the leptons muon and taon) are bosons.

All hadrons, leptons and other elementary particles have electric charges either equal to zero or multiples of the electron charge. Therefore, all bodies formed from them also have charges equal to the charge of an integer number of electrons or protons. This is called quantizing electrical charge. The charge of quarks is proportional to a third of the charge of an electron, but they cannot be met individually. So what is the quantum of electric charge equal to: the charge of an electron or its third? It's a matter of definition, nothing more. Much more interesting is the very fact of quantization.

Figure 6.2 depicts colour combinations (R, G, B), giving anticolours (the same letters with overlap) and white. Naturally, the sum of the three different anticolours is colourless. There are also some diagrams showing the structure of baryons made from quarks, antibaryons from antiquarks and mesons from quarks and antiquarks.

Fundamental particles include not only leptons and quarks, but also quanta of fundamental interactions. Among them are photons, quanta of electromagnetic radiation. And also quanta of strong interaction, of which there are eight and they are called gluons, from the word *glue*. All of them have no rest mass and move at the speed of light in emptiness. The important difference between gluons and photons is that gluons can interact with each other, which greatly complicates the theory of strong interactions. I will talk about the quanta of other fields a little later.

6.2. Strong and Weak Interactions

It's time for me to fulfil my promise and talk about strong and weak interactions and the differences between them. I have already mentioned one of them. A strong interaction acts only on hadrons and quarks, of which they are composed, and a weak one also acts on leptons.

Strong interactions keep nucleons in the nucleus and quarks in the hadron. In both cases, the total rest mass of the composite object turns out to be less than the sum of the masses of its parts. You already know that this is called a mass defect and this nuclear deficit is on the conscience of the strong interaction.

What is the rest energy of elementary particles? The proton has 938.3 MeV, the neutron has a little more — 939.6 MeV — and the electron has only 0.51 MeV. But if a proton and a neutron form a deuteron D or ^{2}H$^+$, then its rest energy is equal to 1875.6 MeV, which is 2.22 MeV less than the sum of the energies of a proton and a neutron. This is the mass defect in the deuteron and is caused by the nuclear forces holding these two particles together.

We also call this quantity the energy yield of the reaction p + n $\rightarrow$ D + γ + 2.22 MeV. In it, a proton and a neutron form a deuteron and a gamma quantum, denoted by the Greek letter γ (gamma). The products of this reaction, i.e. a deuteron and a gamma quantum, carry away the excess energy of 2.22 MeV in the form of the sum of the photon energy and the kinetic energy of the deuteron.

From two deuterons, it is possible to form an alpha particle ^{2}He^{2+} with a rest energy of 3.72 GeV, which is 28.11 MeV less than the mass of the nucleons that form it. From this, it is easy to calculate the energy yield of the reaction in which two deuterons combine to form an alpha particle. It is equal to 28.11 MeV $-$ 2 $\times$ 2.22 MeV = 23.67 MeV. Since it is energetically favourable, such a reaction, if it is possible, will occur, despite the electrical repulsion of two deuterons, or rather protons inside them. And this will be the transmutation of the hydrogen isotope into helium, and more specifically the reaction of thermonuclear fusion.

I wrote "if possible" for a reason. This process will not happen by itself and the deuterium is stable. But if deuterons are accelerated to speeds at which they are able to overcome the so-called Coulomb barrier caused by their electrical repulsion, then they may well merge into a helium-4 nucleus due to the action of nuclear forces. In other words, they must be heated to enormous temperatures and the required concentration of deuterons must be ensured during the time required for the fusion.

There are other reactions, in principle, suitable for thermonuclear power plants, in which the number of protons and neutrons at the input and output is the same. These particles are simply redistributed over atomic nuclei. For example, two nuclei, deuterium and tritium, are converted into a helium-4 nucleus and a neutron. This reaction also takes place in a hydrogen bomb. Or two deuterium nuclei are transformed into a tritium nucleus and a proton or into a helium-3 nucleus and a neutron. At the entrance, there are two protons and three neutrons and two protons and neutrons, respectively; at the output, we have the same nucleons, but in different combinations. In this case, there are no transitions of a proton to a neutron or vice versa, and the energy yield of the reaction is determined only by the strong interaction.

6.2.1. *A weak interaction is capable of converting some quarks into others, and a strong one determines the energy output of this reaction*

But the weak interaction is capable of transformations that the strong one cannot provide, despite the fact that it is much weaker. It can change the flavour of a quark, or, more simply, transform one quark into another. Need an example? A neutron consisting of u-quark and two d-quarks (formula udd) is capable of decaying into a Proton, consisting of two u-quarks and one d-quark (formula uud), an electron, an electron antineutrino and, possibly, a gamma quantum according to the scheme $n \rightarrow p^+ + e^- + \bar{\nu}_e$ (99.7% decays) or $n \rightarrow p^+ + e^- + \bar{\nu}_e + \gamma$ (0.3%). A neutrino is denoted by the Greek letter ν (nu), an antineutrino by it, but with an overline, and the subscript e, μ, or τ indicates its type: electron, muon, or tau neutrino. Electrons and antineutrinos are leptons and do not consist of quarks, like a photon. Thus, in this decay caused by weak interaction, one d-quark turns into a u-quark.

The neutron decay rate is low due to the extremely low energy release, equal to 0.78 MeV. It has a really long lifetime among elementary particles: the lifetime of a neutron is about a billion times longer than that of a muon, the next in this parameter elementary particle. And the lifetimes of other unstable elementary particles are even shorter.

But why don't the neutrons decay inside stable nuclei? What prevents them from transformation? After all, the decay of a neutron is provided by a weak interaction that is not associated with nuclear forces. Why does it stop working inside nuclei?

The answer does not require much knowledge of the details of these two interactions. It's all about energy. Consider the simplest radioactive nucleus in which there is a neutron. This is the tritium nucleus. In it, one of the neutrons can decay into a proton, an electron, and an antineutrino, turning it into a 3He nucleus with an energy release of 18.6 keV = 0.0186 MeV, which is 42 times less than in the decay of a free neutron. There is still a profit, but it is significantly less, so the half-life of tritium is 12.3 years. Let me remind you that the analogous time for a neutron is 10 minutes.

Why has the energy yield of the tritium decay reaction dropped so much? This is the result of a strong interaction in the nucleus. When three nucleons merge into one nucleus, energy is always released, but the merging of one proton and two neutrons by forming the tritium nucleus is energetically more favourable than the merging of two protons and one neutron by forming the helium-3 nucleus. This circumstance significantly reduces the energy yield of the tritium decay reaction, but it is still positive because the neutron mass is slightly larger than that of the proton.

But for the deuterium nucleus, decay is impossible, since this would form a strange nucleus consisting of two protons, which would not be held by nuclear forces. It is clear that the output of this hypothetical decay would be negative. Therefore, it does not occur in nature and deuterium is a stable isotope.

In any other stable isotope, for example gold, ^{197}Au, each of the neutrons could theoretically decay to form an electron, an electron antineutrino, and a proton. This would turn the nucleus into the nucleus of the ^{197}Hg mercury isotope, but purely hypothetically. The ^{197}Hg nucleus itself is unstable; the nuclear binding energy in it is not very high, and so the energy yield in the decay of gold would be negative. Energy would not be released, but absorbed. So a weak interaction in principle could do this transformation, but a strong interaction makes it unprofitable, which means it won't happen on its own.

If you still have some questions, I make the following analogy. Consider a glass-walled aquarium that is divided in half by a vertical partition. The bottom is flat and even, but bottom heights are different for the two halves. A little water is poured into the left half. Will it appear in the right one?

The answer depends on two circumstances. First, let's look at the permeability of the baffle material. If it is metal, glass, or plastic, then water simply will not seep through the partition. And if it is sandstone or even a sieve, then the water will get where it wants, and it wants to flow to where the bottom is lower, this is the second detail.

In this analogy, two parts of an aquarium correspond to two nuclei. The left one is the original nucleus, and the right one is the nucleus in which one neutron turned into a proton. The bottom height characterizes the binding energy of nucleons in the nucleus. The higher the bottom, the smaller it is. The leakage of water is the transformation of one nucleus into another with the emission of an electron and an antineutrino. Weak interaction is what "makes the septum permeable". It gives a fundamental opportunity for the neutron to decay. A strong interaction is what determines "where the water wants to flow", i.e. positive or negative energy yield during decay.

So this type of decay with the transformation of a neutron into a proton and the emission of an electron, called beta decay, is possible only with the participation of a weak interaction and under the condition associated with a strong one. We can say that weak forces allow all nuclei, except for protium, to decay with the emission of an electron and antineutrino. In this process, the atomic number of an element increases by one and the mass remains almost unchanged. The nuclear forces, in turn, forbid this to many isotopes because of the negative energy yield of such a transformation.

I would like to draw your attention to one circumstance. With beta decay, we get elements that have a higher number in the periodic table. Their atomic number is shifted from the beginning to the end of the table. With alpha decay, the opposite is true, we get a nucleus with a lower number, and the emission of gamma quanta does not affect the number in any way.

If it were not for the existence of stable isotopes that are resistant to beta decay, a weak interaction with the help of this process would eventually turn, say, the nucleus of the isotope of iron ^{56}Fe, which has atomic number 26, into the nucleus of zirconium, the element with number 40. Then the transformations would continue, turning it into niobium, then molybdenum, and finally the impossible in nature isotope of barium, with a nucleus consisting of 56 protons only. Or it would transform the nucleus of the lead ^{206}Pb isotope into uranium and even something transuranic, even if this conversion takes a huge amount of time. But this could not happen in nature. It is because of the strong interaction that the chain of element transformations due to the beta decay will stop at some stable nucleus. Or, on the contrary, it could end with a particularly unstable one, which falls apart into pieces. Naturally, the latter possibility could not be realized for light nuclei.

But weak interaction is not reduced to neutron decay only. It can change the flavours of quarks in different ways. And it's great that it is also capable of converting protons into neutrons. Naturally, it will not work simply to convert, because the neutron is slightly more massive than the proton, and the energy yield of this reaction would be clearly negative. But the resulting neutron can immediately find a proton and, through a successful marriage, turn into a deuteron, while receiving more energy than is necessary for the transformation of a proton into a neutron.

This is where the long chain of hydrogen transformations into helium and other elements begins, which provides energy to the stars, including our Sun. Gradually, step by step, the number of neutrons in the nuclei being formed grows. Each step takes many millions of years and requires high temperatures from 10 million degrees and a not very small density of the original particles. All this is called the proton–proton cycle; it has several options for the course.

As I have already mentioned, two protons can turn into a deuteron, emitting a positron and neutrino $p + p \rightarrow D + e^+ + \nu + 0.4\,\mathrm{MeV}$. The characteristic time of such a transformation on the Sun is estimated to be 5.8 billion years. If instead of positron radiation a passing electron is caught, then the energy yield could be increased: $p + p + e^- \rightarrow D + \nu + 1.44\,\mathrm{MeV}$. This process on the Sun takes $2.3 \cdot 10^{12}$ years on average because it requires the meeting of three particles, two protons, and an electron. These two processes provide the main stream of solar neutrinos.

Then, in $3.2 \cdot 10^{12}$ years, the deuteron meets a proton, forming a helium-3 nucleus and dropping the excess energy in the form of a gamma quantum: $D + p \rightarrow {}^3\mathrm{He} + \gamma + 5.49\,\mathrm{MeV}$. Two such nuclei can turn into a helium-4 nucleus and two protons, completing the conversion of hydrogen into helium-4 with the release of 24.7 MeV. The process ${}^3\mathrm{He} + {}^3\mathrm{He} \rightarrow {}^4\mathrm{He} + 2p + 12.86\,\mathrm{MeV}$ takes $1.5 \cdot 10^5$ years. In addition, ${}^3\mathrm{He}$ and ${}^4\mathrm{He}$ nuclei can form the nucleus of the ${}^7\mathrm{Be}$ isotope, which will continue the chain of transformations. I will not tell you about all these options to the very end. I just show the principle by which the Sun and its colleagues work.

Now let's remember our aquarium example. I told you the truth about him, but not all the truth. If the aquarium is peacefully resting on the table, then everything will happen exactly as I described. But there are special cases. For example, a mini-whale was launched into the aquarium, and it released a fountain of water. Or a strong earthquake has started and the aquarium is shaking as if on a vibrating table. Is an earthquake not enough for you? Imagine our aquarium in the hands of a mischievous active child. There is no time to think about where the water will flow. It will be everywhere it can get, and a little where it cannot be.

In Section 1.5, I talked about nucleosynthesis in stars. Elements with numbers higher than those of lead cannot be obtained in such a way. Strong interaction won't allow that, no matter how hard weak interaction tries. But if a star explodes like a supernova, then all considerations related to the minimum energy work poorly, because during the outburst, each atomic nucleus will bathe in the excess energy released during the explosion and in the flux of protons and neutrons. It could quickly increase its atomic number in such an environment. Some nuclear careerists manage to grow to uranium and transuraniums during the not very long time of the outburst.

6.2.2. *Fission of atomic nuclei*

Let's move on from nuclear fusion to another variant of transmutation— fission reactions. They are different, but they have the same goal — to achieve the release of energy. We begin the description with the simplest process of radioactive decay, in which the nucleus, called the parent radionuclide or radioisotope, decays and produces at least one other nucleus, called the daughter nuclide.

The spontaneous decay of atomic nuclei found in nature is called natural radioactivity, and the decay of nuclei obtained artificially, for example, as a result of a previous decay, is called artificial radioactivity. The fact is those daughter nuclei can also be radioactive and become parent nuclei for subsequent decay. As a result, a whole chain of decays occurs, ending with the appearance of stable daughter nuclei. Let's consider one of the links in this chain. Let the nucleus pass into the nucleus of another element or isotope spontaneously, without any external influence, while emitting some particles.

If, for example, a neutron is emitted, then the result of the reaction will be an isotope of the same element, but with a smaller number of neutrons. If the original nucleus already suffered from an excess of neutrons, then this may well give an energy gain. If one or more positively charged particles escape, such as an alpha particle or a proton, then the number of protons in the nucleus decreases and the resulting nucleus will have a lower atomic number, moving towards the beginning of the periodic table. If, say, an electron has escaped, then the number of protons in the nucleus, on the contrary, increases and the daughter nucleus is located further in the periodic table. In the last two cases, the nucleus of one chemical element turns into the nucleus of another.

Gamma radiation by itself does not cause any change in the type of nucleus. It is possible when the nucleus dumps the excess energy in order to go from an excited state to the ground or a less excited state, similar to how an electron in a shell emits a photon when passing to another orbital. So this process can hardly be called decay. But many types of real decay are often accompanied by the emission of gamma quanta, so they are also a product of radioactive decay.

There are also more exotic types of decays with the emission of a proton, two electrons, and a positron. Among them is a reaction reminiscent of the relationship between carnivorous plants and insects, called electron capture.

One of the protons in the nucleus captures an electron from the shell, usually the inner one, and turns into a neutron with the emission of an electron neutrino. If we recall the recent analogy with a company sitting at a table in a cafe, then a predatory table eats one of the customers, while turning into a table of a slightly different design. However, it can, on occasion, eat two clients at once.

In 2019, the participants in an experiment to search for dark matter particles in an Italian laboratory in Gran Sasso recorded that a unique decay occurred in their detector with xenon-124. Xenon is not capable of capturing one electron, but it can capture two electrons from the shell at the same time. This process, called double electron capture, was previously known only for two nuclei: barium-130 and krypton-78, and even then only presumably.

So, the half-life of a xenon atom (an atom, not a nucleus, since a naked nucleus without an electron shell cannot decay in this way — there are no electrons to be stolen from the shell) is $1.8 \cdot 10^{22}$ years. This is about 10^{12} times more than the age of our Universe.

And now I rewrite this number for those who have a poor feel for numbers with powers of 10. This is 1,000,000,000,000 or a thousand billion ages of our world.

And physicists were able to see and identify such a rare phenomenon. This became possible due to the fact that the 3.5 tons of xenon in the installation in Gran Sasso contain a huge number of atoms. Why talk about tons when one gram of xenon-124 contains $6 \cdot 10^{23} : 124 \approx 4 \cdot 10^{21}$ atoms. To obtain this estimate, I divided the number of atoms in one mole of a substance, called the Avogadro constant and which is equal to $6.02 \cdot 10^{23}$, by the mass of a mole of xenon in grams, equal to its atomic mass in Da. So, the double electron capture is a unique event, but it can be observed.

In the process of spontaneous nuclear fission, a massive nucleus is transformed into several lighter ones called fragments. This process is usually accompanied by the emission of several small particles, such as alpha particles, neutrons, or gamma rays. It is considered a type of radioactive decay. So, lighter nuclei can get energy by the synthesis of heavier elements. Heavier nuclei could decay. The nuclei with 50–80 nucleons from the middle of the periodic table are the most stable ones, such as ^{56}Fe, with its 26 protons and 30 neutrons.

However, fission reactions are not always spontaneous. If the nucleus is ready to decay, then it is like a person who is about to fall. They can lose balance or slip. In all cases, they fall more likely if you push them. If a particle hits an unstable nucleus and "pushes" it, then the nucleus can immediately decay.

An arriving neutron is a push for the decay of uranium-235 or plutonium-239 isotopes. The nucleus decays, emitting several neutrons among other decay products. These neutrons can fly out of uranium fragments, be absorbed by impurities, or hit another nuclei and cause fission in them. If the average number of neutrons produced during fission and then hitting other nuclei exceeds 1, then the number of neutrons and fission reactions will grow exponentially, that is, very quickly.

A nuclear chain reaction is started. If you do not take measures to reduce the number of neutrons, the whole process will end with an explosion. By controlling the number of neutrons, it is possible to keep the aforementioned average number of neutrons produced at a level very close to unity. Then the chain reaction will neither attenuate nor intensify. This is what happens in nuclear reactors.

I have told you about the main nuclear reactions. Humanity has been able to use them for military purposes for more than 70 years. Atomic bombs use a fission chain reaction and hydrogen bombs use a fusion reaction.

6.2.3. *Nuclear reactors*

A nuclear reactor, like any technically complex device, has a lot of modifications and nuances. I will tell you about the most general details of how it works. Fission processes take place in the reactor core filled with nuclear fuel components and surrounded by neutron reflectors and radiation shielding. The nuclear reactions occur, and heat is generated inside the core. The released heat warms the coolant, which cools the core. The maintenance of the reaction at the proper level is monitored by the control system, the task of which is to prevent the acceleration and explosion of the reactor and the unplanned attenuation of the chain reaction.

The produced heat is used to generate electricity or for other purposes for which the reactor is designed. To some extent, from the point of view of the ways of using thermal energy, it differs little from other types of devices in which something is heated, from the same boilers of thermal power plants. You can, for example, turn water into steam and make it turn the turbines of an electric generator.

The neutrons formed during the reaction are characterized using their average kinetic energy. If it exceeds 0.1 MeV, then the neutrons are called fast, and if it is less than 0.4 eV, then they are considered slow or thermal, because it is close to the average energy of the thermal motion of gas molecules at room temperature — that is, 0.025 eV.

The world's first nuclear reactor was created by a group of scientists led by Fermi in Chicago in 1942. They used natural uranium and its dioxide. Fast neutrons appearing after the fission of ^{235}U nuclei were slowed down by graphite blocks to thermal energies, so this type of reactors are called thermal neutron reactors. A moderator is required for their operation, and it is not necessarily graphite.

In common types of reactors, either ordinary water or heavy water, deuterium oxide D_2O, is the coolant and at the same time the moderator. But there are also reactors in which the coolant is gas, liquid metal, or something more exotic.

It is clear that nuclear energy has its own peculiarities. For example, after passing through the core, the coolant becomes radioactive. Oxygen nuclei, which are part of water, both ordinary and heavy, partially transform into nuclei of the radioactive isotope of nitrogen ^{18}N when irradiated with neutrons. Its half-life is only 7 s, so this induced radioactivity quickly disappears. But other radioactive isotopes can exist longer, so water that has passed through the core is returned to it.

In the so-called pressurized water reactors, this coolant water circulates in the first loop through the core. Then it heats the water in the second loop, which does not pass through a core zone. The heat exchanger is used for this. Water from the second loop in the form of steam turns turbines, generating electricity at nuclear power plants.

We are so used to nuclear energy that we are surprised that people once denied the very possibility of its existence. Who do you think said, "Any one who says that with the means at present at our disposal and with our present knowledge we can utilize atomic energy is talking moonshine"? It is a quotation from Lord Ernest Rutherford, who was addressing a meeting of the British Association for the Advancement of Science in 1933. There were 9 years until the construction of the first nuclear reactor in Chicago.

6.2.4. *Natural nuclear reactor*

Many readers will be surprised to learn that there are also nuclear reactors not built by man. In 1956, their existence was predicted by the American Paul Kazuo Kuroda. I already wrote that there are radioactive elements in the bowels of the Earth. They not only heat these bowels, but also manage to form a natural nuclear reactor. It happened in the Oklo uranium deposit in Gabon, Africa, in which, about 1.8 billion years ago, a spontaneous chain reaction of uranium fission took place, mainly ^{235}U. The half-life of this isotope is 700 million years, so, due to spontaneous decay and decay during the work of the rector, its concentration managed to decrease and the chain reaction stopped after several hundred thousand years.

Moreover, there were several reactors. These are lenticular formations of uraninite (UO_2) with a diameter of about 10 m and a thickness of 20 to 90 cm. The fraction of uraninite in them ranged from 20 to 80% by weight. In three different parts of the field. 16 single reactors are found. They are united under the general name Oklo Natural Nuclear Reactor. The isotope ratios in its rocks were used to test the constancy of fundamental physical constants over the past 2 billion years. No signs of their change were found.

In Section 3.6, I made some guesses about where the Moon came from. Among those not mentioned there is such an exotic hypothesis as the release of the substance that formed our satellite during the explosion of a natural nuclear reactor on Earth. It is strange that its supporters were not surprised when another moon did not appear in the sky after the explosion at the Chernobyl nuclear power plant. And after the explosion of a particularly powerful thermonuclear bomb nicknamed "Tsar Bomba" and "Kuzkina Mother" in the USSR, they could hope for at least a couple of new satellites.

6.2.5. *A spectre of thermonuclear energy seduces the world*

But with thermonuclear reactors, everything is more complicated. You could get energy by synthesis. Take a giant vat of water and detonate a thermonuclear bomb inside.[a] Use the steam generated in the process to rotate the steam turbine. *Voilà!* You have received thermonuclear energy. Naturally, this is a purely speculative option. What we need is a reactor with a controlled thermonuclear reaction in it, such as going inside the stars, but a little different one, as I will clarify.

---

[a]Don't try this at home.

But I'll start with the most unpleasant feature of the controlled fusion reaction on Earth: it doesn't exist. The atomic reactors appeared even before the creation of the atomic bomb and they are used successfully. The thermonuclear bomb has been known for more than half a century, but there are still no thermonuclear power plants. Why? After thinking a little more, two whole questions arise: why there are no thermonuclear power plants and why mankind persists in trying to create not only them but their laboratory prototype.

I'll start by answering the second. Humanity is like a peasant who stubbornly climbs at the fair up a "butter week" slippery pole smeared with grease or covered with an ice crust. He slides, falls, but continues to climb, because such a wonderful pair of boots hangs above. Or briefly: "I really want to!"

Seriously though, a thermonuclear reactor theoretically has a lot of advantages over a nuclear reactor. Usually, such advantages include the following:

- Energy yield is large, with a small amount of raw materials.
- Hydrogen is available and cheap because its reserves on Earth are practically inexhaustible. No country has a monopoly on it.
- There is very little radioactive waste with relatively short half-lives, and no combustion products or greenhouse gases.
- The reaction cannot get out of control, become uncontrollable, and lead to an explosion.
- Materials used for the production of nuclear weapons are not used or created, so they cannot fall into the hands of terrorists.

So, the boots are really pretty. Are there any hidden flaws in them? There are a few, which I will tell you about. But first, I will answer the question of why there are no thermonuclear power plants. I will start with some rather general considerations and only then answer more specifically.

A problem arises with the implementation of the idea: the synthesis reaction requires high temperatures and concentrations of the starting products. That is, we need a high-temperature hydrogen plasma. In addition, it should be in this state for some time.

What should contain this plasma in, and in which vessel or tank should it be placed? None of the known materials can withstand the required temperatures of millions of degrees. We have no star at hand with a strong gravitational field which holds the substance inside.

Only two of the numerous solutions currently being considered have at least some chances of success. These are magnetic traps and pulsed systems in which small targets are heated with laser pulses or beams of particles or ions. In the first case, the process is quasi-stationary, whereas in the second, it takes the form of microexplosions. Quite figuratively, in the first version, we crack nuts by pressing on them, while in the second, we knock on them with a hammer.

It is worth explaining what kind of thingummy is called magnetic trap or magnetic bottle. This is a device in which the plasma is confined using a magnetic field of a special configuration. We can say that it is poured into an invisible and intangible magnetic bottle, where it is stored for the time it takes to react. Let me remind you that plasma consists of charged particles — in particular, hydrogen plasma consists of a mixture of protons and electrons, and deuterium plasma consists of deuterons and electrons. In a magnetic field, these particles rotate around the lines of force of the magnetic field, in addition to moving along them.

The field is arranged in such a way that the field strength increases towards the edges of the "bottle". In this case, the pitch of the helical line begins to decrease when approaching the edges, and the speed of movement along the line of force decreases, then vanishes and changes sign. In other words, the particle is, as it were, reflected from an invisible obstacle and returns back to the bottle. Charged particles move in the same way in the Earth's magnetosphere. The Polar Regions, with a stronger magnetic field, play the role of edges.

But plasma in a magnetic field of such a configuration is still unstable and bursts out of the bottle. The reason is not in the bottle, but in the growth of instability inside the plasma itself, something similar to the growth of the large-scale structure of the Universe from the initial small fluctuations.

It is necessary that plasma temperature be very high for the nuclear fusion process. How much? For the reaction of deuterium with tritium, this is more than one hundred million degrees. Will the fusion reaction be able to complete during the plasma confinement time while it is still in the "bottle"? And it is necessary that the product of the concentration of plasma particles and the confinement time exceed a certain minimum value equal to 10^{14} cm^{-3} s. This condition is called the Lawson criterion.

The record for plasma confinement time for a magnetic trap is somewhere around 100 seconds. It was set and improved in 2016–2018 at various installations, for example, at the Wendelstein 7-X stellarator of the German Max Planck Institute, at the South Korean tokamak KSTAR (Korea Superconducting Tokamak Advanced Research), in China, etc. Tokamak is the transliteration of the Russian abbreviation for the toroidal chamber with magnetic coils. A tokamak is a torus-shaped or doughnut-shaped magnetic trap. The name stellarator comes from the Latin word *stella*, i.e. star. However, this is not at all a configuration of the field in the form of a star with a certain number of rays, but a reminder that fusion reactions take place inside stars.

And here I must admit that there are many differences between these reactions and controlled thermonuclear fusion on Earth. No, both are nuclear fusion reactions, transmutation, but they are different. The reason for this is in different mechanisms of plasma confinement, different characteristic sizes, and other details. The magnetic trap only remotely replaces confinement due to the gravitational field of a huge star with thermal insulation in the form of a radiant transfer zone.

Protons turn into neutrons inside the Sun and other stars. This requires the transformation of quarks, and only a weak interaction is capable of such a trick. But it is called weak because it works slowly. This process will not go far during the confinement of plasma in terrestrial installations.

Therefore, other processes are used in which only strong interaction is involved. In other words, those in which the initial set of protons and neutrons is rearranged into a different combination with a lower energy. I have already named them a little higher. The already mentioned reaction of deuterium D with tritium T requires a lower temperature than others and looks like this: $D + T \rightarrow {}^4He + n + 17.6\,MeV$.

However, it has a number of disadvantages. Firstly, you will not find ordinary hydrogen among the raw materials, and its use has been declared as a clear advantage of the fusion reaction. But tritium is needed, and it is expensive and decays over time. Secondly, there is also a neutron among the reaction products, in addition to the alpha particle. And this is undesirable.

In a star, the formed neutron and its energy will not go anywhere. But this cannot be said about the reactor. Neutrons are highly penetrating and difficult to shield against. And the radioactivity induced by neutron irradiation (as in the aftermath of a neutron bomb explosion) is undesirable.

By the way, I mentioned a small amount of radioactive waste. Formally, they are totally absent in this reaction, but the elements of the reactor structure that have received induced radioactivity, including due to neutron radiation, are considered waste. Let the reactor be surrounded by a material that absorbs neutrons. The kinetic energy of neutrons is absorbed along with the neutrons themselves. It is spent on unwanted and pointless heating of the protective shell.

Consider the reaction of the collision of a deuteron, a deuterium nucleus, with a tritium nucleus in the system of their centre of mass. It is also the system of the centre of mass of the alpha particle and neutron resulting from the reaction. They move in opposite directions and have the same momentum. The energy of 17.6 MeV released during the reaction manifests itself as the sum of the kinetic energies of these two particles. Let's perform the simplest calculations, requiring only a knowledge of the formula for the kinetic energy $E = mV^2/2 = p^2/(2m)$ of a particle with mass m, velocity V, and momentum $p = mV$.

The mass of an alpha particle is four times the mass of a neutron. Therefore, its energy has to be four times less. So, 80% of the energy goes to the neutron and is used to heat the containment and turn it into radioactive waste. Only 20% of the energy yield can be used. It is clear that something could be done. For example, one could surround everything with a thick layer of water and heat it with neutrons, using their energy. But it is better to use other reactions in which no neutrons are produced.

Examples of such reactions are $D + {}^3He \rightarrow p + {}^4He + 18.4\,MeV$ or ${}^3He + {}^3He \rightarrow 2p + {}^4He + 12.9\,MeV$. They require the isotope ${}^3He$ as fuel, which is quite expensive. There are plans to mine it on the Moon. I talked about them in Section 3.6. But it is difficult to assume that ${}^3He$ transportation to the Earth can be economically viable in the foreseeable future.

There is a synthesis reaction using only deuterium as a raw material. More precisely, a whole cascade of reactions, starting with two: $D + D \rightarrow p + T + 4.0\,MeV$ and $D + D \rightarrow n + {}^3He + 3.3\,MeV$. As you can see, one of them is accompanied by the emission of neutrons. The protons and nuclei of tritium and ${}^3He$, obtained during these transformations, in turn, react with deuterons and with each other, forming, in particular, alpha particles.

In pulsed systems, in which small droplets of fuel are compacted by laser radiation or particle beams, a wider range of reactions can be used. One of the most promising is the reaction of protons with boron isotope nuclei $p + {}^{11}B \rightarrow 3\,{}^4He + 8.7\,MeV$.

So far, in many experimental installations, energy can be obtained by a thermonuclear reaction, but more energy is spent on maintaining its operation than is received. Scientists expect to soon receive more energy than they expend, but the cost of the energy received will be enormous in the beginning. However, the problem of obtaining a controlled synthesis reaction is not in something fundamental, but mainly in technological and economic aspects. It is possible that a solution will be found that will make it one of the most used energy sources on Earth.

In the summer of 2020, in the south of France, 60 km from Marseille, the assembly of the international experimental thermonuclear reactor ITER with a toroidal chamber began. Its cost is 20 billion euros. In 2035, it is planned to begin full-scale experiments with deuterium–tritium plasma, during which the reactor will have to hold the high-temperature plasma for 400 seconds and reach the full thermal power of 500 megawatts. At the beginning of 2022, the largest tokamak Joint European Torus (JET) near Oxford,

UK, set a new record by producing about 59 megajoules of energy in 5 seconds (enough to boil water in about 60 kettles), reaching a power of about 11 MW. JET more than doubles the previous world record set in similar experiments in 1997. However, more energy was expended than received.

6.2.6. *Weak interaction breaks the mirror symmetry of the world*

Weak forces violate the principle of conservation of the number of quarks of the same type, which is fulfilled in the case of strong interaction. Naturally, I'm talking about the number of quarks minus the number of antiquarks of the same kind.

And these are not the only conservation laws that weak interaction manages to break. The 1957 Nobel Prize in Physics was awarded to the Chinese–American physicists Yang Zhenning and Li Zhengdao for their discovery of parity violation in weak interactions. Parity is the property of a physical quantity to preserve its sign (or change it to the opposite) during a mirror reflection, or rather a transformation that changes the direction of all axes of the Cartesian coordinate system to the opposite.

Parity is preserved under strong and electromagnetic interactions. Only an anarchic weak force breaks all the rules that they try to impose on it. The fact that our world changes after a mirroring greatly puzzled physicists. This question deserves a story, but I will defer it until Section 8.3, in which I discuss the role of symmetry in physics.

6.2.7. *Tough life of hadrons*

I told you about the strong interaction that acts on quarks and the hadrons that are made of them. But I did it in the most general terms, greatly simplifying the picture of very complex objects and their interactions. In this section, I give a more detailed overview of the daily life of hadrons. Those who are not very interested or consider it too difficult can skip the text to the end of the chapter at any time.

Nuclear forces are well described by quantum chromodynamics (QCD), but this theory is mathematically very complex and nonlinear, and the principle of superposition does not work in it. Consequently, the result of several consecutive actions in it depends on their order.

When the world got acquainted with the Rubik cube in the 1970s, theoretical physicists working on QCD discovered many parallels between its states and their favourite theory. Some combinations are called mesons, and others heavy mesons. There is even a "heavy meson with cherries". Their image is easy to find in any problem book or reference book on patterns on a cube, sometimes called solitaires. But it is known that the state of the Rubik cube depends not only on the turns, but also on their sequence. It is enough to do them in the wrong order and the result will change.

This section is not about theory, but about facts. It is for those who are ready to delve a little further into the mysterious forest of particles. There are no formulas in it, because there are simply no elementary equations in QCD. But there are details that cannot be found in a popular science description.

For example, consider the abundance of particles that were named after all lowercase and uppercase letters of the Greek alphabet. Aren't there too many of them, although they all consist of only six quarks? And are they particles in the full sense of the word?

Some of them disintegrate so quickly that there is no possibility to observe them as a separate object. It's just that small peaks appear on some tricky chart, sometimes very small ones. Then physicists say that this proves the existence of rapidly decaying particles, called resonances, and assign them names, or rather another Greek letter that has not yet been accidentally used for others. Then they go to celebrate the epoch-making discovery. As a result, particle reference books outstripped Tolstoy's *War and Peace* in the number of pages.

I will not be distracted by resonances and immediately begin my story with the most important mesons, called pi-mesons or simply pions. There are three variants of them, which differ in their electric charge.

Uncharged pion is the so-called truly neutral particle, which coincides with its antiparticle. Among the fundamental particles belonging to this group, you can find several quanta of interactions. There are also constituent truly neutral particles. Usually they write that these are mesons consisting of a quark and an antiquark of the same flavour.

In the physics of elementary particles, they received the names π^0, φ^0, η^0 (pi-mesons, phi-mesons, and eta-mesons, the zero above meaning the electrical neutrality of the particle), etc. As you can see, the π^0-meson is included in this particular group. If a particularly meticulous reader asks which quark and antiquark is a pair, then I will have to clarify that when talking about mesons, I have simplified a little.

A π^0-meson is a specific combination of pairs of a u-quark and its antiquark and a d-quark and its antiquark. For phi- and eta-mesons, a pair of an s-quark and its antiquark in different proportions is added to this cocktail. In general, "mix, but do not stir".

But the other mesons are just a couple. For example, u- and anti-d-quarks form a positively charged particle named π^+-meson. Its antiparticle is called the π^--meson and it consists of d- and anti-u-quarks. Naturally, it is negatively charged. At the same time, the name of both of them does not have the prefix "anti". It is usually applied to leptons, quarks, nucleons, and compound nuclei such as the deuteron. In lists and tables of particles, the π^+-meson is listed as a particle, and the π^--meson as an antiparticle.

Pions are at least three times lighter than any of the other mesons. They do not live long outside the nucleus, but all other mesons live even less. Charged pi-mesons have a rest energy of 139.6 MeV and decay due to the action of a weak interaction on a mu-meson and a muonic neutrino or antineutrino in $2.6 \cdot 10^{-8}$ s. The neutral pi-meson with an energy of 135 MeV decays into two photons due to electromagnetic interaction, and very quickly, in $8.5 \cdot 10^{-17}$ s. Very rarely, they decay on a photon and an electron–positron pair, or into one or two such pairs.

It is the exchange of pi-mesons that prevents the atomic nucleus from fragmentation, despite the electrical repulsion between its protons. This hypothesis was put forward in 1935 by the Japanese

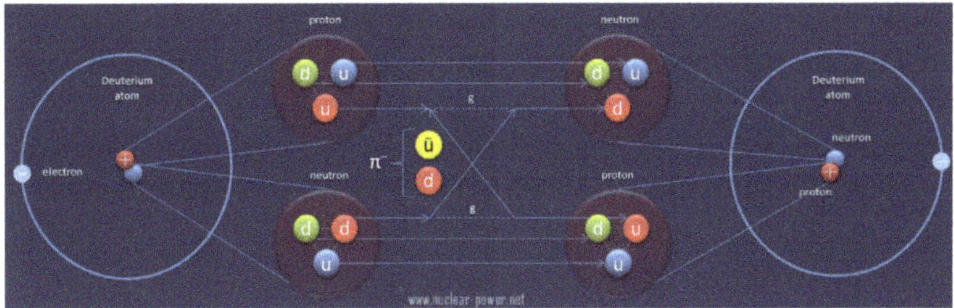

Figure 6.3.   Protons and neutrons in deuterium nuclei consist of three quarks of different colours. Nucleons stay together due to the exchange of pi-mesons, consisting of a quark and an antiquark with a colour and a corresponding anti-colour. Quarks are gathered into nucleons due to the exchange of gluons, particles–carriers of strong interactions. In the figure, the antiquark is indicated by an over-line above its symbol ($\bar{u}$), and gluons by the letter g. The picture is taken from the site www.nuclear-power.net

physicist Hideki Yukawa, who received the Nobel Prize in Physics in 1949 "for his prediction of the existence of mesons on the basis of theoretical work on nuclear forces" after triumphantly confirming this prediction. We can talk about the exchange of virtual pi-mesons or we can talk about the emission and absorption of pions by nucleons. The main thing is that these processes do work. That is why I called pi-mesons the most important in their class. Figure 6.3 shows the quark composition of nucleons and pi-mesons that hold them in the nucleus.

But the interaction between quarks forming a hadron is provided by the exchange of gluons. The reason for all this lies in the peculiarities of the strong interaction. It decreases with decreasing distance between quarks and turns to zero at very small ones. This is called asymptotic freedom.

But with increasing distance, this interaction increases and does not allow quarks to exist other than in the form of colourless hadrons. This is called confinement and limits the distance between individual quarks to about 10^{-15} m. So the strong interaction manifests itself differently on the scale of nuclei and individual hadrons.

Imagine that we have caught a charged π^{+} meson, somehow grabbed its constituent u- and anti-d-quarks, and begun pulling them in opposite directions. First, the force required for stretching will increase. Then another quark–antiquark pair will be born from the vacuum between the quark–antiquark pair as a result of quantum effects.

The newly born quark will fly to the old antiquark, the other two particles will also find each other and we will get two mesons. We ourselves provided the energy for the formation of an additional meson by working on stretching the pair.

So, we learned how to get pi-mesons from the two lightest u and d-quarks and their antiquarks. What about baryons? I have already mentioned the proton and neutron with their composition uud and udd. But there are also other possible combinations.

Simple and seemingly plausible reasoning can lead to the conclusion that a spin 3/2 baryon cannot be obtained from first generation quarks only, i.e. at least another quark is needed in addition to u-quarks and d-quarks.

The reason is that in order to get a hadron with a spin of 3/2, it is necessary that all quarks have the same spin directions, either spin-up or spin-down. But if there are only two types of them, then two quarks with the same flavour will be in the same state, which is forbidden by the Pauli principle.

But this is not the case. The baryon is formed from the "red", "blue", and "green" quarks. And since they have different colour charges, which I discussed in Section 6.1, they are not in the same state and the Pauli principle has nothing to do with this case.

So from two quarks of the first generation, you can form particles with a spin of 3/2, which are called Δ-baryons (delta). There are four types of them with the compositions uuu, uud, udd, and ddd. It is easy to calculate that one of them is neutral, and the electric charges of the other three are equal to the charge of the electron, the charge of the proton, and twice the charge of the proton. These particles were discovered back in 1952 by a group of American scientists led by Enrique Fermi. They live very shortly, only $5.6 \cdot 10^{-24}$ s, and decay into nucleons and pi-mesons.

And I still keep silent about particles with spin 1/2 and quark composition uuu or ddd. Because of the Pauli principle, their ground state should have an orbital rotational moment, in addition to spin, which makes their consideration quite difficult. I will not lead readers into the dense thicket of elementary particle physics and just forget about these particles, which are also called delta-baryons and are included in the official lists of elementary particles.

As you can see, at least six different baryons and their six antiparticles, as well as the triple of pi-mesons, have already turned out from two quarks of the first generation and their antiquarks. I hope that with this example the reader has understood why the number of different elementary particles that can be formed from six quarks turned out to be so huge. Some of them have already been discovered, while others not.

6.3. The Standard Model of Fundamental Particles and Fields and Attempts to Go Beyond It

What we know today about fundamental particles can be brought together in a scheme called the Standard Model. This is a modern theory of the structure and interactions of particles and fields, which is based on a small number of postulates. Its predictions are usually confirmed by experiments, sometimes with extremely high accuracy. It paints us a beautiful picture, which has been confirmed so far in all experiments on accelerators.

The working tool of the Standard Model is quantum field theory. All the "entities" it describes are summarized in the diagram in Fig. 6.4. They consist of fundamental fermions and quanta of fundamental fields, which are related to bosons. The model does not try to describe the gravitational interaction in any way, although in some versions of the scheme, there is a so far undiscovered quantum of gravitational interaction with a spin number of 2, named graviton.

I will at first outline its foundations, sometimes repeating the information presented above, and sometimes simply informing about conclusions obtained by physicists on the basis of the most complex experiments and theories based on them. And then I will explain why many physicists consider it necessary to go beyond this theory.

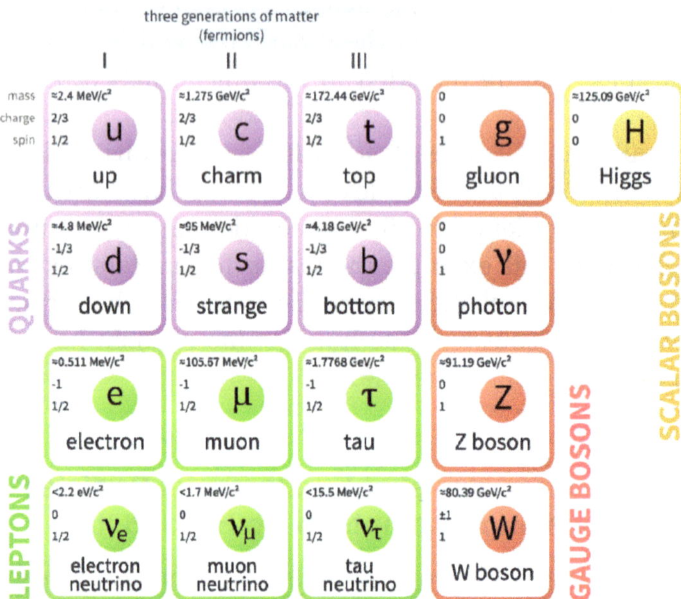

Figure 6.4. Fundamental particles and quanta of the Standard Model. In the public domain.

6.3.1. *Fundamental particles*

In the diagram, we see six varieties (they are also flavours) of quarks, each with its own antiquark. They are divided into three pairs, which are called generations of quarks. Each generation has a quark with a positive electric charge equal to 2/3 of the electron charge and a negative charge equal to 1/3 of this charge. They participate in strong, weak, electromagnetic, and gravitational interactions and have a special charge called colour.

A quark of one kind can turn into a quark of another kind, but only with the help of a weak interaction. Bosons, also known as gluons, carry out strong interactions. There are eight different types of bosons. The colour of the quark changes when interacting with gluons. Gluons themselves participate in the interaction, the quanta of which they are, which greatly complicates their theory. They have no electric charge and do not exist in a free state.

It is possible that in the very early Universe (before the age of one millionth of a second after the Big Bang), there was a stage when it was filled with quark–gluon plasma, a kind of very hot soup of quarks and gluons. And after this stage, quarks do not occur separately, but only in the form of colourless hadrons, as their combinations are called.

Let's add three charged leptons to the quarks: an electron, a muon, and a tau-lepton. They participate only in weak and electromagnetic interactions, but not in strong interactions. Each lepton has its own neutrino, which is involved only in weak interactions. They all have their antiparticles. These six leptons are naturally divided into three generations, each of which contains a charged particle and a neutrino associated with it.

6.3.2. *Field quanta or gauge bosons*

Three types of forces act on fundamental particles. I don't talk about gravity because it is considered separately. There are particles called quanta of these interactions, and all of them are bosons. These are special bosons called gauge bosons, which are defined as force carriers that mediate strong, weak, and electromagnetic fundamental interactions. Each force has its own set of such particles. They all have spin 1, and most of them are electrically neutral.

The gauge bosons of the electromagnetic interaction are well known to you. These are photons. When you look at this page, it is them that the retina of your eye perceives. They bring information about the text that you are reading.

The weak interaction is carried by the W and Z bosons. The W^+ boson is an antiparticle for the W^- boson and vice versa. They have electric charges of opposite signs, equal to the electron charge. The Z boson is electrically neutral. Moreover, it is a truly neutral particle that is its own antiparticle, like a photon. All these particles exist for a very short time, and their average lifetime is about $3 \cdot 10^{-25}$ seconds. Only the W bosons, but not the Z boson, are capable of changing the flavours of quarks.

Photons and gluons that are quanta of strong interactions have no rest mass. This fact is verified experimentally with great precision for photons. The gluon mass is zero in theory, but scientists always check everything. Experiments constrain the value of the gluon mass, which should not exceed 2.5 electron masses.

The zero mass of the quantum makes the interaction long-range. This is true for electromagnetism, but not for nuclear forces. And physicists understand why this is happening. The strong interaction is nonlinear; the principle of superposition, which is very helpful in the study of electricity and magnetism, does not work in it. Because of this, the interaction between two quarks increases with the distance between them.

Gluons hold quarks in hadrons. And nucleons, which are located in the nucleus at distances significantly exceeding their size, exchange not gluons but pi-mesons; this requires much less energy. The masses of pi-mesons are nonzero, and this detail makes nuclear forces short-range.

Zero mass has been expected for a long time from quanta of weak interaction. But then this force would be long-range, which contradicts the experimental facts. In nature, weak interaction has a very small radius of action — about $2 \cdot 10^{-18}$ m, which is approximately 1000 times smaller than the size of the nucleus. For two nuclei located at a distance of 10^{-10} m, it is weaker for not only electromagnetic but also gravitational interaction.

Experiments have shown that the W and Z bosons not only have mass, but are also obscenely heavy; they are one of the most massive particles. They are almost 100 times heavier than a proton and are close to the mass of rubidium and technetium atoms, respectively. This is what causes the weakness of the weak forces.

Both strong and weak interactions are no longer the completely incomprehensible forces that they were for a long time. I have mentioned in this book many Nobel Prize winners in physics who have been honoured by it for their work in this area. I add a few more theorists to the long list. These are Gerard 't Hooft and Martinus Veltman from the Netherlands "for elucidating the quantum structure of electroweak interactions in physics" (1999) and the Americans David Gross, Hugh David Politzer, and Frank Wilczek "for the discovery of asymptotic freedom in the theory of the strong interaction" (2004). They — and not only they — have significantly advanced our understanding of two fundamental interactions.

Naturally, I talked about interactions and their quanta only in the most general terms on our excursion. Readers who want to view the comic about fundamental interactions could find a corresponding book. But it is difficult to count on even a minimal understanding of these issues after reading a comic. Alas, not everything can be explained, but some particularly difficult moments are worth mentioning.

The first of these is something that all three non-gravitational forces have in common. Their quanta are called gauge bosons for a reason. They are somewhat similar. They are all described on the basis of the gauge principle, that is, the symmetry of the theory with respect to certain transformations.

The Nobel laureate Frank Wilczek in his popular books tries to explain something about these transformations and the symmetry hidden behind them, but I cannot afford to devote tens of pages to this. I'll just write that electromagnetic, weak, and strong interactions have common hidden features. And it was this complex gauge symmetry that gave the physicist the key with which they opened the door to the world of strong and weak interactions. It follows from this symmetry that all gauge bosons cannot have mass. So, where did the mass of the W and Z bosons come from, and how can it be explained?

I add one more detail. The 1979 Nobel Prize in Physics was awarded to the Americans Sheldon Lee Glashow and Steven Weinberg, and to the Pakistani Abdus Salam "for their contributions to the theory of the unified weak and electromagnetic interaction between elementary particles, including, inter alia, the prediction of the weak neutral current".

What kind of unified theory is this? The main thing in it is that these three scientists and their numerous colleagues have combined the electromagnetic interaction with the weak one into a unified theory of the electroweak interaction. And this theory not only explained a lot, but was also confirmed experimentally.

The unification took place at the level of subtle details of the interaction of elementary particles. To solve problems on electricity and magnetism, you do not need to know anything about the weak interaction, including the very fact of its existence. Moreover, this knowledge is not necessary to understand the macroscopic phenomena described in this book related to electricity and magnetism, such as lightning, magnets, or electrets.

So, did electromagnetism and weak forces come together? And does this unification manifest itself in the macrocosm? The answer to both questions is the same: "yes and no". Electroweak interaction shows the unity of two forces only at huge energies of interacting particles. Therefore, it can be explored with the most powerful accelerators.

At energies less than a few hundred GeV, it splits into two seemingly different fundamental forces, which we observe when we are not dealing with accelerators and high-energy particles from space.

Let us convert the energy of the particles into degrees, and we get that $100\,\text{GeV} \approx 1.16 \cdot 10^{15}\,\text{K}$. The Universe cooled below this temperature when it was very young and inexperienced, about $10^{-12}\,\text{s}$ of age. Since then, that is, almost 14 billion years, the electromagnetic and weak interactions in it have existed separately from each other. But for the first $10^{-12}\,\text{s}$, the united electroweak force acted.

But what turned it into two practically different forces? Suddenly, it revealed that there is a single answer to two seemingly different questions: this one and about the large mass of W and Z bosons. And it is associated with another quantum, which is present in the table in Fig. 6.4. It is called the Higgs boson, has no spin, and is therefore placed in a separate column titled "scalar bosons".

6.3.3. *Higgs boson and spontaneous symmetry breaking*

How can a massless photon be combined with massive W and Z bosons? The so-called Higgs mechanism comes to the rescue. It requires the existence of one more field in nature. It interacts with other fields and breaks some of their inherent symmetry.

Simply put, the quanta of weak interactions were initially massless in the absence of the Higgs field. But this field feeds these quanta, giving them mass and making the weak force short-range. It also leads to a difference in the masses of quarks of one generation. Like any field, it has a quantum, the so-called Higgs boson or scalar boson. The Standard Model predicts its lifetime of $1.56 \cdot 10^{-22}$ s.

The Higgs boson was discovered in experiments at the Large Hadron Collider at CERN in 2012–2013. It turned out to be a very heavy, short-lived particle with a mass of about 130 protons. This boson in the press began to be called "the particle of God", which hold the readers' interest. In 2013, the Nobel Prize in Physics was awarded to the Briton Peter Higgs and the Belgian François Englert "for the theoretical discovery of a mechanism that contributes to our understanding of the origin of mass of subatomic particles, and which recently was confirmed through the discovery of the predicted fundamental particle, by the ATLAS and CMS experiments at CERN's Large Hadron Collider".

But where does the Higgs field itself come from? Since we observe the results of its action, then this field should not only be able to exist in nature, but also be really present at every point of the Universe. Its value must be nonzero.

This is due to the so-called spontaneous symmetry breaking, for which the Japanese–American physicist Yoichiro Nambu received half of the Nobel Prize in Physics in 2008 "for the discovery of the mechanism of spontaneous broken symmetry in subatomic physics". The second half was shared by the Japanese Makoto Kobayashi and Toshihide Maskawa "for the discovery of the origin of the broken symmetry which predicts the existence of at least three families of quarks in nature".

What is spontaneous symmetry violation? Remember my story about the true and false vacuum and the decay of this false vacuum in Section 5.3? So, the Higgs mechanism is that the true vacuum is a state in which the value of the Higgs field is nonzero. A false vacuum corresponds to a state with a zero Higgs field and the following things: restored electroweak symmetry, massless

W and Z bosons, and the equal masses of quarks of the same generation and hence the approximate equality of the proton and neutron masses. Since a false vacuum always sooner or later turns into a true one, a nonzero Higgs field will arise by itself.

The Standard Model uses about two dozen constants external to it: the masses of fundamental fermions, the numerical values of the constants characterizing the strength of various interactions, and some other quantities. All of them are determined by comparison with experiment.

6.3.4. *Beyond the Standard Model: A way to a New Physics?*

The Standard Model is certainly very beautiful. It explains a lot, almost everything. But the results of some experiments are not consistent with it, and there are some problems that do not fit very well within its framework.

For example, there seems to be much more matter than antimatter in our Universe. It is clear that in the early Universe there could be so many different particles and antiparticles that a very small difference in the percentage of particles and antiparticles formed then could lead to very large differences in our time, after the mass annihilation of particle–antiparticle pairs. Some mechanisms for the emergence of this asymmetry were invented, but they go beyond the Standard Model.

It does not describe gravitational interaction. Don't be confused by pictures like Fig. 6.4, which depicts the graviton as well. No gravitons are included in the model. An equally serious problem is related to the concepts of dark matter and dark energy, which we discussed a little in Section 1.9. Let me briefly recall these entities.

After analysing the spectrum of CMB temperature fluctuations and some other factors, it became clear that most of the DM cannot consist of baryons. Naturally, there is some baryonic dark matter, like wandering planets, intergalactic meteoroids, etc., but there is so little of it that you can ignore it. Thus, most of the DM does not belong to ordinary matter, which stars, planets, and you and I are made of. This DM attracts surrounding bodies, but does not participate in electromagnetic and strong interactions.

Therefore, it is invisible, because it interacts with photons only gravitationally. It can deflect light with its gravitational field, but it cannot emit, absorb, or scatter it. It is not known if DM is involved in weak interactions. It was also shown that its speed is significantly less than the speed of light.

The presence of DM practically does not affect anything on the Solar System scale, but on a galactic scale it is important. Moreover, DM forms a sphere around galaxies, the so-called dark halo, which is many times larger than the galaxy itself. Some galaxies have been found that contain very little of the usual baryonic matter, mostly only dark matter.

DM cannot consist of any particles described by the Standard Model. But there are many theories that extend this model by introducing new particles and interactions. Therefore, physicists have a huge supply of hypothetical particles of every taste and size that have not been discovered. Naturally, they are proposed for the role of DM particles. In addition, not very clear particles were considered, called weakly interacting massive particles, or simply WIMPs.

In order to detect DM particles experimentally, physicists installed a variety of detectors in their underground laboratories, sheltered from cosmic rays, which were supposed to record all the transformations and collisions of the nuclei of the detector's substance. After many years of work, the conclusions are disappointing: although something was recorded in some experiments, according to their authors, this did not convince the scientific community.

Moreover, one device was recently improved, after which the articles published by the group working with it, the only one that continued to insist that the particles of DM were caught by them long ago, changed their tone dramatically. Confidence disappeared, and then timid hope was replaced by a complete disregard for previous statements. In the depths of the article, published in the summer of 2019, the team members admit that traces of radioactive thallium isotopes may have been present in their installation, which led to wrong conclusions.

In 2020, an article appeared claiming that another group with a very reliable reputation, the one that recorded double electron capture, found several suspicious events on a detector containing over a ton of xenon. As serious scientists, they see several possible explanations, from observing interaction with a DM particle to the possibility that in their detector there are three tritium atoms per kilogram of xenon, which have sneaked inside in spite of all the cleaning procedures. So let's wait for their further messages. If it appears that DM quanta have been caught, it would be a breakthrough in the study of the dark side of the world.

Most of the experiments did not reveal any traces of DM particles. This does not mean that they do not exist. It is possible that they interact with matter much weaker than we can record in experiments. Or maybe DM could not participate in weak interactions at all. It is curious that it does not take a long time to look for a candidate for the role of DM particles with such properties. It appears in one of the simplest generalizations of the Standard Model and is called a sterile neutrino.

Experiments have shown that practically all neutrinos, both emitted and recorded in laboratories, have a spin opposite to the direction of the momentum. They are called left-handed neutrinos. In more rigorous scientific terms, these are neutrinos with negative helicity.

What about right-handed ones called sterile neutrinos? Maybe they just don't exist. Or they are not observable. One could assume they have a large mass by the standards of elementary particles or simply do not participate in weak interactions. The idea arose that only left-handed neutrinos with negative helicity can participate in weak interactions, while their right-handed sterile counterparts with positive helicity cannot. But with antiparticles, the opposite is true: right-handed antineutrinos interact weakly, but left-handed ones are sterile. It is these sterile neutrinos or antineutrinos that can be particles of DM.

Experiments on the capture of DM particles are interesting because their results will have a significant impact on the very idea of DM. Naturally, scientists will accept any result of the experiment, whether they like it or not. But it's not hard to predict what to expect in the two possible scenarios.

If the particles of DM are caught red-handed after weak interaction with ordinary matter, then the areas of physics related to DM would immediately receive an incredible impetus. It would be found out what DM consists of, as well as the mass and some other properties of their quanta. This would make it possible to narrow the front of research a hundred times, concentrating enormous forces on it. The information already received would make it possible to quickly put into operation new, more accurate devices, sharpened for the study of precisely those reactions in which DM participates. And the construction of these installations would be financed with great enthusiasm. Theorists, too, would focus their efforts, rather than scattering them into thousands of hypotheses. All this would accelerate the movement towards the goal. Naturally, in this case, there will be a lot of questions, such as whether there are different types of DM in nature. And a lot of new questions will arise related to the results of the experiments.

If no traces of the weak interaction of DM with baryonic matter were reliably detected, then this would not be fatal for the idea itself. Theorists will continue to delve into theories, experimental installations will work for some time and even new ones will be built, but then funding will gradually dry up. The theory that has no experimental evidence stagnates rather quickly, overflowing with thousands of formulas and hypotheses, from which it is impossible to reject the wrong ones. Alas, the choice between these two options is not in our power.

In any case, the existence of DM hints that the Standard Model needs to be supplemented or radically improved. Dark energy does not fit into it either. The simultaneous existence of two mysterious entities, DE and DM, naturally raises the question of their possible connection. Although we know very little about dark matter and even less about dark energy, this is not an obstacle to the flight of fantasy. Naturally, articles considering all kinds of options were published. Options were considered when one of these mysterious phenomena somehow produces another, decays with the release of another, or when there are different types of DM. But such things are typical enough for any revolution in physics, and this is what we are witnessing.

Why not look for an analogy of the forces we are used to on the dark side of the world? If ordinary matter can interact with the help of electromagnetic interaction, and dark matter cannot, could there be an unknown "dark electromagnetism" which acts only on DM but not on baryonic matter? And whose quantum could be called a "dark photon"? In general, theoretical physicists have no problem with how to kill time.

All of the above reasons and some additional considerations, such as the huge spread in the masses of fundamental fermions, led to the idea of modifying the Standard Model. The theory, which had not yet existed, was called New Physics in advance. Many scientists hoped that with the introduction of the latest particle accelerators, primarily at CERN, the effects of the New Physics would pour out as if from a cornucopia. However, the reality turned out to be much more prosaic, and after several years of operation of the accelerator, not a single reliably proven effect was found that required abandoning the Standard Model.

These effects are sought not only in accelerators, but also in astronomical observations and in terrestrial laboratories. For example, some hypotheses and theories suggest that a proton, the lightest baryon, can decay into a positron and some set of neutrinos, but this rarely happens. All attempts to detect this effect were unsuccessful, for example, in a series of looking for observable manifestations of proton decays at the Japanese Super-Kamiokande facility.

If a proton decays, then its lifetime is not less than $2.9 \cdot 10^{29}$ years, which is a huge number of times longer than the lifetime of the Universe. However, given that if there is a possibility of proton decay, then each proton can potentially decay, and there are a huge number of them in each kilogram of matter, this estimate is quite justified.

Be that as it may, physics now has a beautiful and perfectly working Standard Model and is looking forward to seeing the next scientific revolution, this time under the banner of New Physics, if nature will provide us with a reason to start it. After all, the desire of scientists alone is not enough.

6.3.5. *How many different types of forces are there in nature?*

After I have introduced you to all the fundamental interactions, you can once again return to their list, shown in Fig. 1.3. Today we know four, three, or five types of forces. In Section 1.8, I have already listed the first four types of forces in nature. These are gravitational, electromagnetic, strong, and weak interactions. But then, why did I write not only about four, but also about three interactions? The fact is that two of them were combined into a single electroweak interaction.

Are there a couple of additional interactions hidden somewhere? Or at least one? The answer to this question is not simple. Many people call the Higgs field the fifth interaction. Others disagree with this and talk about the field and the Higgs mechanism, not counting this as a special interaction.

So, let's list the existing lists of interactions. Three forces: gravity and strong and electroweak interactions. Or four interactions: gravitational, electromagnetic, strong, and weak. In a variant with five interactions, a Higgs field is added to the last list.

Naturally, for a long time, physicists have been trying to combine all known interactions or part of them and have come up with a lot of theories and hypotheses for this. However, all other theories, which should unite forces of different types, except for the theory of electroweak interaction, have not yet received experimental confirmation and have been either rejected or remain hypotheses.

But maybe this number should be increased? For more than half a century, there has been an intensive search for new types of interactions. Their subject has received the conditional name of the fifth force, although after some people declared the Higgs field as the fifth interaction, this name is no longer so obvious. In this case, the fifth force is understood as any interactions acting at any distance and on any scale, as long as they are not reduced to those already known.

Until now, in increasingly accurate experiments, nothing has been found that is inexplicable within the framework of the existing picture or deviates in magnitude from the predictions of generally accepted theories. Naturally, the search continues, and right now there is a hunt for this hypothetical fifth force, so desired by many.

And I can't help but mention one somewhat discouraging nuance. In the quantum world, which I tell you about in the fifth Chapter, there are some forces at work, caused precisely by quantum properties — for example, an exchange interaction.

It is not customary to mention these forces when listing possible types of interaction, but they do exist. For example, they gather atoms into molecules and prevent them from decaying. As for me, it would be worth considering this as a manifestation of special quantum forces, but historically, they are not included in the list of fundamental interactions.

7   CLOSING THE CIRCLE OR *OUROBOROS*. SCALE: EVERYTHING IS COMPLICATED

As you may have noticed, in the last chapters, we thought about cosmology and astronomy all the time, but did not mention, for example, semiconductors. Speaking of cosmology, we remembered quantum and nuclear processes, elementary particles, and other inhabitants of the microworld. Going down the scale ladder to the very bottom, we suddenly found that it was directly connected to the top. This staircase can be continuously descended and ascended, as in the famous engraving by Maurice Escher (you can view the image at: https://mcescher.com/gallery/impossible-cons tructions/#iLightbox[gallery_image_1]/4). The staircase is looped! And this is not a whim of the authors, but a trend: the smallest scales are combined with the largest. I have already mentioned the ancient alchemical symbol called the *ouroboros*, which depicts a snake biting its own tail, or rather devouring itself, starting from the tail. Its images, both ancient and modern, are easy to find on the Internet. Upon closer inspection, our staircase of scales loops, becoming like a staircase from an Escher engraving or an *ouroboros*.

In physics, a subsection of this science has appeared, called astroparticle physics, which works at the junction of these scales, the smallest and the largest. And the discoveries associated with the world of quanta and the Universe enrich each other.

Specialists in particle physics and quantum field theory find in space a free ultrahigh-energy cosmic ray accelerator, as well as cosmological and astrophysical testing grounds, to test the processes involving their favourite particles. These conditions are often extreme and inaccessible on Earth.

Indeed, in order to understand the world of elementary and fundamental particles, scientists need more and more powerful accelerators that provide ever higher energies of accelerated particles. And in the early universe, all particles were high-energy ones.

If the power of modern accelerators is not enough to test the hypothesis associated with elementary particles and fields, then we can calculate how the assumption of the correctness of this hypothesis affects the properties of the early and modern Universe.

Astrophysicists and cosmologists receive a basis for explaining the observed processes and the consequences of what happened once, say, in the early Universe, in the form of the Standard Model. These two groups of scientists, albeit not suddenly, but gradually, became very necessary to each other, mutually providing ideas and information.

Let me give you two examples. I have told you about thermonuclear processes in the Sun and other stars. They are very important to astronomers. But all the details, for example, the masses of the initial and formed nuclei, and hence the energy yield of the reaction, were learned by atomic physicists.

And here is the opposite example. The three types of neutrinos have a mass that is desirable to know. Astronomers help specialists in elementary particle physics here. The difference in masses of various types is constrained, in particular, by differences in the arrival times of different types of neutrinos from a supernova explosion called SN1987A. The upper limit of the sum of the masses of the three types of neutrinos was obtained from the analysis of the anisotropy of the relict radiation.

In this chapter, I will talk about objects in which different scales are mixed, as well as substances exhibiting quantum properties in the macrocosm, and a little more about the effects of SRT and GRT, manifested at all scales.

7.1. Neutron Stars

Since we have found that our scale ladder is looped, why not adorn this ring with some kind of jewel-like decoration? I mean symbolically. It should be something special, combining the properties of both the macrocosm and the microcosm. So, meet our heroine. This is a neutron star.

As a star, it clearly belongs to the world of astronomical objects and has an enormous mass, up to two solar masses. But its dimensions are not at all astronomical; the radius is about ten or two tens of kilometres. It can safely fly through a hoop the size of the Moscow Ring Road or 6th Ring Road around Beijing, but not through the *Boulevards des Maréchaux* (Boulevards of the Marshals) in Paris. In fact, it is something like a huge atomic nucleus. When I discussed the possibility of obtaining a stable nucleus with a large number of nucleons in Section 5.4, I was referring naturally to ordinary atomic nuclei, not neutron stars.

I wrote about the size and weight. It is easy to understand that everything is fine with the density of matter in a neutron star. It even exceeds the density of ordinary atomic nuclei. After all, a neutron star, as the name implies, consists mainly of neutral neutrons, which do not repel each other by electric forces. In addition, the strong gravitational field forces them to pack more densely than in the nucleus. A standard matchbox with neutron star matter inside would weigh 3 billion tons.

On its surface, the acceleration of gravity is approximately $2 \cdot 10^{11}$ times greater than that of the Earth. No amount of diet will help if you have to weigh yourself on this surface. A big neutron star is not much larger in size than a black hole with minimal mass, so it should not be studied without taking into account the effects of general relativity.

And these stars really exist. Astronomers have discovered at least 2,500 of these curiosities, most of them in our Galaxy. It is clear that telescopes cannot see their details, but something can still be learned from observations. And much can be predicted by theorists who once predicted the existence of such stars, unknown at the time. So I'll tell you something about the life of a typical neutron star.

"What? Are there also atypical ones?" you ask and you are not mistaken. Believe it or not, there are more exotic types of neutron stars, such as magnetars, of which more than twenty have been discovered. They have the strongest magnetic field in nature (from 10^8 to 10^{11} T), hundreds of millions of times greater than the field of any magnet ever made by man. Against this background, the Earth's magnetic field with its 30–70 μT looks like a laughing stock. This strongest field changes the properties of the physical vacuum around the magnetar and leads to many incredible phenomena.

A typical neutron star was born from the remnants of an ordinary star that had exploded like a supernova, from its very central part. The newborn was clearly suffering from fever, as evidenced by its high temperature, about 10^{11} K, which is about 10,000 times higher than that in the centre of the Sun. But soon, just a few minutes later, it felt better and the temperature dropped a hundred times. Then again, the temperature dropped in ten, but already during a hundred years.

Any physicist's head starts spinning from this information. "How? How did it do it? How did it shed this energy so quickly?" he mutters incoherently. And it's easy to understand. Physicists are accustomed to dealing with three mechanisms for the transfer of heat and energy. Two of them — heat transfer and convection — are impossible, because the star is surrounded by the world's best heat insulator, vacuum. Thermal radiation works, but the small surface area of a neutron star does not allow much energy to be emitted so quickly. But here a completely different cooling mechanism works and works very effectively. Energy is carried away along with neutrino flows.

Like every celestial body, the neutron star has an internal structure. The outer crust is several metres thick, and the inner crust is up to several kilometres thick. Both are composed of ions or atomic nuclei, electrons, and neutrons. Inside is a core of neutrons with small impurities (a few percent) of protons and electrons. Everything is surrounded by an atmosphere of a thin (from a few millimetres to tens of centimetres) layer of plasma. Some physicists believe that there may be a core of quark–gluon plasma inside a particularly massive neutron star, which filled the Universe in a brief fraction of a moment after its birth.

7.1.1. *Pulsars*

"If you want to live, you should be able to spin," says a Russian proverb. Its meaning is similar to "It's the squeaky wheel that gets the grease" or "You must hustle". The neutron star is particularly successful in this. The fastest-rotating known star, affectionately nicknamed PSR J1748-2446ad, makes 716 revolutions per second. However, among such objects, there are also lazy stars that appreciate idleness and peace. After all, how else can you rate a miserable half a revolution per second? But when you are spinning all the time, it is difficult to make it so that, say, the tax authorities do not notice it. The same rotation made it possible to learn about the existence of neutron stars.

Indeed, how can they manifest themselves so that we know about them? Their size is small. They shine weakly due to their small surface area. Sometimes they occur in binary systems, but more often alone. But, like illegal scouts on enemy territory, they betray their existence by radio signals.

Neutron stars also have their own magnetic poles, from where a powerful radio emission along the direction of the magnetic axis can reach the Earth and be detected using a radio telescope. Usually, the magnetic poles do not coincide with the geographical ones; that is, the magnetic axis does not coincide with the axis of rotation and rotates rapidly together with the neutron star. The radio signal directed along it behaves like the light of a rotating searchlight, for example, on a lighthouse.

It hits the Earth, but sometimes with the rotation frequency of a neutron star. Therefore, we observe unusual objects in the sky that send us periodic radio signals. They are called pulsars. All pulsars are neutron stars, but not all neutron stars emit so that their radio beams hit the Earth and we learn about their existence.

In 2017, scientists reported that they were able to record the merger of two neutron stars. In addition to electromagnetic waves of different lengths, the signal of this event came in the form of gravitational waves. It was the gravitational waves detected by the LIGO detector that attracted the attention of astronomers to this event, and with the help of the most powerful optical telescopes, including the Hubble Space Telescope, they were able to quickly observe the section of the sky where the merger took place.

I hope that our *ouroboros* ring has received a worthy decoration. And we got acquainted with an example of an object that violates many of the previously formulated rules. It is something like an atomic nucleus, but a dozen kilometres in size. In it, both the gravitational forces prevailing in the macrocosm and the nuclear forces ruling in the microcosm are important. To describe it, you need to use both general relativity and quantum mechanics, not to mention the theory of strong interaction. And some theorists talk about superconductivity in neutron stars. Think about how difficult it is even to approach the creation of a model of such a chimera.

7.1.2. *Space valentine*

Our journey is nearing its end. From the microcosm, we returned to the vastness of space. Travellers love to show photos taken during their trips. So I want to present a beautiful photo in memory of all kinds of cosmic exoticism. It's not a shame to show something like this on Instagram.

Figure 7.1. "Cosmic Valentine". Photo of interacting galaxies Arp 147. *Source*: NASA/CXC/MIT/S.

This photo, which some call a space valentine, is shown in Fig. 7.1. I must say right away that this is a combined photo, a superposition of photographs obtained by different American astronomical satellites in different ranges of electromagnetic waves. So the colours in the image are not real but artificial. For example, pink spots show the location of X-ray sources according to the data from the Chandra satellite. Also shown are images in three colours of the visible range from the Hubble Space Telescope.

On the right, two interacting galaxies are visible, both of which are included in the Halton Arp's catalogue of galaxies under number 147 and are called Arp 147. They are located at a distance of 430 million light-years from us in the constellation *Cetus*. The word "interacting" does not fully reflect their complex relationship. They just collided in the not very distant astronomical past.

The elliptical galaxy seen on the left flew through a spiral galaxy, the remnants of which are visible on the right in the form of a ring. This gave rise to a wave of disturbances, which led to a sharp increase in star formation, i.e. the appearance of new stars from gas and dust in a spiral galaxy. The shock wave was followed by a wave of bursts of star formation in the galaxy, which left behind many young, massive, hot stars that emit a lot of ultraviolet radiation (blue colour corresponds to it in the image).

Some of the most massive stars managed to complete their life cycle by supernova explosions in several million years, i.e. very quickly by the standards of astronomy. After these outbursts, there were black holes, which I will talk about in Section 7.3, and neutron stars, which are exactly the objects we are discussing. They can be detected by their X-ray radiation, marked in pink in Fig. 7.1.

What can I say about the elliptical galaxy on the left? In such galaxies, the processes of star formation are much weaker than in spiral ones, so there are not many new stars in it. I'll add that there seems to be a supermassive BH at its centre, but it emits weakly due to poor accretion keeping it on a starvation diet. In the leftmost part, you can see a star at the bottom and a distant quasar at the top. A quasar is also a supermassive black hole.

By searching the name Arp 147 on the Internet, you can find not only a photo as in Fig. 7.1, but in different colours, but also a video in English with a story about this curious Valentine card. Note that this is not the only one that is so beautiful in space. More than one is so full of holes after passing one galaxy through another. Look for an image of the galaxy AM 0644-741, for example. It is no less beautiful.

I'll take advantage of this situation and explain how to distinguish between stars and galaxies in astronomical images. Images of stars have rays. This is simply the result of light diffraction on the telescope mirror mount. Only a few telescopes do not give such rays due to design features.

Why do galaxies, even the smallest ones, have no rays? From the point of view of optics, stars can be considered practically point sources, but galaxies cannot. Therefore, the picture observed during the diffraction of their light by the stretch marks is not so clearly expressed. Roughly speaking, the cross is blurred and almost invisible.

Now you can close your photo album with the air of a seasoned astronomer and admit that galaxies are very photogenic. It is a pity that the album does not contain photos of neutron stars and pulsars due to their remoteness and small size. But there are even more mysterious monsters in nature that scientists have managed to photograph. These are black holes. But before describing some of them, I will tell you about the effects of general relativity.

7.2. Some Effects of General and Special Relativity

7.2.1. *Three classical general relativity effects*

General relativity is generally too complex for a popular story, but some of the effects it predicts are quite understandable. I have already mentioned one of them in Section 2.3; it is the perihelion shift of Mercury and other planets.

I will tell you about two more so-called classical effects of general relativity, which have been tested and confirmed a long time ago. The first is the deflection of light in the gravitational field of a massive object, such as the Sun. There is a same effect in the framework of Newtonian gravity, but in Einstein's theory it is twice as large. The fidelity of general relativity was confirmed back in 1919 by the results of measurements carried out during a total solar eclipse by an expedition led by Sir Arthur Stanley Eddington. The deflection of light in a gravitational field leads to the observed phenomenon of gravitational lensing, which I discussed in Section 4.4.

The second effect is called gravitational redshift. This is the difference in the speed of the passage of time at points with different gravitational potential. For example, it flows a little faster on the roof of a skyscraper than on the ground floor of this building. And this can be verified by direct measurement. Tokyo has the tallest TV tower in the world called the Sky Tree. It was built in 2012 and has a height of 634 m. It was used in 2020 to verify the conclusions of general relativity.

A group of scientists led by a professor at the University of Tokyo Hidetoshi Katori compared the course of a compact ultra-precise atomic clock placed at the foot of this TV tower and on an observation deck at an altitude of 450 metres. The clock above was running about five hundred millionths faster than the clock below. During the day, they ran ahead by about 4.3 nanoseconds. Over the year, the difference would be 1.6 microseconds. The experiment confirmed the formula for the gravitational redshift with an accuracy of 99.9986%.

The first experiments to test this effect were carried out half a century ago by Robert Vivian Pound and Glen Rebka. It has been confirmed by numerous experiments in space, the first of which was the 1976 launch of NASA's Gravity Probe A mission. A rocket launched from Earth with an accurate time standard rose to an altitude of 10,000 km and fell into the Atlantic Ocean in 1 hour 55 minutes. The speed of the passage of time on the rocket corresponded to that calculated on the basis of the effects of general relativity and special relativity.

Astronomers are well aware of the gravitational redshift of the spectrum of white dwarfs. These massive and small stars have a greater gravitational potential on the surface than a terrestrial observer. Therefore, time flows there more slowly than on Earth, and the gap between the two maxima of the light wave is larger for us than for the atoms emitting them. As a result, the astronomer observes the emission spectrum of these atoms, slightly shifted towards the red end in comparison with the spectrum emitted by the same elements on Earth.

General relativity predictions are not limited to the three classical effects. I have already talked about the creation of the science of cosmology and about the prediction of the existence of BHs, which I will discuss in Section 7.3. The influence of general relativity effects is observed not only when observing neutron stars and other exotic astronomical objects, but also in subtle effects on the motion of planets and even on GPS navigation.

7.2.2. *Gravitational waves*

But there was one prediction of general relativity that did not have experimental confirmation for a long time: the existence of gravitational waves. What are these waves? These are small deviations in geometric properties and curvature of space-time, which propagate in all directions from their source at the speed of light in a vacuum (at least in weak gravitational fields, but we do not deal with strong ones). When they reach some objects, they slightly change the distance between them, which makes it possible to record their arrival.

Electromagnetic waves are generated by the acceleration of charges. Gravitational waves are generated due to acceleration of masses. That is, their source is any accelerated motion. Wave your hand and you will emit gravitational waves. Naturally, they are too weak to be registered. Stronger gravitational waves arise from cosmic cataclysms, such as the merger of two massive black holes. Their source is far away, but it is very powerful. It is clear that the further such an event is, the more sensitive the setup must be in order to register the wave.

Everyone understood that detection of gravitational waves is very difficult for technical reasons, but experts had no doubt that sooner or later it would be possible to achieve this. Moreover, there was indirect but convincing evidence in favour of their existence.

In 1974, Joseph Taylor Jr and Russell Hulse discovered a binary system of neutron stars PSR B1913 + 16. The stars revolved around a common centre of mass. At the same time, each of them was moving with acceleration, while emitting gravitational waves in all directions and losing the energy carried away by them. The loss of energy by a pair of stars means their approach, which leads to a change in all parameters of the system's motion, including the period of rotation of the pair. And its change is not difficult to detect during radio astronomy observations.

Scientists not only recorded the predicted change in the period, but also showed that it coincides with the change calculated according to the formulas of general relativity due to the radiation of gravitational waves by stars. Hulse and Taylor received the Nobel Prize in Physics in 1993 "for the discovery of a new type of pulsar, a discovery that has opened up new possibilities for the study of gravitation". Since then, detecting gravitational waves has been viewed as a matter of time. And so it turned out.

A large group of scientists managed to build the installation, register gravitational waves on it and prove that it is there. Three of them were awarded the Nobel Prize in Physics in 2017 "for decisive contributions to the LIGO detector and the observation of gravitational waves". These are the Americans Rainer Weiss, Barry Barish, and Kip Thorne. On the LIGO detector, they confirmed the existence of gravitational waves and black hole mergers.

7.2.3. *General relativity and quantum mechanics*

Scientists have successfully quantized electromagnetism, creating quantum electrodynamics. But there are problems with gravity. General relativity and quantum mechanics are not very fond of each other, because they are based on different principles, often not very compatible. Therefore, there is now no real quantum theory of gravity, although there are several approaches that pretend to be called this. However, by and large, these are not full-fledged theories, but just their ersatz versions. Creating a quantum theory of gravity, or quantizing gravity, is one of the most difficult challenges nature presents to physicists. All scientists know this, but they cannot even find the path along which we would come nearer to the creation of this much desired theory.

The situation is greatly worsened due to the fact that experiments in this area are practically impossible, so you have to choose among numerous approaches and hypotheses, focusing exclusively on your preferences, from philosophical to aesthetic. In science, this situation is very dangerous. It has been met more than once before. Throughout the end of the 19th century, physicists developed beautiful, but completely incorrect theories of *aether*, sometimes compressible, sometimes incompressible, which are now of value primarily for the history of science. As soon as experimental tests of hypotheses became possible, all the developments of several decades went to the trash can.

The point is that physics is based on experiments. The criterion for the correctness of any theory is its ability to explain the observed picture of the world, including the results of all experiments. The contradiction between theory and even one experiment, reliably confirmed, is the reason to show the theory the "black mark", the symbol of revolution on a pirate ship, which is the author's find of the writer Robert Stevenson. Then either it will be possible to reasonably explain the contradiction, or the theory must be discarded or improved. We will discuss this in detail in Section 8.1. And if there are no experiments, then there is no need to discard something, and therefore no possibility of narrowing the options for choosing the correct theory.

7.2.4. *Relativistic shortening of length, slowing down of time and mass increase: Fallacies and their unmasking*

As everyone who has read Mikhail Bulgakov's *The Master and Margarita* knows, an integral part of a black magic session is its exposure. And any journalist and writer would agree that the scandal strongly fuels interest in what is published. I will also try to step on this slippery slope and expose something generally accepted. Before that, I had debunked only the theory of the thermal death of the Universe, and this is clearly not enough.

We'll have to take a closer look at the effects of the special theory of relativity (SRT), namely, slowing down time, shortening length, and increasing mass at tremendous speeds close to the speed of light. They have entered popular culture, and now almost every person with a secondary education knows about them. One bad thing is that they don't exist. More precisely, they do not exist in the form in which they are known by 99.9% of people who have heard about relativistic effects and in general about SRT.

Quite a long time ago, I saw a popular science film about these effects. The author took an unremarkable footage from a train carriage: passengers walking down the aisle, compartments, bored children, and a fallen glass of tea. Then the author slowed down the movie speed three times and squeezed the picture horizontally three times. The passengers got out narrower and moved slowly. This, according to the narration, is exactly what it would look like if the train were travelling at near-light speed.

Many people believe that relativistic effects look like this. But in reality, everything is different. Moreover, to understand that something is wrong in the film, it is enough to ask yourself why its authors squeezed frames horizontally, not vertically or diagonally.

But I will play a little for some time to keep the intrigue and tell you how the concept of relativistic effects entered physics. The first successes of physics, which has become a real science, are associated with mechanics. So, later, when physicists came to understand heat, electricity, and magnetism, they applied with might and main the laws and concepts that came from more advanced mechanics at that time. They included the idea of the *luminiferous aether.* This medium is supposedly necessary for the propagation of light in it.

But time passed and Maxwell derived a system of his equations for electrodynamics. The equations passed all tests by experiments, predicted the existence of radio waves and other electromagnetic waves and entered the golden fund of achievements in physics. And everything was fine with them, but they turned out to be not very compatible with the classical mechanics that settled in this fund much earlier.

Let's imagine that a physical process occurs and a physicist describes it using equations. The second physicist does the same thing, but he is sitting in a car that is passing at a constant speed on a straight highway past the first physicist. If both descriptions are correct, then there should be a simple way to link them together. It is called coordinate and time system transformation.

Moreover, of these two descriptions, one cannot be more important than the other. This could only be in the opinion of Aristotle, for whom each body performs the prescribed movement and the moving body is fundamentally different from the body at rest.

In classical mechanics, very simple Galilean transformations are known that connect these descriptions of the two physicists. But Maxwell's equations did not obey them. To put it in a more scientific language, Maxwell's equations were not invariant under the Galilean transformations. And it was incomprehensible and unpleasant.

While some physicists tried to persuade Maxwell's equations to respect Galilean transformations, others figured out which transformations fit these equations and with respect to which transformations they are invariant.

These transformations were obtained almost simultaneously by a large number of physicists and mathematicians, so different people in different countries call them differently. I'm talking about the Lorentz transformations, named after the Dutchman Hendrik Antoon Lorentz. In France, they will tell you about the Poincaré transformations in honour of the French mathematician Jules Henri Poincaré. And in Great Britain, they remember the FitzGerald contraction and its discoverers, the Irishman George Francis FitzGerald and the other Irishman Joseph Larmor. You guessed it! These are the same transformations.

They are the basis of Einstein's SRT. Why is Albert Einstein called the author of SRT? Because he understood their meaning and suggested that we take the next important step, which is to change the mechanics so that it becomes invariant under the Lorentz transformations. None of the discoverers of these transformations dared to do this.

As a result, classical mechanics turned into an approximate theory, into which relativistic mechanics transforms if moving bodies' velocities are significantly lower than the speed of light in a vacuum. You can jokingly say that electrodynamics took revenge on mechanics for half a century of humiliation.

The discoverers of transformations perceived them as a consequence of the existence of the aether, for example, as a reduction in the length of an object due to the force of pressure produced by the aetheric wind. Or they gave other explanations that differed from the modern views, formulated on the basis of SRT. What did Einstein claim? There is no ether and it is not needed. All relativistic effects can be easily deduced if we proceed from the fact that light in a vacuum always propagates with the same speed c, regardless of the speed of its source and the speed of the observer, frequency, and other factors. Quite simply, in emptiness, light cannot overtake light. It's enough to build a theory that unambiguously leads to the Lorentz transformation.

However, since we are talking about something related to length and time intervals, it is necessary to somehow define these concepts. More precisely, one needs a procedure to measure them.

If there is a picture on the wall, then its dimensions can be measured using a ruler. And how can you measure the size of a painting that is being moved at high speed near you?

It is not difficult to determine and even measure the differences in the speed of the hands in the clock if one sets the same initial time on them and after a certain period compares their readings. The physicist refers to this process of comparing readings as clock synchronization.

You can check your watch once "on the go", when one of the compared chronometers rushes by, but if it continues to move evenly and in a straight line, then there will be no second meeting.

Nevertheless, using the constancy of the speed of light in the vacuum, you can figure out how to determine distances, and therefore lengths, as well as how to synchronize the clock. When the radar sends out a radio signal that reaches the aircraft, bounces off it, and returns back, the distance to the plane can be found by multiplying the half of the time interval between the emission and the return of the signal by the speed of light c (the signal travels twice the distance to the aircraft).

The midpoint between the moment of signal emission and its return is considered synchronized with the moment of signal reflection from the aircraft. The length of an airplane flying directly toward the radar is equal to the difference between the distances to its far and near ends, measured simultaneously.

It is not difficult to turn all this into formulas and get the same relativistic contraction and time slowing down associated with the Lorentz transformation. But I won't do that. First, I promised to avoid formulas in this book, and I intend to keep that promise by making a small exception for the final two sections of this chapter.

Secondly, all this has already been done in Hermann Bondi's wonderful book *Assumption and Myth in Physical Theory* (Cambridge University Press, 1967). With pictures, formulas and detailed explanations, it is quite clear. I recommend reading it. This book by an Austrian-born British mathematician and cosmologist and a fellow of the Royal Society of London is written for a wide range of readers. For example, the paradox of twins and some other SRT issues that puzzle unsophisticated people are perfectly considered there.

The main thing to understand is that it is not about what you can see with your own eyes, but about the measurements performed according to the procedure described above. And it is certainly impossible to see or measure changes in any dimensions and speeds, moving along with the measured object, as in the popular science film I mentioned.

Note that Bondi, like most other authors, is limited to the situation of one-dimensional motion, when a thin rod moves along its axis, and the observer is located on the extension of this axis, so that the rod moves directly to or from the observer. All other motion options are not considered. Therefore, a little later I will have to do it for him, right in this book.

And what about the relativistic increase in mass? Is it there or is it not? The problem is what exactly is meant by mass. And it's not easy at all. There are questions about the well-known formula $E = mc^2$. What kind of energy is indicated by the letter E in the left part? What mass is indicated by the letter m on the right?

To completely confuse the question, I will add that I constantly explain to students that a formula that contains a vector on the left and a scalar on the right cannot be correct. And in the formula $E = mc^2$, on the left is energy, a component of the four-dimensional vector of energy and momentum, and on the right is a scalar. This does not mean that the formula is incorrect. It means that it needs a detailed explanation, which is often missing.

It is good that the relevant work has already been done by someone else to whom you can refer. Lev Okun, a well-known theoretical physicist and expert on elementary particles, has published an article[a] in which he analysed this issue in detail. Its text is not very difficult, although some details are not easy for a non-physicist. Here are a few quotes from there, several of them abridged by me:

> In the popular scientific literature, school textbooks and the overwhelming proportion of university textbooks, the formula $E = mc^2$ dominates, which is usually read from right to left and

---

[a]L. B. Okun, "The concept of mass (mass, energy, relativity)", *Sov. Phys. Usp.* **32**, 629–638 (1989). You can find it at http://www.itep.ru/science/doctors/okun/publishing_eng/em_5.pdf.

interpreted as follows: The mass of a body increases with its energy, both internal and kinetic.

The overwhelming majority of serious monographs and scientific papers on theoretical physics, particularly on theoretical physics of elementary particles, for which the special theory of relativity is a working tool, does not contain the formula $E = mc^2$ at all. According to these books the mass of a body m does not change when it is in motion and, apart from the factor c^2, is equal to the energy contained in the body at rest.

The formula $E = mc^2$ has long been an element of popular culture. This gives it a special vitality. Sitting down to write about the theory of relativity, many authors proceed from the fact that the reader is already familiar with this formula, and try to use this acquaintance. This is how a self-sustaining process arises.

"Does mass really depend on velocity, dad?" Such is the title of a paper published by C. Adler in the *American Journal of Physics* in 1987. The question posed in the title was put to the author by his son. The answer was: "No!" "Well, yes ...", "Actually, no, but don't tell your teacher." The next day his son dropped physics.

It so happened that in the same year 1987 in which Adler's paper appeared I had to work in a commission created by the former Ministry of Education of the USSR to determine the winners of the competition for the best textbooks on physics. Having become acquainted with about 20 submitted textbooks, I was struck by the fact that they all treated a velocity-dependent mass as one of the central points of the theory of relativity. My surprise increased still more when I found that the majority of the members of the commission — pedagogues and specialists on methods of teaching — had not even heard that a different point of view existed.

The conclusion to which Okun comes is simple. Physicists in this formula understand mass m as the mass of a particle at rest, and energy E as the total energy of this particle at rest. This allows us to talk about the mass defect and much more, but not about the increase in the mass of a moving particle. This opinion is shared by Frank Wilczek in his book *The Lightness of Being: Mass, Ether, and the Unification of Forces*.

When trying to use velocity-dependent mass, there are a lot of different problems. For example, one has to introduce a "longitudinal" mass, used if the acceleration is directed along the velocity, and a different "perpendicular" or "transverse" mass if the acceleration is orthogonal to the velocity.

A mass that changes with a change in speed is a relic of attempts to understand the meaning of formulae made in the early years of the SRT's existence. This interpretation has long been abandoned in science, but it is still flourishing in the mass consciousness and among schoolteachers. I completely agree with this point of view.

7.2.5. *Photographing a body moving at near-light speed*

"Okay," the reader could say. "What prevents us from photographing a rapidly flying body with a camera or filming a video and then seeing what shape it has in the picture? So we can easily understand how it is with the relativistic shortening of the length." Indeed, among the common men, there is the conviction that a moving body contracts along the direction of its motion, regardless of where the observer is.

For example, if we look at a spherical planet moving at near-light speed, it turns into an oblate spheroid flattened in the direction of motion. And if we look at it from a direction perpendicular to its speed, we will see an ellipse. This is all wrong. Since in this case I am not able to hide behind Bondi's back, I have to explain everything on my own.

Looking ahead, I will let you know that the body in the photo will look completely different from what many are sure. Instead of being compressed in the direction of travel, it will be distorted by a shear deformation, and sometimes there is elongation instead of compression in the picture.

I will start with a warning. Alas, in this and the next subsections, I will use formulas. Moreover, I will derive them. The fact is that I am approaching those questions that are not described in textbooks and popular science books. Moreover, I am trying to dispel the misconceptions that some of the readers probably have.

Therefore, the argument "I swear by my mother!" will still be insufficient. I have to prove the statements in an adult way, on a more serious level. The readers can skip these sections without any harm to them or still read, having received my assurances that there will be few formulas; they will be simple and illustrated with drawings.

I'm going to discuss the details of the photography process first, then go through a few examples. But why discuss the process? Press the button and that's it! Nevertheless, you will have to discuss everything in advance. Everything is not so simple even if we are not in the quantum world, where the result is influenced by the measurement process itself.

Look at Fig. 7.2. The rocket flies in a straight line, and the photographer takes pictures of it at three different moments. Is everything obvious? No. The photographer can rotate the lens so that he follows the rocket, which would be in the centre of the frame all the time. Or he can fix the camera on a tripod and the rocket will be shot first in one corner of the frame, then in another, and between them in the middle. And without any relativistic effects, the image of the object will be different.

Let it be not a rocket, but a steam locomotive moving at a ridiculously low speed compared to the speed of light, practically crawling. If it is driving on straight rails, the distance between it and the photographer is changing. Closer objects in the photograph are larger in size.

Therefore, if you rotate the lens, following the subject, as the photographer on the left does in Fig. 7.2 (there are not three of them, but one, but the position of the camera changes), then the picture of a rocket or a steam locomotive in position 2 will occupy more pixels on the camera matrix than in positions 1 and 3. In addition, in position 2, the rocket will be taken strictly face-on, and in positions 1 and 3 at an angle, so the proportions in the picture will already be different.

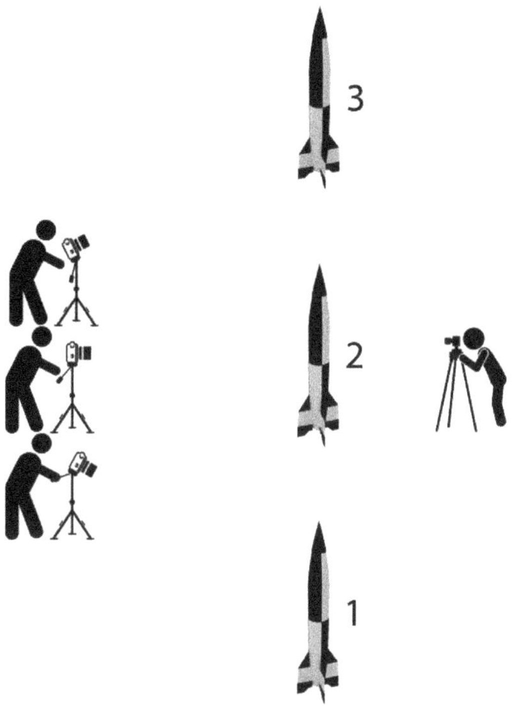

Figure 7.2.   You can take pictures of moving objects in different ways: by turning the camera or by not turning it.

And if you fix the camera so that its matrix is parallel to the rails or the trajectory of the rocket, as the photographer did on the right in Fig. 7.2, then in three frames the image of a steam locomotive or a rocket will be the same in size.

Now let's discuss the issue of the exposure time. Let's remember that we don't get anything good by trying to photograph a very fast moving body. The picture will be blurred. It is better to photograph it, setting the shutter speed as low as possible and lighting it brightly. Then the photo will not be very blurry from the movement. Imagine that our camera can take a photo with a shutter speed, if not infinitely small, then very small, say, in one billionth of a second. During this time, the light will pass only 30 cm, and the size of the object is much larger.

I will try to prove to you that we can use a camera with a flash, and a very short one. Let's say the same one billionth of a second. Then the subject should not be illuminated by anything other than the flash, and the shutter should be constantly open. This corresponds to endless exposure. As I will demonstrate below, these two options are actually very similar. But there is still a slight difference between them: the option with a flash, it seems to me, is easier to explain and easier to understand. Therefore, after I prove the similarity of the two options, I will make the flash shot the main one for the further story.

I will assume that the size of the subject is significantly less than the distance between it and the photographer and that the camera does not distort the image, and everything else, which will make it easier for us to solve the problem. All other details of the shooting are not important to us. For example, we do not care about the colour of the image, although due to the Doppler effect, the spectrum of light from an approaching rocket at position 1 will shift towards the blue end, and at position 3, the spectrum of the receding rocket will shift towards the red end of the visible range.

After a long preliminary discussion of the details, we can proceed. First, let's understand the reason for the distortion of the shape in the image. It is the finiteness of the speed of light denoted by the letter c. The object is moving upward from bottom to top with speed v. We will apply a trick popular in SRT and express it as a percentage of the speed of light: $v = \beta c$.

Here β (read beta) is a dimensionless number. Since it is impossible to move with a speed greater than the speed of light, then $\beta = 1$ for light and for massless particles and $\beta < 1$ for any body having a nonzero rest mass.

First, we will consider the simplest option, similar to shooting a rocket in position 2 in Fig. 7.2. With it, the direction of shooting is perpendicular to the speed of body movement. If the subject is a rod perpendicular to the direction of shooting, then in the photograph it will have the same length as the stationary one. But the rod oriented along the shooting line will look completely different. This can be seen from Fig. 7.3.

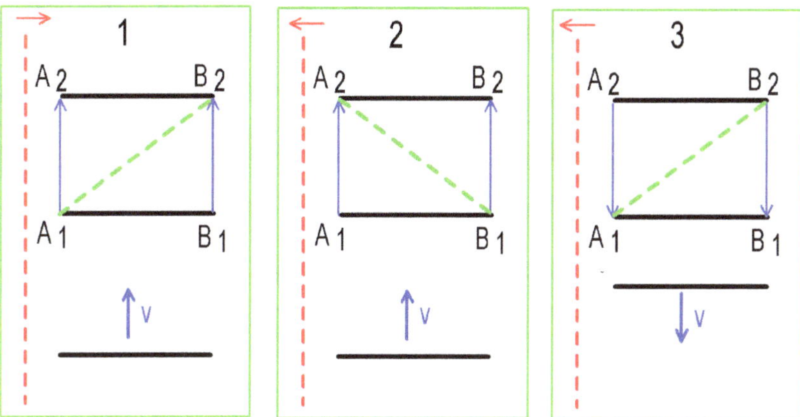

Figure 7.3. The horizontal rod flies upward at a speed close to the speed of light. The black bars show its position at different points in time, and the dotted green line corresponds to how it will look in the photograph.

This horizontal rod of length L moves vertically upward at a constant speed $V = \beta c$. In Fig. 7.3, it is drawn at different points in time as black bars. The trajectories of movement of its ends are vertical. To the left of the picture is a photographer who takes a picture of this rod in motion. It is located quite far away, so that the light rays from the rod to the camera lens are almost horizontal. The last detail was introduced to simplify reasoning; it will not affect the qualitative conclusions in any way. The red dotted lines are the instantaneous position of the light front, which I will discuss in more detail below, the arrow above it shows the direction of its movement at a speed of c.

In the picture on the left with the number 1 on top, the photographer is using a very fast flash and infinite exposure time. The light from the flash at every moment of time is located along the red line and moves to the right at a speed of c. It reaches the left end of the rod closest to the photographer at point A_1. In this moment, the light has not yet managed to reach the second end of the rod at point B_1. This will happen after a period of time equal to L/c. During this time, the rod will still have time to fly the distance $VL/c = \beta L$, which is shown in the figure by blue arrows.

The flashlight will illuminate the far end when it will be at point B_2. In this case, the near end at point A_2 will be in the dark, like the rest of the rod. In the photo we will see the ends of the bar at points A_1 and B_2. The bar itself on it will look like a segment A_1B_2, shown in the left figure with a green dotted line. It is not horizontal, but inclined. It is easy to calculate that the tangent of the shear angle is $\beta = V/c < 1$.

And what is shown in Fig. 7.3 with number 2 above? With the rod, everything is unchanged, but now the photographer uses a fast shutter speed, and the rod either glows or is illuminated all the time. The red line on it shows where at a certain moment the light is located, which will then enter the camera during the shutter opening.

The photons that hit the film or the camera sensor are not only on this red line, but also move to the left at the speed of light. The photons to the left of the red dotted line will reach the camera before the shutter opens, and to the right after it closes. All of them will not reach the film or matrix. Only the photons on the red line will leave their mark on the picture.

Before that, they were emitted or reflected by the surface of the rod. First, this was done by the far end at point B_1, and then by the near end at point A_2. In general, the photograph will show the rod in the position shown by the green dashed line A_2B_1. It is sheared by the same angle as in the left picture, but in the opposite direction.

In the drawing numbered 3, we see the same situation as in Fig. 7.3 with number 2, but the rod is now flying downward at the same speed. It can be seen that it will be photographed in the same position as in the figure with number 1.

With a little thought, you can understand that the similarity of the two options in panels 1 and 3 is not accidental. If we follow the photons that enter the camera during the exposure time in the figure on the right, and then change the direction of time, we will get the movement of the photons from the flash in the left figure. It will be clearer to talk about two films that show the movement of these photons in time. The reversal of time will be reduced to the fact that the film is shown backwards.

But when time reverses, the rod no longer moves down, but up, as in panel 1. So if you change the direction of the flow of time, then what is depicted in one picture or film will turn into what is depicted in another.

This important conclusion applies to any position of the object. Therefore, further, we will discuss only a shot with a flash, realizing that the option with a fast shutter speed will give a similar shot, in which you need to reverse the direction of the object's speed.

But can you think of some other options to photograph the subject? For example, use both flash and fast shutter speed at the same time. Alas, as you might guess, in this case there will be either darkness in the photo or one point of the rod.

From the rod, let's move on to a square with vertical and horizontal sides. Each of the horizontal sides moves like the rod discussed above. At the moment of shooting, they are directed at the camera. Repeating the reasoning applied for the rod, we understand that in the picture we will see an image of a trapezoid. One side of it is vertical, and the other deviates from the horizontal by the angle $\arctan(V/c) = \arctan(\beta)$.

Now we can consider the volumetric body. Let's take the skeleton of a cube and understand that the image in the photo will be distorted as shown in Fig. 7.4. This transformation is called shear deformation.

But the photographer can photograph an object that is not moving perpendicular to the shooting line like a rocket in positions 1 or 3 in Fig. 7.2. What will be in the pictures in these cases? I'll start with the option when it is not a rocket that flies upward, but a square of four rods or a flat sheet with a drawing of the black square from

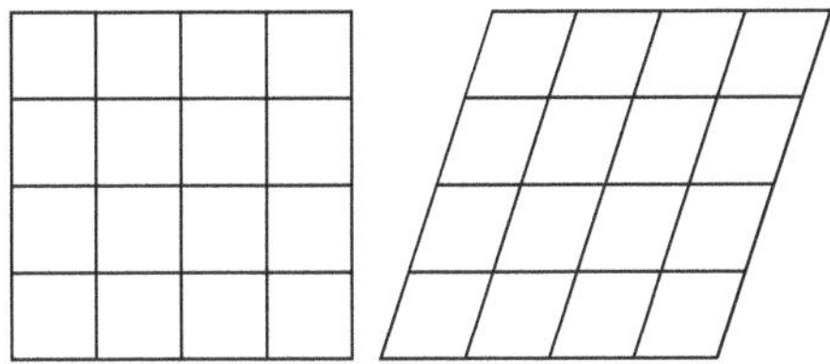

Figure 7.4.   The cube before and after the shear deformation.

Malevich. In position 2 in Fig. 7.2 the plane of the square or sheet is perpendicular to the direction to the photographer and this orientation is maintained during flight. It is not hard to guess that there will be a rectangle in the photo, but what can you say about the ratio of the lengths of its sides?

We figured out everything about the shot in position 2, the square will remain a square. And in positions 1 or 3, two effects will work. The first one is a change in proportions due to the fact that the shooting is not done perpendicular to the drawing's plane. The second effect is associated with the finiteness of the speed of light.

It will take longer for the light from the flash to reach the more distant end of the square. During this time, the far side will have time to fly a little along its path and will be illuminated in a position shifted upward compared to its position at the moment when the flashlight illuminated the near edge. But in position 1 in Fig. 7.2, the farthest end will be the bottom end and this effect will reduce the height in the photo of the square. And in position 3, the far end will be the upper end and the effect will work in the other direction, lengthening the photo.

I derived formulas for the aspect ratio of the image of a flying square for this more general case and they are not very difficult. The reader can do the same or take advantage of the fruits of my work and simply believe that I am not deceiving him.

The conclusion is simple: in position 1 of the photo, the square for both photographers will be noticeably compressed vertically at the speed and angle for which I made the calculations. But in posi tion 3, they will stretch in the direction of travel! A bit unusual for the relativistic length contraction, isn't it? This example alone is enough to understand that relativistic effects are not at all as simple as they seem to many people.

And if the object has depth, and a cube flies, not a square, in the photo it is additionally distorted by a shear, as shown in Fig. 7.4. The conclusion from all this is simple: in any case, the body in the picture will be distorted by shear deformation and compression or stretching along the direction of movement. And it is not at all simply flattened in the direction of movement, as many are sure. This does not mean that the Lorentz transformation formulas are incorrect. They simply are not at all about distorting a photograph.

And the general conclusion is: in reality there is a relativistic shortening of length and slowing down of time, but you need to know the procedure by which the distances and lengths of moving objects are measured and the clocks are synchronized to understand the SRT approach — in general, all the details that popularizers usually miss.

7.2.6. *Superluminal speeds*

Let's discuss the possibility of motion with superluminal speed. I don't mean hypothetical particles called tachyons, which have not yet been found in any experiment, but something more specific. I will use examples to demonstrate the possibility of superluminal velocities that do not contradict SRT and superluminal velocities, which, when checked, turn out to be a trick or a mistake.

Let me remind you that SRT does not permit bodies, energy, or information that are moving faster than the speed of light in a vacuum, and that's all. It does not limit the speed of a spot of light, for example. Let's do an experiment. On the day when the Moon is low above the horizon, let's take the laser pointer, turn it on, and place it on the spinning disc of a turntable that makes 33 and 1/3 rpm. We set it so that the beam from the pointer passes through the disc of the Moon.

And now let's calculate the speed of movement of the spot of light from the pointer along the surface of our satellite using the simple formula $v = \omega r$. Here ω is the angular speed of rotation, equal to $2\pi \cdot 33.3/60 \, \text{s} = 3.5 \, \text{Hz}$, and $r = 384{,}400 \, \text{km}$ is the distance to the Moon; hence the speed of the spot is $v = 1{,}345{,}400 \, \text{km/s}$, which is more than four times the speed of light. The point is that neither energy nor information is transferred with a spot of light. Both of them come from your laser pointer and move at the speed of light.

But this spot can influence the processes on the lunar surface. How? I'll start from afar. The refractive index of water is about 1.33, so the speed of light in water is about $300{,}000/1.33 \approx 225{,}000\,\mathrm{km/s}$. SRT does not prohibit a particle from moving in water faster than this speed, but slower than light in emptiness.

Such a particle leaves behind a conical wave front. It is called the Vavilov–Cherenkov radiation in honour of the Soviet physicists Sergei Vavilov and Pavel Cherenkov, who discovered it in 1934. Because of this effect, blue light is observed in the liquid that cools a nuclear reactor. In 1958, Cherenkov and two theorists who explained the effect, Igor Tamm and Ilya Frank, received the Nobel Prize in Physics for this discovery. Vavilov did not live to see this moment, and the Nobel Prize is not awarded posthumously.

The spot of light on the Moon, moving faster than the speed of light in a vacuum, becomes a source of radiation similar to the Vavilov–Cherenkov radiation. Naturally, no one can detect the radiation from a laser pointer spot, but if you use a particularly powerful laser, this is quite possible, at least theoretically. Naturally, such experiments have not yet been performed.

Another example of supposedly superluminal speed is associated with a very long chain of soldiers, for example, from Earth to Neptune. They stand one after another. The first soldier takes a step forward. Each is given an order: step forward and stop immediately when the front neighbour has taken a step forward. The spot where the soldiers are stepping forward will move along the column. It is not hard to guess that its speed is clearly less than the speed of light. It is because the light takes time to cover the distance from one soldier to the eyes of the next one, among other things.

No surprises so far. But then Mr Wisenheimer appears and offers to equip each soldier with an accurate chronometer and an order to step forward exactly at noon. The soldiers are disciplined and trained and will carry out orders perfectly. Therefore, exactly at noon, the "step forward" signal will instantly travel the distance to Neptune.

And, as I wrote in Section 2.10, the light will take 4 hours to make such a journey. Superluminal motion is evident! But this spot transmits nothing, neither energy nor information. Moreover, if a middle soldier in the column falls asleep or escapes, this will not in any way affect the behaviour of the latter one; he will simply not know about it. Information about the order "to take a step forward at noon" came in advance and reached Neptune no faster than radio waves.

In these two examples, the essence of the trick is almost not hidden, so it is easy to expose it. But here's a real-world example that is much more difficult to deal with. Two jets are often ejected from so-called active galactic nuclei (AGN) at the centre of many galaxies. They are sources of powerful radiation in different ranges of electromagnetic waves, from radio to X-rays. The jets are rather narrowly directed. They eject matter in opposite directions and sometimes the speed of the outflow approaches the light one.

Astronomers calculated the velocities of outflow that often exceeded the speed of light. This happens when one of the jets is accidentally pointed almost directly at Earth. Should we already begin to doubt the special theory of relativity?

To explain the reason for the apparent superluminal motion, I need formulas. Figure 7.5 shows picture of the jet emitted by AGN. This is not a photograph, just an artist's impression. Let's mentally add a couple of designations to it. Let the jet be in the xy plane, where the x coordinate is directed toward the Earth and the y coordinate is perpendicular to this direction, and their origin is in the centre of the active galactic nucleus. The jet makes a small angle α with the direction to the Earth. The velocity of ejected matter is $v = \beta c$.

Astronomers observe how the distance changes over time between the AGN and some characteristic jet detail, for example, a clump or a separate cloud. After a time interval t after the start of the jet ejection, the substance, for example, the already mentioned clump, will move away from the active galactic nucleus at a distance vt. Its coordinates will be $x = vt \cos \alpha$, $y = vt \sin \alpha$.

Suppose that a flash had occurred in the clump. After a time interval Δt another flash had occurred. Flashes are needed simply to indicate a specific place and time, i.e. just like two labels referring to

Figure 7.5.   A picture of the jet emitted by AGN. We assume that the direction to the Earth is horizontal and oriented to the right. *Source*: NASA/ESA/G.Bacon, STScI.

the same part of the jet, they don't affect anything. You can consider two clouds flying out after a period of time Δt instead of two flashes in one clump, if you like it more.

Since the second flash will be closer to the Earth at a distance $\Delta x = v\Delta t \cos \alpha$, its light will have to cover a shorter path to the telescope, which is located on our planet and is directed at these flares. Light on its way to the telescope will win the time $\Delta x/c$.

The astronomer will see that the time interval between the arrival of light from two flares is equal to $\Delta \tau = \Delta t - v\Delta t \cos \alpha/c = \Delta t(1 - \beta \cos \alpha)$. These two flares occurred at different points in the sky; the distance between them, projected onto the celestial sphere, is equal to $\Delta y = v\Delta t \sin \alpha$.

So, the source of the flash has moved in the sky at a distance Δy in the time $\Delta \tau$ according to the astronomer's clock. It seems logical to find its speed as

$$u = \frac{\Delta y}{\Delta \tau} = c\frac{\beta \sin \alpha}{1 - \beta \cos \alpha}.$$

This expression becomes maximum if $\cos \alpha = \beta$. Then it is equal to

$$u = \frac{c\beta}{\sqrt{1 - \beta^2}}.$$

and can exceed the speed of light c. For example, for $\beta = 0.99$ (quite admissible speed in a jet) and $\alpha = 8°$, we get $u \approx 7c$. So, an astronomer on Earth may erroneously decide that the speed of the matter outflow exceeds the speed of light by no less than seven times, although it is not at all superluminal. The finiteness of the speed of light will play a cruel joke on this person.

By the way, in this example, $\Delta t = 50.25\Delta\tau$, which is the source of all calculation errors. All displacements that occurred during the time period Δt are erroneously attributed to a significantly shorter period $\Delta\tau$. Displacements which happened in two days are attributed as reaching one hour.

It is clear that professional astronomers know this trap and try not to fall into it. But the person who read the article in the media titled "Astronomers put Einstein to shame by discovering superluminal speeds in space" is completely unfamiliar with the nuances that I have just described.

So the conclusion of this section is simple: in nature there is no movement faster than light in vacuum. All examples of violation of this rule are incorrect. Both energy and information are transmitted at sublight speed. Probably there is an error in determining the speed. Any sensation in the media should be treated with a grain of salt.

7.3. Black Holes and Naked Singularities

Since we were talking about unusual objects in space, moreover relativistic effects, it is difficult not to recall other exotic objects predicted by general relativity — black holes (BHs). The solution describing the simplest BH was obtained by the German Karl Schwarzschild back in 1915, but it took decades to understand the nature of the physical object it describes. Even the name black hole this object received only at the end of 1967 in a lecture by the American John Archibald Wheeler, who became famous for his apt nicknames and the terms he proposed for things related to general relativity. For a long time, the corresponding decisions of general relativity were perceived as something purely speculative.

After all, there was no certainty that besides the name there were BHs themselves.

Now astronomers have discovered a lot of black holes in space. However, it is still possible, although very unlikely that what astronomers observe is not BH, but much more unusual objects called naked singularities. We'll talk about them a little later, but for now let's temporarily cast aside doubts about the nature of unusual space objects, just in order not to clutter up the book with constant reservations like "if these are not naked singularities or something about which we have no idea yet".

Figure 7.6 is an image taken by NASA's Chandra X-ray satellite. To obtain this image, it took more than 7 million seconds of exposure, because the image shows very distant and weak X-ray sources. The concentration of BHs on it is enormous: about 5000 BHs would fit in the area of the sky occupied by the Moon on a full moon.

Figure 7.6.   Black holes in a deep X-ray image of the sky taken by the Chandra satellite. *Source*: NASA.

So nowadays it is difficult to remain sceptical and deny the existence of black holes. However, in the world where supporters of the flat Earth theory live, you can find many such sceptics, but they have practically no idea of BH as a rule.

It's time for us to describe the most basic properties of the objects under discussion. There are several types of black holes in general relativity. If the BH rotates or it has an electric charge, then this circumstance greatly changes its properties. But the main ones remain unchanged for all isolated BH without massive neighbours. I will describe these general properties using the example of the simplest Schwarzschild BH, which does not rotate and does not have a charge.

7.3.1. *Properties of black holes*

The main feature of BH is the space-time singularity surrounded by the event horizon. I think that nothing became clearer after this explanation. But let's try to find out what is hidden behind these words. A singularity is a place where space-time goes crazy. Its curvature goes off scale, tending to infinity, as well as tidal forces near a singularity. Nobody knows if the laws of physics work in this strange place. Mathematicians really do not like singularities, physicists also prefer not to deal with them, but nature itself forces them to do this.

However, the second part of the sentence, the one about the event horizon, greatly softens the bad impression of the first. The event horizon is like a one-way street or a semi-permeable membrane that lets anything in one direction and doesn't let anything in the other. In the case of a BH, it lets everything into the horizon and does not let anything out. This is a kind of BH boundary, which determines its spatial dimensions.

A little about why the event horizon is one-way. If a ball or a dog or your beloved alligator or any other object moves, then it can change its position in space, as far as the laws of physics and the surrounding bodies allow. But time always flows in one direction, from the past to the future. We can say that the movement in time is one-way. A movement in the opposite direction of time, from the future to the past is found only in fiction, but not in nature.

The Schwarzschild BH is spherically symmetric and we can move along its radius. As long as we are outside the event horizon, this movement is no different from movement along the other two spatial coordinates. It can be arbitrary, the coordinate can change in any direction, but in time we move only towards the future.

Inside the event horizon, there is a kind of change of two coordinates (in general relativity, this is quite easy to establish). The radial coordinate becomes temporal, but the coordinate that was temporal outside the BH suddenly becomes spatial. The location of this transition determines exactly where the event horizon passes. As a result, the radial coordinate inside the BH becomes analogous to our time coordinate and can change only in a certain direction, while the former time coordinate becomes spatial and can change as desired. For details, I refer the reader to other books, including my book *How the Universe Works*.

In other words, within the event horizon of the Schwarzschild BH, everything (matter, radiation, information) must change its radial coordinate in a certain direction. If it is only obliged to decrease, then this is a black hole, if only to increase, then we are talking about another object called a white hole. We'll talk about white holes later.

For now, let's focus on BH. Everything inside them must move towards the centre, from the event horizon to the singularity located in this very centre. Once in the black hole, you must keep falling toward a singularity. Attempts to throw something or just shine a flashlight outside BH are doomed to failure. Nothing can leave the black hole, not even light. Strictly speaking, it is for this reason that the BH got its name. Note that we certainly cannot see the singularity inside the BH, because its radiation cannot even approach the horizon, let alone cross it.

Any BH, like its simplest Schwarzschild variety, has an event horizon that lets everything in, but does not let anything out. There must be a singularity inside this horizon. This follows from a theorem proved half a century ago by the English mathematician

Roger Penrose, who was awarded the Nobel Prize in Physics in 2020 "for the discovery that black hole formation is a robust prediction of the general theory of relativity". Singularity may differ from that which is in the Schwarzschild black hole, but it must be inside the horizon. For now, we will restrict ourselves to this information.

What are the dimensions of the simplest Schwarzschild BH? The radius of its horizon, called the Schwardschild radius, is proportional to the BH mass M and is equal to $2GM/c^2$. This formula uses two fundamental constants, namely the gravitational constant $G = 6.674 \cdot 10^{-11}\,\mathrm{m}^3 \cdot \mathrm{s}^{-2} \cdot \mathrm{kg}^{-1}$ and the speed of light c. It is easier not to multiply the numbers, but to remember that the Schwarzschild radius of a BH with the same mass as that of the Sun is approximately 3 km. The dimensions of such a BH with a diameter of 6 km do not greatly exceed the Garden Ring in Moscow with its length of 15.6 km.

The radius of a BH with the mass of N solar masses is equal to $3 \cdot N$ km. And if we estimate the size of the BH of the Earth's mass, we get a radius of about 9 mm. However, astronomers do not expect to find BHs of either terrestrial or solar masses in space.

Why? To do this, from the question of what a black hole is, one must move on to the question of how it could appear. It is clear that BH, like anything else, could theoretically arise along with the rest of the Universe in the Big Bang. However, at the era of inflation, all exotic objects were carried away very far from us in a fraction of a second. It is unlikely that the part of the Universe that we, in principle, can observe, there is at least one black or white hole, born in the Big Bang.

7.3.2. *The birth of a black hole in the collapse of stars*

So a mechanism is needed to ensure the appearance of new BHs after the end of cosmological inflation. And it was invented long before the concept of inflation itself was proposed. According to this hypothesis, BHs are formed during the collapse of compact massive objects, such as supermassive stars. Collapse is compression, something like an explosion in reverse, when the resulting explosion video is scrolled in the opposite direction.

To start the collapse process, a very large mass must be in a not very large volume. Then it will begin to shrink and can form a black hole. But not a white one, into which nothing can enter from the outside, but only fly out of it. In practice, this means that we can meet a BH in the Universe, but we will not meet a white one.

Ordinary stars can be small in size, such as the so-called white dwarfs. But if the mass of these stars exceeds a certain critical mass, equal to approximately $2.8 \cdot 10^{30}$ kg or 1.4 solar masses, then the gas pressure is no longer able to prevent the collapse of the star and it begins to shrink. It can drop some of its mass or form a neutron star or BH.

The maximum possible mass of a white dwarf is called the Chandrasekhar limit after the British–Indian–Pakistani physicist Subrahmanyan Chandrasekhar, who set this theoretical limit. For this and other achievements, he received the 1983 Nobel Prize in Physics "for his theoretical studies of the physical processes of importance to the structure and evolution of the stars". Neutron stars have an even higher density, but their mass is also limited by another limit, called the Oppenheimer–Volkov limit and equal to 2–3 solar masses.

However, I have already mentioned the star Betelgeuse in this book. It is located near the Solar System and has a mass of about 17 solar masses. Scientists have determined its size and have even been able to get an image of this star. This red supergiant is larger than the orbit of Mars. So this is definitely not a black hole. What prevents it from collapsing? Astronomers know many stars with a mass of about 10–50 solar masses that do not collapse at all. The fact is that thermonuclear reactions take place in them and they actively emit light and other electromagnetic radiation, which creates additional pressure that prevents collapse. As soon as hydrogen and other raw materials for the reaction of thermonuclear fusion in the interior of such a star burn out, the radiation pressure will drop and the process of collapse and explosion of a nova or supernova will begin.

Every massive and very compact celestial body (say, more than 5 times the mass of the Sun) which does not look like a massive star is a candidate for black holes or even more exotic relativistic objects like naked singularities.

7.3.3. *The accretion of the surrounding matter onto a black hole*

So, we figured out what astronomers need to look for. But this raises a natural question: how can we even see a BH if it does not emit light and other electromagnetic radiation? The answer is simple. Yes, the BH cannot do this, but nothing prevents the matter around the BH from taking an active part in the radiation process. It is outside the event horizon and therefore the emitted radiation is able to reach earthly astronomers.

What kind of matter are we talking about? Many BHs (but not necessarily all) are surrounded by a so-called accretion disc. The term *accretion* comes from the Latin word *accretio*, increase. This is the process of increasing the mass of a celestial body by the falling of matter on it. And this matter is always taken from the surrounding space. Usually it is simply stolen from neighbouring bodies. Most often, this matter is gaseous. However, sometimes these bodies themselves or their fragments fall on a supermassive black hole.

Falling matter usually has some moment of rotation relative to the BH. Given the small size of the latter, we can draw an analogy with the Solar System, in which the planets do not fall on the Sun because they are orbiting it. But the planets are doing this for billions of years in almost the same orbits, and gas, dust and particles when falling on a BH with its strong gravitational field lose energy due to the friction, radiation of electromagnetic and gravitational waves, and other processes.

Therefore, they begin to move along a spiral trajectory, gradually approaching the central black hole. A great many of falling matter forms an accretion disc. This disc does not reach the BH. Near it, motion becomes unstable and matter simply falls on the hole by leaps and bounds, increasing its mass and emitting electromagnetic and gravitational waves when falling. The accretion disc gets very hot and starts emitting in the X-ray range. It is this radiation that X-ray and gamma-telescopes observe.

I want to mention such a remarkable object as the system known as OJ 287. The first photograph of it dates back to 1891, when even the concept of a BH did not exist. Now astronomers are confident that these are two BHs, one of which, with a mass of 18 billion solar masses, occupies the second line in the current list of BH masses. The smaller companion is only 10^8 times more massive than the Sun. They revolve around their centre of mass with a period of 12 years at a distance of 3.5 billion light-years from us. Each BH has its own accretion disc. Therefore, from time to time, one of them crashes into the disc of the other and passes through it, providing a periodically recurring flash, or rather a sharp increase in radiation.

So, we do not see directly the BHs themselves, but we can observe the accretion disc around them, primarily by its X-ray and radio emission. Are there other ways to detect these objects? Oh, sure. Gamma-ray bursts can be observed that rapidly change their intensity. They occur when large portions of matter fall on a BH. Gravitational wave detectors recorded waves generated by the merger of two BH and BH with a neutron star. Some stars revolve around supermassive BHs. One can very accurately estimate the masses of these black holes by observing the stars' trajectories. Astronomers have long discovered quasars and active galactic nuclei, which are supermassive BHs, according to modern concepts.

7.3.1. Black holes in the Galaxy and supermassive black holes in other galaxies

Astronomers believe that there are many BHs of different masses in nature. Small BHs with a mass of up to 20 solar masses can be observed only in our Galaxy and then if they are part of a binary system and are a source of X-ray radiation. Naturally, something similar exists in other galaxies or in the distant regions of the Milky Way, but we cannot observe them due to the remoteness. We simply can't find out about the existence of such an object in another galaxy. And if it is close to us, it can go unnoticed if it is not a powerful source of electromagnetic radiation, primarily in the X-ray or gamma range.

The most famous candidate for this type of BH is the object called Cygnus X-1. This one of the most powerful X-ray sources was discovered in 1964. Its mass is 14.8 times the solar mass. It is located about 6,070 light-years from the Sun and has a companion in the form of a star — a blue supergiant.

Twice closer to us is another BH candidate, called A0620-00 or V616 Unicorn, with a mass of 3 to 5 solar masses. The closest possible candidate for BH is located at a distance of about 1000 light-years in the triple star system HR 6819 in the southern constellation *Telescopium*.

We can say much more confidently about the existence of super-massive BHs with masses of more than 10^5 solar masses, which cannot be confused with stars. A hole called Sagittarius A* is located in the centre of the Milky Way* at a distance of about 25,900 light-years. It has a mass of about $4.3 \cdot 10^6$ solar masses, which is significantly less than the mass of the entire Galaxy. Astronomers observe its emission in the radio, X-ray and IR ranges. Sagittarius A* emits electromagnetic waves in all ranges, but visible radiation is absorbed by dust and gas on its way.

Astronomers can observe the motion of stars surrounding it. Some of them are orbiting in elliptical orbits around it, making it possible to determine the mass of the BH. In 2020, two astronomers received half of the Nobel Prize in Physics for the discovery of this hole: Reinhard Genzel from Germany and Andrea Mia Ghez from the United States "for the discovery of a supermassive compact object at the centre of our galaxy."

I talked about the shift of the perihelion of Mercury. The gravitational field in the Solar System should be considered weak from the point of view of general relativity. Astronomers were eager to measure the shift of the pericentre of the body orbiting a black hole to test this effect in a really strong gravitational field. This was successfully implemented in early 2020.

The star, called S2, orbits BH Sagittarius A* in a very elongated orbit with an eccentricity of 0.88 and a period of about 16 years.

With the help of the Very Large Telescope, installed in the Atacama Desert, Chile, astronomers measured the displacement of the periapsis of its orbit, which turned out to be equal to 12′ for the orbital period. This fits perfectly with the prediction of general relativity.

Some BHs known to astronomers are not only large, but also located at great distances. The light from a quasar with an unremarkable name, SMSS J215728.21-360215.1, has travelled to us for about 12.5 billion years. During the trip, the Universe managed to expand 5.75 times. The mass of this supermassive BH is estimated at 34 billion solar masses.

And every day it absorbs from its surroundings the mass of an entire star, increasing its own. It is known that good nutrition is a guarantee of health and strength. And this BH demonstrates its strength to everyone who sees its radiation. After all, its accretion disc is emitting light 10,000 times brighter than the galaxy in which it is located.

7.3.5. *A snapshot of a black hole in the Messier 87 galaxy*

One of the main scientific sensations of the spring of 2019 was a press conference at which an image of a BH in the Messier 87 galaxy was shown. Some journalists were quick to declare it a photograph of a black hole. This is not entirely true.

To begin with, this is not a visible light image. This is the result of a joint observation of ten of the largest radio telescopes on Earth. The colours are artificial and only show the distribution of the intensity of the radio emission from this object. Moreover, the source of this radiation is not at all a BH, but an accretion disc around it.

Observations on an array of radio telescopes were carried out in the spring of 2017. It took two years to process a huge amount of information recorded during the observation process. All radio telescopes participating in the project, dubbed the Event Horizon Telescope, worked together as a single instrument called the radio interferometer. This is to some extent an analogue of a huge radio telescope with an antenna the size of the distance between individual instruments, i.e. almost the size of the Earth. If one of the antennas is launched into space, as in the Russian device Radioastron, then these dimensions can be significantly increased.

The Event Horizon Telescope observed two supermassive BHs, the closest Sagittarius A* and the hole in the Messier 87 galaxy. Why were the observations carried out in the radio range? Supermassive BHs are located in the centres of galaxies, where there is a lot of gas and dust, which makes it impossible to see them in visible light or in the neighbouring ranges of electromagnetic waves (UV or IR). But radio waves are scattered by dust particles much weaker. X-ray and gamma radiation from the accretion disc also reaches us, but we cannot yet use it to work together in the interferometer mode. So we observed what nature allowed us.

And what did these observations give us? For starters, the image turned out to be exactly as expected, similar in appearance to a doughnut. This is due, among other things, to the deflection of rays (in our case, radio waves) in the gravitational field of the BH. The next check of general relativity was successful.

Where did the dark spot in the middle come from? Imagine that you are looking at the section of the accretion disc beyond the BH. The emission that misses the BH is deflected, but its brightness does not change much. But the waves that got inside the horizon cannot leave the BH, trapped inside. The brightness decreases at this area, forming a "torus hole".

For clarity, we can consider the radiation going in the opposite direction from the terrestrial radio telescope toward the Messier 87 galaxy. Some radio wave beams that hit directly the horizon will not leave the BH. Accordingly, in the opposite direction, the astronomer receives a signal only from the part of the accretion disc located along the way of radio waves.

But if we look a little to the side, then the radio beams pass by the horizon and reach the more distant part of the accretion disc behind the BH. This means that in the opposite direction, light from this part will come to an observer on Earth, forming a picture in the form of a torus. Why is the doughnut brighter at the bottom? This is due to the rotation of the accretion disc. Those parts of the disc that we see at the bottom of the torus rotate towards us, and at the top — away from us.

Galaxy M87 is also associated with fake superluminal velocities, which I discussed in Section 7.2. The supposedly superluminal motion was studied in detail in early 2020 using two examples, one of which was the galaxy M87. A jet takes off from it, directed towards us. By measuring the velocities of two knots or bunches of this jet, which have moved away from the centre by 900 and 2500 light-years, astronomers have obtained values that exceed the speed of light by 6.3 and 2.4 times, respectively.

This is exactly the case that I described in Section 7.2. More detailed studies have made it possible to determine the speed of movement of matter in a jet. It turned out that it is moving away at more than 99% of the speed of light, but still slower than light. Such speeds are called sublight and do not contradict SRT.

7.3.6. *Falling into a black hole*

Now it is clear why the story of black holes of different masses fell off the general scale of scales. Objects with sizes that differ billions of times are not easy to fit into it. And by themselves, they are too unusual to be described, say, together with planets of similar sizes. Moreover, in order to fully appreciate this unusualness, one must discuss what will happen if you fall into a black hole.

Let's start not with a fall, but with a slow approach. Imagine that the BH is surrounded by a spherical shell with a large radius, significantly exceeding the radius of the hole itself. On the surface of this shell is the base of the laboratory for the study of black holes with a group of physicists and their instruments. There are hatches in the shell that can be opened, and through them the instruments can be lowered to the BH event horizon with a winch, such as is usually seen at the old-time village well, with a rope wound around it. Have you imagined? Now let's describe two thought experiments for which this simple equipment is enough.

In the first of them, an ultra-precise chronometer is tied to a rope. It can be atomic, quartz, mechanical, etc. The main thing is that it does not use gravity in any form. So a pendulum clock or hourglass will not work. The laboratory assistant lowers the watch closer to the BH and looks at how fast their readings change. In other words, it measures the speed of the passage of time depending on the distance of the clock from the horizon of the hole. General relativity asserts that this speed will fall when approaching the event horizon. I have already talked about experiments that confirm the existence of gravitational redshift.

What does this mean for an object falling into a black hole? Let a very accurate clock fall into the BH and emit a radio signal every microsecond. This signal is received by a receiver located very far from the BH: on the laboratory envelope around the hole, on Earth, Sirius, distant galaxies, etc. Due to the difference in the flow of time, the intervals between the signals received there will be more than a microsecond. At the beginning, the difference will be very small, but as the BH is approached, it will increase.

If the watch with the transmitter almost reaches the horizon, then the intervals between the signals will reach seconds, minutes, hours, years, centuries. If we calculate what time interval the clocks on Earth will count down to the moment when the signal emitted exactly at the moment of crossing the event horizon reaches them, the result will be quite unexpected, for it will turn out to be endless.

Let's imagine that the falling body has emitted a signal near the event horizon. While this signal is on its way, the Sun ends its life cycle, other stars do the same, galaxies disintegrate or merge, all hydrogen in the synthesis processes turns into heavier elements, and protons decay if this hypothetical process has an arbitrarily small nonzero probability. But the signal keeps going where once there was the Earth.

This formal result is often interpreted inadequately. Some argue that a distant observer will constantly see light and other electromagnetic radiation from a body falling into a BH. Simply put, this person will forever see an object falling into the black hole, which will seem to him to be frozen over the horizon.

In fact, a completely different picture will be observed. To begin with, the radiation will be observed at a frequency significantly lower than that at which it was emitted. There are two reasons for this. One is the already mentioned gravitational redshift. The second is also redshift, but caused by the Doppler effect.

But the magnitude of the signal will also change. If the light emitted during a microsecond of falling into a black hole stretches for a year for a terrestrial observer, then its power or brightness should drop a huge number of times because of the conservation of energy. The calculation shows that the brightness of the observed object when falling into the BH will decrease very quickly (for those who want to know the details: exponentially). But this is not the worst part.

Let's remember that light is a stream of photons. And the last photon emitted by the body before crossing the event horizon will sooner or later reach the terrestrial observer. And the next one will not go beyond the horizon, because it will be radiated inside the BH, from where there is no exit. This means that after the arrival of this last photon, the observer simply will not see the falling object.

Let's go back to our experiments on lowering bodies to the surface of a hole using the simple device described above. The bucket of water is trying to go down, twisting the winch. The load tends to do the same because of the attraction to the BH. Let a laboratory assistant attach a dynamo machine (ideal, with an efficiency of 100%) to the winch and measure how much energy was generated by lowering the load to the very horizon. If we eliminate all energy losses for friction and so on, then the generated energy will be equal to that described by Einstein's formula $E = mc^2$.

This is the total energy contained in a body of this mass. Usually, the gravitational mass defect releases only a small part of it. And here each gram of mass will provide us with 90 TJ of energy, which corresponds to approximately 25 GWh — the daily output of a nuclear power plant.

Even a tenth of a percent of this is enough to make any supporter of green energy ecstatic. And if you remember that you can throw away anything in the black hole, for example, used plastic bags and other garbage, then you can solve the problem of waste disposal at the same time, including poisonous or radioactive wastes. But in our vicinity there is not a single BH that can be used in such a way by proponents of green energy.

Having discussed the speed of time and energy production along the way, we are ready to move on to the main question: what will happen if we jump into a black hole? Let's start with the simplest Schwarzschild BH, which is without rotation or electric charge. Any body crossing its horizon has to fall further, continuously decreasing the radius and approaching the central singularity with infinitely increasing tidal forces near it.

Sooner or later, it will be torn apart by them. The same thing will happen to a person. Tidal forces will stretch the human body or any falling object in the radial direction and compress in all perpendicular ones. In short, they will try to stretch it and turn it into spaghetti. Therefore, this process is often called spaghettification.

The pieces will continue to fall onto the singularity and will be torn apart by ever-increasing tidal forces into smaller and smaller pieces, etc. Soon, the molecules will be torn apart into atoms and then into elementary particles. The office shredder and kitchen waste shredder would be shocked after observing the hellish efficiency of this process.

An article was published in 2020 in which astronomers from observatories of different countries (Great Britain, USA, Denmark, Italy, Germany, Sweden, Israel, Poland, the Netherlands, Australia, Chile, France, Spain, Portugal, and India) wrote about the observation of emission caused by the fall of parts of a star of approximately solar mass after its spaghettification by tidal forces onto a BH with a mass of one million solar masses. This confirms what was written above.

So it is clear how the process will end. Where exactly will this happen? Let's roughly estimate how the only parameter that fully characterizes the Schwarzschild BH — its mass — affects the tidal forces. In a weak gravitational field, tidal forces are proportional to the mass of the body that causes them and inversely proportional to the cube of the distance to it. Let us extend this formula to the strong field of the BH to estimate the force value and remember that the size of the horizon is proportional to its mass. It turns out that the tidal force near the BH horizon is inversely proportional to the square of its mass.

This means that when somebody falls on a BH with a small mass, this person will be torn apart outside the horizon and only pieces will cross the event horizon. But you can cross the horizon being alive when falling on a supermassive BH. The moment of crossing the horizon itself does not stand out in any way, and for the falling one, it is not accompanied by any special effects or impressions. No more than for a motorist passing by a city limits sign. This person has crossed the border of the city, but the highway is still the same and there are no orchestras or fireworks. Everything is similar to BH. The falling person will keep falling and will be able to observe the process for some time, knowing full well that it will not last long.

How much exactly? It is clear that the moment at which the falling one will be torn is determined, among other things, by the strength of the body, posture and a number of other details that are foolish to discuss. But it is obvious that this will certainly happen before the falling person or object reaches the singularity with its unlimited tearing tidal forces.

But the time interval between crossing the event horizon and hitting the singularity is limited. It is surprising that it can be estimated without knowing general relativity, but using dimension analysis, one of the techniques widely used in physics.

The fall time cannot depend on the mass of the falling body, because all bodies fall equally quickly. It can only depend on the parameters of the black hole. A Schwarzschild black hole has only one parameter — its mass M. There are also two fundamental constants — the gravitational constant G and the speed of light in vacuum c, which can be included in the formula for the limiting fall time. The only combination of these three quantities, which is measured in seconds, is GM/c^3. Thus, the time of falling into a black hole will be equal to $k\,GM/c^3 \approx 4.93\,kM/M_{\text{Sun}}\,\mu s$, where k is a certain dimensionless coefficient and M/M_{Sun} is the mass of the BH, expressed in solar masses.

We got the answer without calculating anything or using any formulas. This is the beauty of dimension analysis. Here we could add that the value of k is unlikely to be very different from unity in order of magnitude. To find its exact value, you will need both formulas and calculations. General relativity claims that it cannot exceed π in any case,[b] even if the body is a rocket and it turns on its engine, trying with all its might to slow down the fall.

Thus, the maximum time of falling into a black hole is equal to $\pi GM/c^3 \approx 15.5 \, M/M_{\text{Sun}} \, \mu s$. For a black hole in the centre of our Galaxy, this is somewhere around a minute. For the most massive black hole known in the galaxy NGC 4889, with a mass of about $21 \cdot 10^9$ solar masses, it would be about 90 hours.

Let us add that the characteristic time of decrease in brightness with which a remote observer sees an object falling into a BH has the dimension of time and can be obtained from the analysis of the dimension. Naturally, he cannot give anything different from the above estimate. So this time will also lie in the range from tens of microseconds for stellar mass BHs to tens of hours for supermassive BHs. It's a little different from the supposedly observed eternal fall, as told by those who do not understand this issue, isn't it?

However, a person falling into a black hole may die even before he is torn apart by tidal forces. If there is an accretion disc around the hole, then it emits electromagnetic waves in all ranges, including hard X-rays and gamma rays. It also accelerates charged particles to high energies. In addition to radiation, there is also mechanical damage from collisions with objects falling into the hole. But for now, let's forget about an additional source of threats from an accretion disc.

7.3.7. *Rotating or electrically charged black holes*

What changes if a black hole is charged or spinning? It is odd enough to give such an ambiguous answer as both everything and

---

[b]Details could be found in A. P. Lightman, W. H. Press, R. H. Price, and S. A. Teukolsky, *Problem Book in Relativity and Gravitation* (Princeton, NJ: Princeton University, 1975).

nothing. So I'll have to explain it. The fact is that if a BH has an electric charge or rotates, then its internal structure completely changes.

How exactly? It is much easier to talk about charged black holes, which are called Reisner–Nordström holes in honour of two physicists, the German Hans Jacob Reissner and the Finn Gunnar Nordström, who discovered and studied this solution of the general relativity equations a century ago.

An electrically charged BH is centrally symmetric. It differs from the Schwarzschild hole in that there is another, inner horizon inside the event horizon. The radial coordinate in the interval between them corresponds to time, and there any object must move to the centre of the BH or from the centre of the white hole, which cannot form during collapse. But outside this interval, it will be the usual spatial coordinate, and stationary bodies can exist there. The area outside the outer region of its event horizon is our world. We can approach or cannot approach a BH, especially if it is located in a distant galaxy.

Inside the inner horizon, there is another region with a spatial radial coordinate that surrounds the central singularity. Theoretically, a sufficiently powerful rocket can fall into the Reisner–Nordström BH, fly through both horizons, and then turn on the engine, slow down the fall on the singularity and begin to move away from it. On the way back, it crosses the inner horizon again, after which it must move away from the centre. Yes, you are not mistaken; it flies out of the white hole through the event horizon from the inside and finds itself in the Universe.

But which one? Where and when will it fly out? No one can give a reliable answer to this question. Since the rocket in our mental example flies out of a white hole, and there cannot be many of them for the reasons I mentioned; perhaps it will fly out somewhere far beyond the cosmological horizon — the boundary of the region of the Universe available for astronomical observations. Therefore, a person who has made such a journey will not have a single chance to report his success to Earth.

If the BH rotates, then this greatly complicates its structure. A rotating black hole without charge is called a Kerr hole after

Roy Patrick Kerr, who discovered this general relativity solution. You can also fall into it and, theoretically, you can fly out in an unknown direction, avoiding the singularity. At this point, the reader can become optimistic and decide that jumping into the black hole is a particularly dangerous sport for extreme travellers, even if without the opportunity to return and brag. Alas, this is not the case.

It's time to recall the second part of the controversial answer. In theory, rotation or charge completely changes the internal structure of a BH, but almost all travellers who get there will be torn to elementary particles. The fact is that for a BH with a mass large enough to be formed by the collapse of a star or something even more massive, the inner horizon should be located so close to the singularity that a body will be torn apart before it reaches it. So our cunning plan to travel through the Universe is doomed to failure.

But thought experiments, in which no one can really get hurt, ask a lot of unanswered questions. Unfortunately, in order for someone to simply agree to listen to your answer and take it seriously, you must first study GR and its solutions describing BH for a long time. Without that, it will be empty words.

7.3.8. *Let's summarize what we know about BH*

So, BHs can exist and most likely exist in the Universe. We can observe objects around them or the radiation from their accretion disc, and so learn about their presence and estimate a number of parameters of the BH including its mass.

Sometimes the black hole engulfs large objects, including nearby stars. Black holes can merge into one (but cannot divide). All these processes are accompanied by an especially powerful radiation of gravitational waves. As a matter of fact, many gravitational waves recorded by physicists were born in such cosmic catastrophes.

I hope that my explanations have helped you to find out the basic details of the modern concept of BH. First of all, this will help you weed out the flow of all sorts of nonsense about black holes which comes from the media, films, and fiction. After all, even without them, these are truly wonderful objects. Let's add a little romance and they can be safely called a one-way portal to other worlds and mystery star gates leading into the unknown. And although at best, only the elementary particles that were once parts of our body will reach the notorious "other worlds", such little things are not important for true adventurers.

7.3.9. *Naked singularities*

Well, for those who truly love extreme sports, I have even more mysterious objects. These are the above-mentioned naked singularities (NS) with which I have already frightened naughty readers. It's time to get to know them better. In a nutshell, they have a singularity, but no horizon. Therefore, the singularity is not hidden under the horizon, but impudently exposed for all to see. There is a hypothesis that naked singularities cannot form upon collapse. Its author, Roger Penrose, called it the Principle of Cosmic Censorship.

Some readers have probably already come up with some ideas about the properties of naked singularities. Since there is no horizon, if you fall on them, you do not have to reach them necessarily. In theory, at any moment you can break, turn, and fly away from the singularity. In addition, matter, light, or other radiation can be emitted from the naked singularity. In this way, it could influence what is happening in our Universe, and affect it completely unpredictably. To prevent exactly this possibility, Penrose proposed his hypothetical Principle of Cosmic Censorship.

Could giant tentacles or a flock of devils with pitchforks crawl out of there? Who knows, who knows ... But you should not be afraid, because near the naked singularity the tidal forces become infinite; that's why it is called a singularity. Therefore, they will immediately tear apart the tentacles, devils, and pitchforks, like everything else that tries to get to us.

So NS are just like a two-way portal to another world, but you can't go through it alive or intact in any direction. What does such a singularity look like? The choice of options is much richer than that of BH. By the way, objects such as BHs with too much electric charge or rotating too fast also turn out to be a naked singularity.

In general relativity, there are a lot of solutions describing naked singularities, but it is not clear whether they have any physical meaning and whether the objects they describe can exist in nature. Taking into account the ability of inflation to remove all kinds of exoticism, "sweeping" it beyond the cosmological horizon, the question can be concretized: Can naked singularities be forming during a collapse? Or is Penrose right with his Principle of Cosmic Censorship?

An accretion disc may exist around such a singularity. Stars can orbit around NS. So how can you distinguish it from BH? It is possible that some of the massive compact objects in space are naked singularities. However, today, no one can know for sure what we are dealing with.

One of the hypothetical possibilities to enforce the Principle of Space Censorship is as follows. For the formation of a naked singularity, a very asymmetric collapse, or a collapse of rotating or charged matter is needed, because the usual, almost symmetric collapse of non-rotating uncharged matter will lead to the formation of a BH. If some radiation or other factor, classical or quantum, can slow down the collapse so much that they have time to go through the processes of isotropization, loss of charge or angular momentum, then in this case the collapse will be able to create only a BH.

From objects about which we know a lot or almost everything, such as our planet, rainbow, or lightning, we moved on to black holes, and then to naked singularities that may or may not exist somewhere in the nooks and crannies of our huge and still mysterious world.

7.4. Quantum Phenomena in the Macrocosm

The quantum properties of particles naturally manifest themselves in the microcosm. But this is not only happening there. Outside it, they also affect substances and phenomena. And these are not only neutron stars. In the section devoted to objects and phenomena associated simultaneously with different scales of the world, it is worth talking about this, or at least mentioning it.

7.4.1. *Bose condensation*

At school, we are told about the properties of an ideal gas placed in a vessel, and about the relationship between its temperature, pressure, and volume. But let us remember that the particles of any real gas have features that distinguish them from ideal molecules of such a gas that interact according to the laws of classical mechanics. For example, they can be bosons or fermions and their energy distribution is described by other formulas than for abstract ideal gas molecules. For the sake of brevity, I refer to the latter as classical molecules.

I'll start with a gas made of bosons. If it is cooled to very low temperatures, then some of the bosons will be in the identical quantum state with minimal energy. Bosons love to do it and can afford this behaviour. The quantum state I mentioned is called the Bose–Einstein condensate, or Bose condensate for short. Its energy is often taken as the zero one.

The rest of the bosons will have nonzero energies. Their energy distribution will be different from that of classical molecules, but this is less important than the existence of two different phases: condensate and gas. A little further I will tell you how this manifests itself in the phenomenon of superfluidity of helium-4, whose atoms are bosons.

Condensation exists only when the gas temperature is less than a certain critical one. There is a second-order phase transition when it is passing through the critical temperature. At temperatures above the phase transition point, the existence of a Bose condensate becomes energetically disadvantageous. There is only a gas of bosons, the pressure of which is slightly lower than that of the gas of classical molecules. This can be explained at the elementary level by the fact that the bosons at the edge of the vessel are not averse to be in the same state as the bosons in its centre, which provides something like their mutual attraction. This slightly reduces the pressure on the vessel walls.

But if the temperature is only slightly higher than the phase transition temperature, then droplets of Bose condensate appear in the gas. They should be considered as fluctuations that exist temporarily. They affect the property of the gas in the same way as the properties of non-existent virtual particles affect the property of the physical vacuum, which I talked about in Section 5.3. So the existence of fluctuations manifests itself in the results of experiments carried out at such temperatures.

The state of the condensate was experimentally obtained and demonstrated for a gas from rubidium atoms in 1995. It was 70 years after the papers of Bose and Einstein. Only then did the technologies necessary for cooling the atoms to such low temperatures appear. Eric Allin Cornell and Carl Wieman of the University of Colorado at Boulder, USA, were able to combine 2,000 atoms into a Bose–Einstein condensate that was large enough to be seen through a microscope. Cornell, Wieman, and the German Wolfgang Ketterle received the Nobel Prize in 2001 "for the achievement of Bose–Einstein condensation in dilute gases of alkali atoms, and for early fundamental studies of the properties of the condensates".

As you can imagine, physicists did not confine themselves to hanging the head of yet another exotic prey on the wall, but, inspired by the first trophies, continued their research. By 2012, their collection already contained Bose condensates from isotopes of atoms ^{7}Li, ^{23}Na, ^{39}K, ^{41}K, ^{85}Rb, ^{87}Rb, ^{133}Cs, ^{52}Cr, ^{40}Ca, ^{84}Sr, ^{86}Sr, ^{88}Sr, ^{174}Yb, ^{164}Dy, and ^{168}Er. But in the laboratories there was a factor that prevented further success. This is gravity, which naturally affected the sample. Its influence limited the lifetime of the resulting condensate. With more time to experiment, scientists could cool the condensate even further by causing gas to expand and do work. But if the gravitational field interferes with this, then you can get rid of it by sending the device into space, to the International Space Station (ISS) with its microgravity.

And so they did. The ground prototype of the experimental instrument for the production of the Bose condensate started working in 2014, in the summer of 2018 the apparatus the size of a dishwasher was sent to the ISS, and in 2020 it received the first product in orbit. A bubble of condensate about 30 microns in size with a temperature of 200 trillionth of a degree lasted more than a second.

It is assumed that in the future the temperature will reduce by a factor of 10, and the lifetime will increase fivefold.

In 2010, physicists from the University of Bonn prepared a Bose condensate from light, more precisely, from photons rushing between two parallel mirrors. With your permission, I limit myself to only mentioning this fact, without giving details.

Other properties of the Bose–Einstein condensate are also interesting. In 1998, the Dane Lene Vestergaard Hau was able to slow down laser light to 60 kilometres per hour by passing it through a cigar-shaped sample of such a condensate of sodium atoms. In later experiments, Howe's group succeeded in completely stopping the light in the Bose condensate by turning off the laser as the light passed through the sample. I will not even try to explain to you the physics of this process.

7.4.2. *Superfluidity*

The existence of the Bose condensate is closely related to the phenomenon of superfluidity discovered by Pyotr Kapitsa and John Frank Allen in 1937 for helium cooled below 2.17 K. Half a century later, in 1978, Kapitsa received half of the Nobel Prize in Physics "for his basic inventions and discoveries in the area of low-temperature physics", and the second half went to Penzias and Wilson. This wording included the discovery of a superfluid state, into which helium is converted by a second-order phase transition.

Superfluidity exists, in particular, in certain isotopes of helium, rubidium and lithium, cooled to almost absolute zero. Superfluids have no intrinsic viscosity. Placed in a test tube, they begin to crawl up the sides to the outside and then down. Liquid helium leaks easily as it can slip through even microscopic holes.

The phenomenon is associated with the fact that atoms of helium or another element are in the ground energy state due to Bose condensation. Let them move together with the superfluid liquid, meeting some obstacles on their way. Since the energy of states is discrete, each atom can receive not any portion of energy, but only the one that is not less than the energy gap between neighbouring energy levels.

But at low temperatures, the collision energy in the course of the flow may turn out to be less than this value, as a result of which the energy dissipation will simply not occur. In other words, the fluid will flow without friction.

7.4.3. *Degenerate Fermi gas*

And how does the gas consisting of fermions, which is also called Fermi gas, behave? At temperatures close to absolute zero, they occupy all possible levels with the lowest energy. Due to the Pauli principle, they cannot accumulate together at the lowest level, this is the privilege of bosons. Fermions, like electrons of the atomic shell, sequentially fill the levels available to them with energies from the minimum to a certain maximum, called the Fermi energy. At absolute zero temperature, all lower levels are completely filled and all higher energy levels are empty.

Physicists call this a degenerate Fermi gas. Its pressure is greater than that of an ideal gas and it does not vanish even at a temperature equal to absolute zero. At a little higher temperature, some fermions move from levels slightly below the Fermi energy to levels a bit above it, but the energy distribution of particles is very different from the distribution of a classical gas.

So, the properties of a degenerate Fermi gas are different from the properties of a classical gas at low temperatures. But the word *low* here can be misleading. I mean temperatures that are significantly lower than the Fermi energy (remember that temperature is energy measured in degrees). And this energy can be enormous.

In stars, matter is in a state of plasma with a high degree of ionization. Practically, there is a gas of electrons and a gas of protons, deuterons, helium nuclei, and other elements. Electrons are fermions and they form a degenerate Fermi gas even at the enormous temperatures that are characteristic of stellar interiors. And in a neutron star we are dealing with a degenerate gas of neutrons.

7.4.4. *White dwarfs*

Taking into account the properties of a degenerate electron gas is very important for understanding the properties of some types of stars. For example, white dwarfs. These stars appear after the explosions of new stars and supernovae. The brightest star in the night sky, Sirius, is actually a double star. Its companion is a white dwarf invisible to the naked eye.

The mass of Sirius B (the components of the stars are named in capital Latin letters) is approximately half the mass of Sirius A and is close to the mass of the Sun. But its radius is close to the size of the Earth. Its surface area is significantly less than that of the companion, which provides a lower luminosity of the white dwarf. It is the closest star of this type to us, only 8.6 light-years away.

No thermonuclear reactions take place in a white dwarf. It slowly cools down, lowering its initially high temperature due to radiation losses. It can support itself against gravitational collapse only by electron degeneracy pressure. As a result, it is extremely dense. The average density of matter in white dwarfs is almost a million times that of ordinary stars. In our Galaxy, from 3% to 10% of stars are white dwarfs.

As you can see, the very important quantity from the microworld — the spin of a particle — determines whether it refers to bosons or fermions. This circumstance affects the property of matter, consisting of these particles. So quantum effects are showing up even in stars. This once again confirms that the microcosm and the macrocosm are connected and nature is arranged like a snake biting its own tail.

Incidentally, the electron gas exists inside metals. And without taking into account its quantum properties, it is impossible to explain many properties of solids, in particular metals and semiconductors.

7.4.5. *Superconductivity and other consequences of the formation of fermionic pairs*

Surprisingly, the phenomenon of superfluidity is also observed in helium-3, whose atoms are fermions. Its phase transition to a superfluid state at temperatures below 2.6 mK and at a pressure of 34 atm was discovered in 1972. The Americans David Morris Lee, Douglas Dean Osheroff, and Robert Coleman Richardson received the 1996 Nobel Prize in Physics "for their discovery of superfluidity in helium-3". And this superfluidity of fermions had to be explained somehow. However, the answer did not take long.

In the same 1972, the American physicists John Bardeen, Leon Neil Cooper, and John Robert Schrieffer received the Nobel Prize in Physics for their theory of superconductivity, known by the authors' initials as the BCS theory. By the way, Bardeen is the only person to have received the Nobel Prize in Physics twice. In 1956 he received it for the first time for the discovery of the transistor.

Superconductivity is sometimes called electron superfluidity. Indeed, both effects, superconductivity and superfluidity of helium-3, are caused by a similar cause. It is the existence of pairs of fermions that together form something like a composite boson. They are called Cooper pairs.

What is the mechanism behind this phenomenon, which physicists call pairing? Fermions begin to interact with each other, sometimes using very complex mechanisms. This interaction often leads to the fact that two fermions can go into a state of lower energy by forming Cooper pairs.

Such composite bosons in the form of a pair of fermions can be accumulated at the lower energy level for pairs, so the pairing of fermions can lower their energy in comparison with the energy of a degenerate fermion gas. The reason for the superfluidity of helium-3 is Bose condensation of Cooper pairs of atoms of this isotope.

The resulting special state of matter is called a fermionic condensate. It was demonstrated in 2003 by Deborah Shiu-lan Jin and her colleagues when 500,000 potassium atoms were cooled to a temperature of $5 \cdot 10^{-8}$ K in an alternating magnetic field.

What are Cooper pairs? These are not two particles spinning side by side around a common centre. They have opposite spins and move with the same speed in opposite directions. There may be other particles in between that are not in a pair.

An analogy can be given: a criminal is moving in the crowd, and the detective is watching him at some distance. They do not seem to stand out in the crowd, do not get close, there are random passers-by between them, but the movement of the detective is determined by the movements of the criminal.

Keep in mind that this example is only meant to illustrate how the motion of two particles can be correlated. It does not describe the behaviour of real electrons in a Cooper pair. In my example, the detective is tracking the criminal and this makes their roles very different. Closer to reality, there would be an example with two detectives spying on each other in the crowd.

7.5. Problems. Scale: Everything

We are completing our journey along the closed circle of scales. I think that during it, you had a lot of questions, some of which I was able to answer. But obviously not all questions were answered. For some of the answers, you have to go to the Internet or read books, and for other questions, no one will give you a generally accepted answer. And there are quite a lot of them, which show that science does not know everything. But this demonstrates that it continues to do its job. Science discovers and explains new phenomena, creates theories, and applies all of the above in the interest of humankind. I mean not only technical and utilitarian purposes, but also an understanding of the laws of the world around us.

So, let us list the main problems of modern physics. There are many of them, because with the growth of what humankind knows, the number of very important questions is increasing. I mean the ones which scientists have already realized deserve to be asked, but which do not yet have robust answers. I took the liberty of highlighting some of the most important issues from this whole vast mass. I'll simply list them:

- What is dark matter, what does it consist of, is it capable of weak interaction?
- What is dark energy? Is it reduced to a cosmological constant? What determines its density?
- Is it possible to create a theory of gravity that is compatible with quantum concepts, and if so, what is it?
- Do so-called naked singularities exist in outer space?

- Are there situations in which the known physical laws stop working, for example, near singularities?
- Do space and time quantize?
- Are there additional dimensions?
- What caused cosmological inflation?
- What causes the asymmetry between matter and antimatter?
- Are there other types of Higgs boson besides the one already discovered?
- Does the Standard Model of elementary particles need to be improved, and how exactly?
- Does a proton decay?
- Are there magnetic charges called monopoles?
- Are there sterile neutrinos?
- What determines the masses of fundamental particles?
- Why does strong interaction have three colours, why do quarks and leptons have three generations, and is this somehow related to the dimensions of space?
- Are quarks and leptons made of something even more fundamental?

Naturally, I did not include some particular questions such as the nature of ball lightning or problems that are mainly of applied importance, for example, obtaining high-temperature superconductivity at room temperature or cheap energy from fusion reactors.

More detailed lists can be found on the Internet, for example, in the "List of unsolved problems in physics" on Wikipedia. Some of the issues mentioned there were not even mentioned during our trip or in preparation for it. Well, one cannot but agree with the proverb "No living man all things can".

8 BONUS MILES

Many reputable airlines provide their customers with so-called bonus miles. Having made a particularly long trip, you can accumulate enough of them to fly somewhere for free. So, after our journey down the scale ladder, we have bonus miles accumulated for several small additional flights. Their destinations are mostly linked with our main route, but not so much as to be included directly in it. Therefore, you can read or skip any of the sections in this part. Moreover, some of them will focus more on history or philosophy, even if they are related to physics and science in general.

8.1. What Is Science?

It is time to take a closer look not only at the world around us, but also at what studying it is. That is, to science — more precisely, to the natural sciences: physics, chemistry, biology, astronomy, geology, and other disciplines, besides the adjacent mathematics, which does not belong to the natural sciences for the reason described further in the text.

What is science? What is the scientific method it uses? How is science different from non-science, including philosophy? Sometimes it's not very easy. If you read encyclopaedias, you can find out that scientists own a kind of magical tool called the scientific method of cognition. Unfortunately, its description is sometimes so vague that it is impossible to understand what it is. The meaning is hidden behind a cloud of words. As a result, the scientific method (and this is the main achievement of humankind in science), so obvious to scientists, is usually completely incomprehensible and unknown to the majority of the general public. More precisely, the term itself is well known, but not what is hidden behind it. I will try to describe the scientific method as a procedure and give many examples from physics that illustrate the nuances of its application. Consider this chapter as a simplified guide to applying science.

If someone draws you a triangle, the corners of which are labelled "God", "Man", and "Soul", and says that this is a diagram of the Universe, then it is very easy to understand that this is not science at all. What if it says "Matter", "Energy", and "Information"? Or "Particles", "Fields", and "Interactions"? "Bosons", "Fermions", and "Condensate"? Some may be tempted to classify this as a science. In fact, a diagram of this kind with any inscriptions cannot relate to science unless it is accompanied by a lot of additional details. More precisely, the scheme can play a very auxiliary role at most in explaining scientific ideas. I will not teach you how to recognize pseudoscience, but I will try to tell you what features and habits real scientific ideas have.

I will begin, as is customary in science, with the definition of certain concepts. The point is that it is very important that different people mean the same thing by using the same terms. Back in the early 17th century, Sir Francis Bacon, Lord Keeper of the Seal, Lord Chancellor, Baron Verulam and Viscount of St Alban, and part-time philosopher, wrote about the problem, which he called the Idols of the market (*Idola fori*). This is one of the "idols" or typical mistakes that stand in the way of knowledge. "For men associate through conversation, but words are applied according to the capacity of ordinary people. Therefore shoddy and inept application of words lays siege to the intellect in wondrous ways," he wrote in his *Novum Organum*. Incorrect use of terms is a serious fault of the scientist. And if he or she has to introduce a new term or use the old one in a slightly different or expanded sense, this must necessarily be stipulated and, if necessary, explained.

Let's give a few preliminary definitions, albeit not very strict. Philosophy creates a conceptual picture of the world and operates with concepts. Science creates a factual picture of the world and operates with knowledge. Mathematics is a discipline that creates an abstract model of the world and operates with models. For comparison, art documents and translates emotional impressions of the world.

"Idols of the theatre" (*Idola theatri*) also frightened Bacon. They are false ideas about the structure of reality that a person learns from other people:

> Lastly, there are the Idols which have misguided into men's souls from the dogmas of the philosophers and misguided laws of demonstration as well; I call these Idols of the Theatre, for in my eyes the philosophies received and discovered are so many stories made up and acted out stories which have created sham worlds worth of the stage. [. . .] Neither again do I mean this only of entire systems, but also of many principles and axioms in science, which by tradition, credulity, and negligence have come to be received.

Idols of the theatre are related to science. Popular science literature is designed to counteract them. However, sadly, some of it, on the contrary, serves these idols, promoting outdated, dubious, and even completely unscientific ideas.

8.1.1. *Scientific method*

I will briefly describe the scientific method as I understand it. How does science work? It probably starts with a question that can be specific, such as "Why is the sky blue?", "Why is the plane flying and not flapping its wings?", or "What happens if I put that green powder in this bottle with a bubbling transparent liquid?", or it can be general, such as "How to cure cancer?" or "What is everything made of?"

To answer all these questions, ideas are needed that are somehow related to science or just the world around us. An idea may be very vague at first, but when it ceases to be only an idea for internal use and is communicated to others, it must already satisfy a number of requirements.

To begin with, it must be formulated, at least in rough terms. Further, it must pass the selection for elementary compliance with our knowledge of nature and for compliance with the scientific criterion, which will be discussed a little further. Once an idea has overcome these initial barriers, it can be considered a scientific hypothesis.

Then the scientific hypothesis begins to fight for the right to become a theory. The theory is at a higher level on the conventional scale of authority and importance. At the same time, no theory can be considered recognized forever. But it can easily be relegated to outdated theories, such as the theory of *aether* or phlogiston. The fact is that in science, by and large, there is only one criterion for correctness. This is a coincidence with practice, observations, and experiments.

Thousands, even millions, of experiments are performed to test theories, and each of them can end the career of a particular theory. A real theory must satisfy all existing observations and experiments, as well as consequences from known facts and other theories. However, if there are contradictions, they can be resolved with the help of additional theories or hypotheses.

For example, Euclid and I hypothesized that light rays are straight. Everything is fine, but someone says that he saw an inverted ship in the sky. We can doubt this fact, introduce an additional hypothesis that ships sometimes fly in the sky, and upside down, or about mass hallucinations among sailors caused by excessive use of rum and opium. We can improve the theory by assuming that the rays are almost straight, but sometimes they bend so much that they give rise to mirages, which we discussed in Section 4.4.

Moreover, each theory usually has a predecessor, theory, or hypothesis. And a recognized theory must be better than its predecessor — either more general, explaining a greater number of facts, or more accurate, that is, its predictions must correspond to some experiment clearly better than that of the previous theory, or more general and more accurate at the same time.

If both theories are not very consistent with the experiment, but the new one deviates less, this, of course, is an argument in its favour, but it is clear that the theory is not final and must be improved or replaced by a completely different theory.

Note that compliance with facts and experiments is the only criterion for correctness. Popular literature likes to quote Niels Bohr's statement: "We are all agreed that your theory is crazy. The question that divides us is whether it is crazy enough to have a chance of being correct." But "craziness" has never been regarded as a criterion for the correctness of the theory. Moreover, these words were said by Bohr to Wolfgang Pauli regarding the idea of introducing an electron spin, which turned out to be correct. And certainly, Niels Bohr perfectly understood and accepted the traditional scientific criterion of correctness in physics.

But some additional considerations are still taken into account, especially the beauty of the theory. A beautiful physical theory receives a lot of correct, that is, consistent with the experiment, conclusions from simple assumptions, and these presuppositions are usually few or even one. In theory, there are no fitting or adjustable parameters or very few of them.[a]

It is easy to refute it in experiments, but experiments only confirm it. It is often mathematically simple or elegant. Physicists get purely aesthetic pleasure from working with such theories. These include, for example, the general theory of relativity or the idea of inflationary inflation of the Universe and, to some extent, the Standard Model.

Note that beauty alone is not enough. Many rejected or incorrect theories are also beautiful in their own way. And the recognized beautiful theory, the same general relativity, according to physicists, needs to be improved, by the creation of a quantum theory of gravity.

Another additional criterion is purely utilitarian. It is the simplicity of the calculation. In geocentric models of the Solar System,

---

[a]These are quantities, unknown in advance, whose values are determined from the condition of better agreement between theoretical predictions and experimental results. An example is the value of Planck's constant obtained by Max Planck himself. If we use fixed values obtained in other experiments, then they are not adjustable parameters.

the Sun and planets revolved around the Earth along very complex trajectories, described by superimposing many circular motions along the so-called epicycles. At the same time, the movement along the epicycles occurred at a variable speed. All this required complex calculations, performed manually, since in those days, there was nothing like a computer or even an adding machine.

When Nikolai Copernicus (Mikołaj Kopernik) proposed the heliocentric model, it also required epicycles, but there were significantly fewer of them than for the geocentric model, with the same accuracy of correspondence to astronomical observations. If the calculations were performed by a computer, it would not be so important, but they were performed by the astronomer himself and he perfectly understood the difference in calculation times for these two models.

And after the discovery by Johannes Kepler of the laws of planetary motion, which were also beautiful and did without any epicycles, the coincidence of theory and observations became obvious. Astronomers threw away the old geocentric model not yet having any evidence that the Earth revolves around the Sun, and not vice versa.

The first direct proof was obtained a century later by the Englishman James Bradley, who discovered the so-called aberration of light caused by the movement of the Earth in orbit around the Sun and the finiteness of the speed of light.

And what is practically not taken into account as a supplement to the conformity of the experiments? These are the opinions of individual scientists, including prominent ones.[b] In ancient times, including the Middle Ages, in some not quite real sciences, such an appeal to authority (*Argumentum ad verecundiam*) was considered irrefutable proof. But you rarely find quotes in physics articles, except in popular ones. Even collective opinions, expressed in

---

[b]Let me remind you that I am talking about discussions within the scientific community. When it comes to recommendations for politicians, economists, decision makers, or society in general, the influence of the authority of a person or organization is very significant.

the form of an expert opinion or a vote, are not very important. The decision of the Académie des sciences (French Academy of Sciences) in 1775 to consider no projects of the perpetual motion machine because of the impossibility of its creation was an exception. However, I would like to emphasize that this decision, which played an important role in the development of science, was still organizational.

What changes scientists' views about the nature of phenomena, generally accepted theories, and paradigms? Mostly new facts, including those obtained through observations and experiments. Naturally, there are scientists who are not ready to change their ideas. They fight for old views to the best of their ability. But their opponents, often young, at this time develop new theories, conduct experiments to test and tune them, and write books and articles elaborating and promoting new scientific ideas. However, more than once I have come across articles in which famous scientists wrote something like: "All my life I fought against the idea of BH (DM, etc.) existence, but now, under the pressure of new results, I cannot but admit that I was wrong."

Over time, the bearers of old views leave science in a natural manner, and their opponents come to the forefront. The only way to prevent this process is to either dramatically improve an old theory or find an experiment that is better described by it than by the competing ones, but this is very difficult.

But in totalitarian countries, with the support of the authorities, you can simply ban the opposite point of view in science, education, and even in the media and literature. Examples are the persecution of general relativity, quantum mechanics, genetics or cybernetics in the USSR, and the so-called Aryan physics (*Arische Physik*) in the Third Reich. As a rule, the country experienced a huge degradation in this field of science and disciplines related to it, which required decades of effort to overcome, and even then success was not always possible.

Theories and paradigms in physics and other natural sciences can never feel safe. None of them is recognized as forever correct, as proven theorems in mathematics are. They are always in danger of not coinciding with the results of more and more accurate experiments. They can be easily refuted, but they can't be proven. Therefore, when critics of science talk about its rigidity, nothing can be more wrong. On the contrary, scientists, especially theorists, adore revolutions, a glorious time when they make great articles, win Nobel Prizes, and write their name in the history of science. But their fantasies, calling for revolution, are still kept on a leash. And this leash is in the hands of experimental physicists.

So maybe they are the real kings of physics? Not really. And the point is not even that they constantly have to check the theories and predictions that theoretical physicists give out. Often they choose what to investigate, rather than just testing theories. Sometimes they manage to conduct a wide variety of studies, sometimes rather strange ones. We are not talking about honestly mistaken people or professional fraudsters trying to find a buyer for another perpetual motion machine, low-temperature thermonuclear fusion, inertia-free propulsion, a quantum engine, and other miraculous developments. Even real scientists from time to time conduct quite unexpected experiments and sometimes, even if very rarely, these random searches bring something interesting.

Experimental physicists also have specific limitations: in their experiments, they must fulfil several conditions necessary for the scientific community to recognize the results obtained. Among them are the obvious requirements of honesty, openness of all the details of the installation, correct processing of results, and taking into account the influence of all factors that can affect the result.

Experiments must be repeatable, that is, the results should be stable. Naturally, if we are talking about some experiments, where a tendency manifests itself against the background of huge random factors, then the results should be repeated not in every single experiment, but statistically, they should be repeated.

It is even better if the repetition of the results is checked by scientists at different experimental facilities. If the installation is unique, then they try to involve external specialists to monitor the progress of the experiment and control the processing. Sometimes sensational results are the result of a mistake, often an elementary one.

So, the report of superluminal neutrinos turned out to be the result of a poorly plugged plug-in the installation.[c] Therefore, after the first message about the discovery of something especially unexpected, physicists are in no hurry to recognize it, but wait for independent confirmation. At the same time, sometimes errors in the experiment or unaccounted factors are discovered decades after the experiment itself.

Perhaps you have already seen for yourself that everything written here is poorly applicable to mathematics. Despite the fact that all physics speaks a mathematical language, and the combination of physics and mathematics has long been steadily widespread, these are very different sciences. And there is a regular debate about whether mathematics can be considered a science. Physics is the top of the pyramid of natural sciences. Mathematics certainly does not belong to these sciences, although this does not mean that it should be called an unnatural science.

Indeed, there is no experiment in mathematics and no one checks the theorem for compliance with it. Something, proved once, remains true in mathematics forever, except that the formulation is somewhat refined due to the appearance of new terms and concepts.

So what is math? Some consider it a special branch of science, but perhaps we should raise it to a meta-science that is on the same level as philosophy or the natural sciences in general. However, the wording may vary depending on the personal preferences of their authors. But no matter how you look at it, mathematics is something special. It is not one of the main characters in this book, but they can't do without this character behind the scenes.

I want to make it clear that experiments were once used in mathematics. First and foremost, this applies to the emerging theory

---

[c]Only an anecdote remained from this story, maybe not very clear to the uninitiated: "... and the bartender says: 'we do not serve a neutrino moving faster than the speed of light!' A neutrino enters the bar ...".

of probability and mathematical statistics. To test hypotheses and experimentally obtain the form of various probability distributions, mathematicians were not afraid to use random physical processes. For example, in 1894, Walter Frank Raphael Weldon, the creator of biometrics based on the use of statistical methods, threw a set of 12 dice 26,306 times and recorded what exactly fell. These results were used by Karl Pearson to develop his method that scientists still use today, known as the chi-square (χ^2) method. Francis Galton constructed a machine in which beans fell down and collided with a number of pins, showing the limiting form of the binomial distribution to demonstrate the so-called limit theorem.

So let's leave the axioms to mathematicians; in physics, chemistry, biology, and geology, there are no axioms and cannot be. But there are paradigms that change from time to time in the course of scientific revolutions.

8.1.2. *Science as a procedure*

I have sketched for you the basics of the scientific method of knowing. This is not some trick known only to scientists who have passed it on in secret to their very best students. These are scientifically validated criteria for correctness that make up half of the success. The scientific procedure depicted in Fig. 8.1 is often perceived as a series of iterations to get closer to the truth. This is not entirely true, because scientists have to run around in this hamster wheel forever, never being sure that they know the final answer or are close to it. Scientific revolutions, paradigm shifts, and the emergence of previously unknown objects of study are all evidence of this.

If the above seems too abstract to you, then I will explain it again with examples from detective stories, as well as law and grammar. Maybe it will help one of the readers. In many detective stories or blockbusters, heroes are faced with riddles that need to be solved, or with the need to penetrate the secret plans of the enemy. The main character or the smartest guy in the company of heroes suddenly solves them, and everyone immediately takes action based on this guess and does not doubt its correctness.

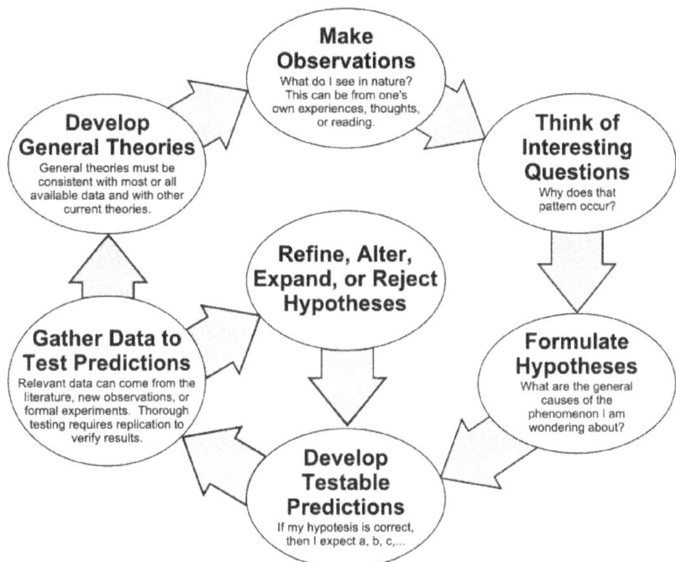

Figure 8.1.   Scheme of the scientific process. Extracted from Wikimedia Commons (CC BY-SA 4.0).

If the secondary characters start asking for details, then the author of the idea simply tells what happened or what the enemy is up to, without burdening themselves with any evidence. At this point, they are very different from Sherlock Holmes, who explained in detail to Dr Watson the reasons and the evidence that led him to this particular guess. However, a faith in guesses is possible for a police officer when searching for a criminal, just as it is possible for a scientist at the stage of thinking over a hypothesis.

The blockbuster decision-making method "it fits, therefore it's true" can be likened to the traditions of a time when science had not yet become what it is now. During the period when it was still part of natural philosophy, conjecture replaced experiments or proofs. Ideas were not proved, but were lambasted or praised.

But this is far from the final stage. It is in the movies that evidence comes in the course of the action, or the main enemy suddenly begins to utter long pathos or penitential speeches with a story about his crimes and plans. In real life, there is usually an investigation and a trial. And they need proof of guilt, taking into account a reasonable doubt. The prosecutor cannot just say, "I swear by my mother!" and win the case. He needs to provide evidence, refute doubts and alternative interpretations, false alibis, and even the testimony of dishonest witnesses. Only then will the accusation turn into a verdict. In the case of science, a hypothesis becomes a theory.

It is curious that from 1587 to 1983, the so-called Promoter of the Faith (*promotor fidei*) took part in the process of beatification and canonization in the Catholic Church. He had the unofficial title Devil's advocate (*advocatus diaboli*).

Its function was to collect all possible arguments that could prevent the canonization or beatification of a righteous person, which could only take place if the Promoter of the Faith did not find arguments of sufficient importance to cancel the procedure. The similarity of this procedure to that accepted in science has long been noticed. Physicists are sometimes even called "devil's advocates".

Are there any significant differences between the scientific method and the judicial process? Yes, the principle of "once in jeopardy", which is used in the legal system of some countries, does not work in science. It consists in the fact that a person who has been found not guilty of a criminal offence by a court cannot be re-charged in the same case. In science, any hypothesis, theory, or even paradigm is constantly under threat and must constantly confirm its "innocence".

How do scientific rules differ from, say, grammar rules? They do not and cannot have exceptions, which are so common in some languages. If in the course of the research there is evidence that in some situations the recognized rule does not work, then this first attracts increased attention of scientists, the number of studies of the alleged exception increases. Further, either the rule changes, so that the former exception fits into it (the rule "Lead cannot turn into gold" is supplemented with "... in chemical reactions, but some chemical elements are converted into others in radioactive decays"), or it turns out that there is simply no exception and the phenomenon gets its explanation.

In rare cases, the situation remains in limbo for a long time. This was the case with the idea of an atomic nucleus and electrons orbiting around it, which followed from experiments, but contradicted the idea of emitting electromagnetic radiation, which would lead to their fall on the nucleus. So, in science, exceptions to the rules are impossible, their existence means that we just haven't figured out the rules completely yet. As an anecdote, I mention the answer received by the author during an exam. The student said that he read on the Internet that the skin effect[d] is observed for any alternating current, except for that received by a certain electric generator. Note that after the question of how the currents differ from different generators, the student himself realized that he was repeating someone's stupidity.

It can be added that the scientific environment has preserved something like the atmosphere of a medieval guild. There are written and unwritten laws that must be observed. A scientist can make mistakes and even persist in doing so, but a mistake can only be accidental. If it becomes known about deliberate falsification of results, forgery, or manipulation of original data, this completely destroys the reputation of the culprit.

Summing up, let's focus on which question science answers. Surprisingly, this is not an inquisitive child's typical question "Why?" Science could be a bit boring and its adepts can spend hours explaining why the water is wet and the dog doesn't talk. But if we persistently repeat the question "Why?" after each answer, we will quickly make even the most boring of physicists surrender. Let's imagine a dialogue between a boring physicist and a scientific troll.

"Why did the apple fall on Newton's head?" asks the scientific troll.
"Because all bodies fall to the ground," answers the boring physicist.
"Why?"

---

[d]A decrease in the amplitude of the alternating current with distance from the surface of the wire, akin to the attenuation of the alternating electromagnetic field inside the conductor, described in Section 4.2.

"Under the influence of their weight, that is, a downward force."
"Why?"
"The force of gravitational attraction acts on each body. We call the force of attraction of an apple to the Earth its weight."
"Why does the force of gravitational attraction work?"
"Einstein believes it's all about the curvature of space-time."
"Why does space-time curve?"
"Nobody knows; it follows from observations and experiments."

No matter how the series of questions begins, it will quickly end with this answer. The point is that science studies nature, not the reasons why nature is the way it is. It is often claimed that science answers the question "How?" How does an apple fall? How does its speed change? How does the weight strength depend on the mass of the apple? How does the acceleration of gravity change with altitude? Finally, how exactly is space-time curved?

Or it answers the question "What?" What determines the apparent colour of the sky? What keeps us from falling through the floor? What is formed during this reaction? What's in the bowels of Jupiter?

However, we must admit that the answer to the question "Why?" is important. A difference between real theory and empirical laws is hidden in the answer. The theory finds the reason in some more fundamental theory, the rules of thumb are not explained in any way, and they just describe the experimental data well. For centuries, astronomy has been concerned with describing the apparent movement of stars, planets, and the Moon across the sky and predicting eclipses and other important events. Even in antiquity, complex methods were developed that made it possible to calculate all this quite accurately. These were empirical rules. Ancient astronomers were not very interested in the reasons why they worked.

Johannes Kepler was able to significantly clarify and at the same time simplify these methods, but the laws he discovered remained empirical. However, it could not be otherwise, because the reasons for the observed motion of celestial bodies were studied by another science — physics. And only after Newton discovered the law of universal gravitation, Kepler's laws received their explanation. It became possible to answer the question of why they are carried out, referring to a more fundamental theory.

But there was no answer to the question of why bodies are gravitationally attracted according to the law of inverse squares of distances. This means that we have reached the most fundamental theory at that time moving along the ladder of questions. From time to time, new, even more fundamental theories arise in science, which make it possible to answer the question "Why?" about what was previously considered the most fundamental. These imply breakthroughs in science, and sometimes, even scientific revolutions.

8.2. The First Steps of Science

At an early stage in the development of philosophy, it was based on guesswork, fantasy, logic, and a minimum of observation of life around us. The question of the criterion of truth was not raised; it was successfully replaced by the question of how to win an argument or convince the audience. In fact, the idea that there is an objective truth is not so obvious to philosophy, in which idealism and materialism have been fighting for several millennia and no one sees anything unusual in this. Science, on the contrary, proceeds from the fact that the truth exists and it must be established, at least approximately.

Science originated in the depths of philosophy. It inherited the idea of finding causes that produce consequences and appreciated the predictive power of ideas. In the beginning, philosophy and science were dominated by a qualitative approach, but as science developed, it was replaced by a quantitative one. The more advanced the science, the further this process has gone. And the philosophy has remained qualitative.

The first real science was mathematics, although it does not belong to the natural sciences, because it is based not on experiment but on logic, calculations, and transformations. Then, progress affected astronomy, which in ancient times stood on three pillars, namely observation, empirical laws, and mathematics ("a harmony of spheres").

Measurements in physics and astronomy came later. Note some particularly important discoveries. The law of hydrostatics received the name of Archimedes (Ἀρχιμήδης) who discovered it by the third century BC. It was used to solve a purely applied problem of how to find out how much gold a dishonest jeweller replaced with silver when making the crown of king Hieron. There were also the measurement of the Earth's radius by Eratosthenes (Ἐρατοσθένης ὁ Κυρηναῖο) and the experiments of Claudius Ptolemy (Κλαύδιος Πτολεμαῖος) on the study of the law of refraction of light, not to mention his famous astronomical observations.

Note that the very name "physics" comes from the ancient Greek word φύσις — nature. This is the title of one of the writings of one of the greatest philosophers of antiquity Aristotle (Ἀριστοτέλης) from Stagira (384–322 BC). What was written by Aristotle after this work was called metaphysics, i.e. "after physics". These are basically philosophical treatises collected and edited by Andronicus of Rhodes (Ἀνδρόνικος ὁ Ῥόδιος).

I will give an example of natural-philosophical reasoning. Aristotle proposed the postulates of physics, which can be summarized as follows:

- Natural place: Each element gravitates towards its natural place, somehow located relative to the centre of the Earth, which is the centre of the Universe.
- Gravitation and levitation: A force acts on objects to move those objects to their natural place. Fire and air rise upward from the centre of the Universe, while earth and water fall to it.
- Rectilinear motion: In response to this force, the body moves in a straight line at a constant speed. This speed is inversely proportional to the density of the medium. This implies the impossibility of a vacuum, since the speed of movement in a vacuum would be infinitely high, which was unacceptable to Aristotle. Therefore, every point in space is filled with matter. This is the so-called all-pervading *aether* (αἰθήρ).

- Bodies cannot consist of atoms; otherwise there would be a vacuum between the atoms, which contradicts the previous conclusion.

All this reasoning is not devoid of a peculiar grace. They are beautiful, logical, but absolutely wrong. They contradict the observed properties of our world.

However, this did not bother Aristotle much. In his writings, there are details that are incorrect for anyone with any knowledge of the issue. For example, he wrote that a fly has eight legs, wolves and lions have only one bone without vertebrae, some animals (horse, mule, donkey, deer, camel) do not have bile, etc.

Now some more global postulates:

- Finite Universe: the world is finite, i.e. completed, therefore, perfect; the world does not embrace anything, from which it follows that the world has no place defined as the innermost motionless boundary of what contains it. Celestial bodies above the Moon sphere are made of other matter than terrestrial ones, for example, from *aether*, they are eternal, unchanging, perfect. In the sublunary world (the region of the geocentric cosmos below the Moon, consisting of the four classical elements: earth, water, air, and fire), including the Earth, everything is perishable, temporary, transient, imperfect.
- Perfect cosmos: The Sun and planets are perfect, immutable spheres. Circular motion: The planets are in perfect circular motion.

This example clearly shows how speculative philosophy differs from science. Even more exotic ideas about the world of other famous ancient philosophers are described in John Dreyer's book *A History of Astronomy from Thales to Kepler*. By the way, understanding that the laws of nature operating on Earth also operate in heaven, i.e. in space, came much later. So Newton's idea that one and the same force of attraction underlie the apple fall and the laws of motion of celestial bodies was bold enough for his time.

8.2.1. *Educational institutions*

What about education? Let me remind you that in the Hellenistic world there were temples of the muses called *Musaeums* or *Mouseions*, a kind of centre of arts and sciences, which were considered as something related. The muses patronized music, dance, poetry, history, and theatre. Urania was the muse of astronomy, one of the natural sciences. It is no coincidence that the main scientific achievements of antiquity are associated with mathematics and astronomy.

Plato insisted that first one should take up the art of music, and then the art of gymnastics, i.e. train first the mind and senses, and only then the body. Philosophy was also included in the so-called musical arts and science was a part of it, according to Plato and others. "I would teach children music, physics, and philosophy; but most importantly music, for the patterns in music and all the arts are the keys to learning," he wrote. To study science and other musical wisdom, you need a teacher and mentor, and this is already a profession. So the link between education and science has more than two thousand years of history.

Everyone knows that schools and teachers are different. What played the role of modern Oxford or Harvard in the ancient world? It was undoubtedly the *Mouseion* or Museum of Alexandria (Μουσεῖον τῆς Ἀλεξανδρείας). I will tell you a little further about its first leader Ctesibius or Ktesibios (Κτησίβιος). But it was not he who founded this educational, scientific, and religious institution at the beginning of the third century BC. The founder was the satrap and king of Egypt Ptolemy 1 Soter on the initiative of Demetrius of Phaler, the absolute ruler (*epimelet*) of Athens and also a part-time philosopher. *Museion* included the famous Alexandria Library, an observatory, and an anatomical theatre, and was supported by the state. Perhaps it was then that the first government-paid scientists appeared.

Ctesibius himself was an engineer and inventor. Mechanical devices of that time were quite advanced and made life easier for those who used them. Some of them still amaze with their perfection. Like Archimedes, Ctesibius graduated from the Alexandria Museum. Among the graduates were many mathematicians and astronomers: Euclid, Eratosthenes, Aristarchus of Samos, Pappus, Hipparchus, as well as medics. During its heyday, more than a thousand students studied there at the same time.

Museion lost its significance and was closed during the Roman period, in the third century AD. There have been attempts to revive it, including elsewhere, but not very successful. It is believed that one of them is associated with the name of the mathematician Theon of Alexandria (Θέων ὁ Ἀλεξανδρεύς). Better known is his daughter Hypatia of Alexandria (Ὑπᾰτία ἡ Ἀλεξάνδρεῖα), one of the first women scientists. Her interests included philosophy, mathematics, astronomy, mechanics, and, of course, teaching. She was the head or *skolarch* of the Theon–Hypatia school in Alexandria until her death at the hands of a crowd of Christian fanatics.

Hypatia was more a teacher and commentator or interpreter of the works of more famous predecessors than a scientist. Nevertheless, she contributed to science and technology. In particular, she invented the hydrometer, a device for determining the density of liquids using the buoyancy of floats with a load inside. On the basis of it, Galileo's thermometer (see Fig. 8.2) and an alcoholmeter were created to determine the strength of a mixture of alcohol and water.

Ancient scientists did not particularly distinguish between philosophy, natural sciences, in particular astronomical measurements and empirical laws derived from them, and mathematical theorems and technology represented by some mechanical, pneumatic, and optical devices. With all due respect to them, I believe that they did not use the scientific method in its modern sense.

The first universities arose in the Middle Ages. These important events took place in 1088 in Bologna, Italy, in 1096 in Oxford, England, and in 1134 in Salamanca, Spain. The latter was awarded the status of a university in 1254 by King Alfonso X. The Paris University was founded in the middle of the XII century and was officially recognized as the university by the French king Philip-Augustus in 1200 and by Pope Innocent III in 1215. Universities were also founded in Montpellier, France, and Cambridge, England. But the first universities appeared in the Muslim world:

Al-Karaouin, Morocco, in 859, and Al-Azhar, Egypt, 988. However, they were and still are rather religious centres and schools.

So, more than eight centuries after the closure of the Museum of Alexandria, a new source of educated people appeared in Europe. And this could not but bear fruit.

8.2.2. *Galileo Galilei*

Galileo Galilei and some of his contemporaries can be considered the fathers of modern physics. Galileo made many discoveries in mechanics. Dropping objects of different masses from the Leaning Tower of Pisa, he experimentally proved that they fall in the same time. Moreover, he established the law of this fall: the speed increases in proportion to the time, and the path — in proportion to the square of the time.

For this, he studied the rolling of bodies from an inclined plane and established its laws. Galileo proved that any body thrown at an angle to the horizon flies in a parabola, and the maximum flight range is achieved for a throw angle of 45°.

He found that the period of oscillation of a pendulum depends on its length, but not on the maximum deflection angle. He formulated the law of inertia, which directly contradicted the works of Aristotle. In statics, he introduced the fundamental concept of the moment of force.

But Galileo did not confine himself to mechanics. In 1609, he built his first telescope with a convex lens and a concave eyepiece, shown in Fig. 8.2. The spyglass had appeared several years earlier in Holland, independently from different craftsmen. In 1608, the first patent application was filed by a spectacle maker, Johann Lippershey. The merit of Galileo is that he applied this invention in astronomy and was the first to publish new results. With the help of a telescope and its improved version, he made many astronomical discoveries. Galileo's primitive microscope was used to study insects.

An unusual thermometer with floating and sinking balls is named after Galileo. However, it was invented later. Galileo Galilei himself invented the first thermometer. It used the increase in

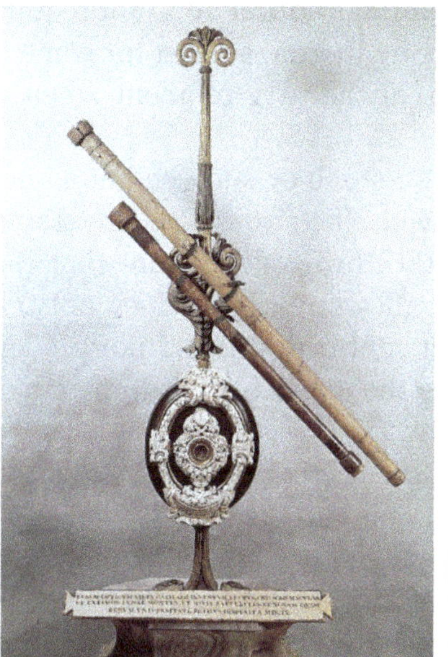

Figure 8.2. Telescope made by Galileo (left panel). Many astronomical discoveries were made with its help. The Galileo thermometer (right panel) is used nowadays mainly as a souvenir and decoration.

pressure of gases when they were heated and in appearance did not resemble at all the Galileo thermometer shown in the right panel in Fig. 8.2. But the main thing is that he did not reason about what should have happened based on some theories or principles, but set up experiments to find out what was really happening.

Much has been written about Galileo's achievements. I will not list them all, but immediately formulate their main distinguishing feature. The scientist entered into direct confrontation with admiration for the wisdom of the ancient Greeks and other thinkers of antiquity, which was previously the basis of philosophy. First of all, he questioned the indisputable authority of Aristotle and his works.

I told you about the main basis of Aristotelian physics. And among them, there was not a single one which was not refuted or questioned by the brilliant Italian. I'll start with the part related to the properties of the world around us, or the sublunary one in the terminology of Aristotle.

First of all, Galileo proved that motion does not require applying a force. A body can move by inertia, at least for some time. On the contrary, it does not stop at the very moment when the force that pushes it disappears.

Anyone who has thrown a stone or snowball at something or someone knows this detail well. The stone does not stop after the hand has released it. And the thrower is aware that the movement of the stone is not at all straight and does not occur at a constant speed. However, for centuries this did not prevent clever stone throwers from admiring the unearthly wisdom of the bearded philosopher from Stagira.

Galileo realized that the force acting on a body determines not its speed, but its acceleration. This laid the foundation for Newton's first two laws. Galileo changed the acting force by varying the inclination of the plane along which the body moved. Since then, this experience has been included in the list of simple experiments for school and student laboratory studies.

Aristotle assured that the body can participate in only one movement. Galileo not only allowed the movement of a rotating body (the same Earth) in space, but in his calculations he decomposed the movement into horizontal and vertical movement. This allowed him to calculate, for example, the trajectory of a cannonball. Now this trick is taught to every student, as well as the addition of velocities based on the so-called Galilean transformation.

Does this seem obvious to you? Galileo's contemporaries were surprised that the bodies falling on the rotating Earth are not carried far to the west. After all, they fall to the centre of the Earth, and its surface is rapidly moving towards the east. Some even experimented with dropping weight from the mast of a fast-moving ship. It fell at the foot of the mast and not behind the stern of the ship, in full accordance with Galileo's reasoning.

The law discovered by Galileo, linking the period of oscillation of a pendulum with its length, later allowed Huygens to design a very accurate chronometer. I have just described Galileo's thermometer, telescope, and microscope. A clock, a thermometer, and optical instruments are the most basic instruments in the arsenal of a physics laboratory. And they are all based on the ideas of one person. Now, of course, we are using their distant descendants. The clock has become quartz or atomic, the thermometer does not use the buoyancy of balls for a long time, etc. But the goal remained the same; the design just improved. So Galileo's contribution to the development of physical devices is invaluable. You will learn further about the contribution of him and his students to obtaining a vacuum, directly forbidden by Aristotle, and to measuring the pressure of gases.

Aristotle assured that everything in the above-moon world is unchanged and ideal. I have already mentioned in Section 1.9 about how difficult it was for him to explain the appearance of comets. Tycho Brahe in 1577 determined the distance to the comet that appeared then and proved that it is not at all in the atmosphere, but very far, farther than the Moon. For a long time, supporters and followers of Aristotle also ignored supernova explosions.

I add a little about variable stars, the brightness of which clearly changed over time, usually periodically. Sometimes it is quite perceptible. But don't look for a mention of it in ancient or medieval astronomers. However, one star was clearly associated with something demonic, decreasing its brightness more than three times every 69 hours. The Greeks placed it in the constellation Perseus, where it matched the eye of the severed head of Medusa the Gorgon. The Arabs called it Al-gul (hence its modern name Algol), which means werewolf or ghoul.

We now know that this is a triple star that contains a close pair of stars. When rotating around a common centre of mass, one of the components periodically obscures the other, providing a fall in the light flux. European astronomers learned about the variability of Algol only in 1669, after the report of the Italian astronomer Geminiano Montanari.

Naturally, Galileo Galilei, the person who first used the telescope for astronomical observations, could not but upset the supporters of Aristotle's ideas about the world above the Moon. For starters, he did what a person does when they get binoculars, or other instruments that magnify distant objects, for the first time.

That is, he directed it toward the Moon and discovered that it is not perfect at all, but has mountains and craters.

Moreover, its surface turned out to be uneven, more uneven than the Earth. A twenty-fold increase was quite enough for seeing this. And on the Sun, Galileo saw spots that, in addition, clearly rotated with its surface. So the luminaries turned out to be not only imperfect, but also rotated in the most inappropriate way.

In addition to all this, on January 7, 1610, Galileo discovered four large satellites of Jupiter. Astronomers now officially recognize 79 of its moons, not counting the rings that not only Saturn has, but also Jupiter. The newly discovered celestial bodies are clearly orbiting around Jupiter, not the Earth, as it should be for all their brethren in the Ptolemy system. Galileo Galilei also discovered the phases of Venus and saw that the Milky Way consists of many stars. In general, in the history of astronomy, he was noted no less than in the history of physics.

It is not surprising that he was an ardent admirer of Copernicus's ideas and promoted them in his books. At first he got away with it, but in 1616 the Catholic Church officially proclaimed itself a supporter of the Ptolemy system. The belief in the heliocentric system of Copernicus or in the rotation of the Earth around its axis was declared heretical. The prohibition of Copernicanism was lifted only by Pope Pius VII two centuries later, in 1822, and the works of the heliocentrists were excluded from the Index of Forbidden Books 13 years later.

Other times came and Galileo did not publish new results for 15 years. In 1632, for some reason, he decided that the position of the church had softened and released his famous work *Dialogue Concerning the Two Chief World Systems*.

Censorship allowed the publication, but Pope Urban VIII declared the book heretical. Galileo was tried by the Inquisition and, in 1633, was forced to publicly recant his views. He spent the rest of his life under house arrest.

From 1979 to 1981, on the initiative of Pope John Paul II, a commission for the rehabilitation of Galileo worked. On October 31, 1992, the Pope officially recognized that the Inquisition made a mistake in 1633, forcing the scientist to renounce Copernicus's theory.

Maybe Galileo Galilei was a genius who was ahead of his time and showed the way to others. Yes, he was a genius, but at the same time, some other scientists working in other branches of physics came to a similar approach to nature. This shows that the point is not in the brilliant insight of a loner, but in the fact that humanity is ripe for understanding the foundations of the scientific method.

8.2.3. *Johannes Kepler*

Fast forward from Italy in the south of Europe to Prague in its centre. The outstanding German astronomer Johannes Kepler worked there at the same time. He not only improved the telescope in 1611 and obtained important results in mathematics and optics, but discovered the laws of motion of celestial bodies. He did it not by speculative considerations such as world harmony and beauty, but by processing numerous astronomical observations. He used data obtained by himself and his teacher Tycho (Tyge Ottesen) Brahe, as well as the results of many astronomers' sleepless nights since antiquity.

Kepler was glorified by the laws of planetary motion that he discovered. There are three of them and they are so simply formulated that they are included in the school course. Nevertheless, it took more than two millennia for astronomers to discover them.

Everyone is familiar with the apparent movement of the Sun, Moon and planets from Mercury to Saturn across the sky. Mars draws loops in the sky, changing the direction of its movement. But the philosophers of ancient Greece assumed that the planets must move perfectly, that is, in a circle and with a constant angular velocity. A different movement would offend their aesthetic feelings. Opinions about the centre of the circle were divided, but most believed that the Sun, the Moon, and planets move around the Earth.

But astronomical observations had to be explained somehow, and the models of motion changed. At first, the planets began to revolve around the Earth, attached to the celestial spheres, and the axis of this rotation itself turned at a constant angular velocity. Then they came up with a system of epicycles, in which the centres of rotation themselves moved in circular orbits. Then the circles became displaced, and the movement was uneven. All this is described in detail in John Dreyer's already mentioned book *A History of Astronomy from Thales to Kepler*. However, I warn you that dealing with all the equants, eccentricities, and apse lines is not easy and requires effort and perseverance.

The result was a system named after the Alexandrian astronomer Claudius Ptolemy (Κλαύδιος Πτολεμαῖος), which has been in use for over a thousand years. It more or less accurately described the trajectories of the planets in the sky, but not the distances to them. The distance to the Moon changes several times according to this model. Everybody knows that angular dimensions of the Moon change slightly, which clearly does not correspond to the essential changes in distance. Ptolemy's system was sometimes slightly modified and refined, but not significantly.

It was fundamentally changed by the Polish astronomer Copernicus, introducing the motion of the planets and the Earth around the Sun. At the same time, the Earth made several movements, including revolution around the Sun and rotation around its axis, which was an incredible innovation for that time. And not only the Earth, but any body since the time of Aristotle was considered capable of only one movement, either in a straight line or in a circle. In fact, the Earth participated simultaneously in four movements according to the Copernican heliocentric model, two of which were later abandoned as unnecessary. But the main thing is that Copernicus "stopped the Sun and set the Earth in motion".

Nevertheless, the same mechanism of epicycles and other celestial machinery remained in the Copernican system. "And so altogether, Mercury moves on seven circles, Venus on five, the Earth on three and the Moon about it on four, and finally Mars, Jupiter, and Saturn on five each. Therefore, taken as a whole, 34 circles are sufficient to represent the entire structure of the heavens and the entire choric dance of the planets," Copernicus himself wrote at the end of the book *Commentariolus*. This was significantly less than 79 that was required in Ptolemy's model and partially solved the problem of distances, but the model was still not accurate enough.

It was Kepler who gave the final form to the heliocentric model with the planets orbiting around the Sun. To do this, he had to establish three important details by processing observational data. That is, what is the shape of the trajectory of the planets, how the speeds of their movement change depending on the position on this trajectory and how they relate to each other? As a result, three Kepler's laws appeared, which are formulated as follows:

1. Each planet moves along an ellipse, in one of the foci of which is the Sun.
2. The radius vector of the planet sweeps out equal areas during equal time intervals. A planet's area speed is constant.
3. The squares of the orbital periods of the planets around the Sun are proportional to the cubes of the semi-major axes of their elliptical orbits.

Note that it is much more difficult to discover each of these laws without knowing the other two than to make the same discovery with information about the other two. To some extent, this is similar to a crossword puzzle, where each guessed word helps to guess the neighbouring ones, and the wrong word stops the whole process.

Kepler had to solve a crossword puzzle in which his predecessors had already written several erroneous words and therefore he had to first erase them, and only then start to enter the words again. Although it must be admitted that the previously inscribed words were close to the correct ones, the wrong letters were often placed at the intersections of words.

Be that as it may, but Kepler completed the revolution in astronomy begun by Copernicus. And after half a century, Newton was able to explain Kepler's laws from the point of view of physics. It is curious that Kepler did not use in his explanations of the essence of the model the properties of motion discovered by Galileo, namely motion by inertia. He was sure that the planets needed to be pushed along their path all the time. And after reading Gilbert's book on the magnet, he tried to connect the ellipticity of the trajectories with the action of the Sun on the planet magnets.

On the other hand, Galileo, who corresponded with Kepler, in his books on the heliocentric system, always wrote about circular orbits, but not about the elliptical ones discovered by Kepler. Most likely, he did not consider the shape of the orbit to be the most important detail against the background of the more fundamental question of what revolves around what. And the orbits of the planets were not very different from the circles. Significant eccentricity was observed only in the orbit of Mercury among the planets known at that time.

Astronomy did not bring a very stable income and Kepler had to earn money by drawing up horoscopes. He wrote:

> Of course, this astrology is a stupid daughter, but my God, where would have become of the wise old mother, astronomy, if she had not had a stupid little daughter. For the world is much more stupid and so stupid that the daughter should chatter and lie for the sake of its wise old mother. And the salaries of mathematicians are so negligible that the mother probably would have starved if the daughter had not earned anything.

The fact that Kepler, who moved to Germany at the end of his life, had to defend his mother in court, who was accused of witchcraft, speaks volumes about the state of science in Europe. The charge contained 49 counts: connection with the devil, blasphemy, necromancy, etc. The investigation lasted five years, during which time six women were burned on similar charges in the small town of Leonberg, where Katharina Kepler lived.

8.2.4. *Magnet properties: William Gilbert and others*

Now let's go to the north of Europe, to the foggy Albion. More precisely, to the court physician of the English Queen Elizabeth I and King James I, William Gilbert. He was 10 years older than Galileo and 20 years older than William Shakespeare. Why did I choose him? Let us recall the forces existing in nature. The concept of strong and weak interactions appeared in physics only in the 20th century and that of radioactive decay and the atomic nucleus at the very end of the 19th century. At the time of Galileo and Gilbert, it was possible to study only gravity and electromagnetic interaction. I talked about the discovery of the law of universal gravitation in Section 1.8.

But at the time of Gilbert, nothing was found out about electromagnetic forces. Nevertheless, many mysterious phenomena related to electricity and magnetism were known at that time, which were later combined into a single electromagnetic interaction in the 19th century. And here Gilbert was able to prove himself.

In about 1600, he invented the *versorium*, the simplest version of an electroscope for detecting the presence of an electric charge. It looked like an iron compass needle, only not magnetized. The arrow showed the direction of the electric field. Gilbert studied electrification of bodies by friction, and then these studies were continued by many, including von Guericke, du Fay and Franklin. I'll tell you more about them later.

Note that Gilbert dealt with issues that were practically not written about by his ancient predecessors, including philosophers. Therefore, unlike Copernicus, Galileo, or Kepler, he did not have to contend with opinions sanctified by time and the authority of the church. Studying science did not interfere with his life and career, and his works did not fall into the lists of prohibited books.

Gilbert's main achievements are associated with magnetism. It is not without reason that the unit of measurement of the magnetomotive force in the CGS system is named in his honour. Visual manifestations of magnetism in the form of permanent magnets and a magnetic compass were well known in those days. Each of these two items is able to make an indelible impression at the first meeting.

What was known about the magnet in those days? The word itself has existed since ancient times and apparently comes from the name of the city of Magnesia. It was named after the Macedonian tribe of magnets. Their legendary progenitor was the son of Zeus and Phia named Magnet. There were seven cities with this name, for example, a city in Asia Minor, now the Turkish city of Manisa. It was founded in the 8th century BC on the banks of the Meander River by Greek colonists from Magnesia in Thessaly, Greece.

Residents found stones that attracted light iron objects near one of these Magnesias. These stones were composed of various minerals, one of which we now call magnesite ($FeO \cdot Fe_2O_3$).

Very little has been learned about the magnet one and a half thousand years after this discovery, but many stupid legends and superstitions have been invented. Gilbert's contemporaries were well aware that:

- Taking a magnet inside prolongs youth.
- If you put a magnet under the head of a sleeping woman, it will throw the adulteress out of bed.
- The magnet opens locks.
- The magnet attracts more strongly by day than at night.
- If you rub a magnet with garlic or put diamonds next to it, its power disappears. If you anoint such a damaged magnet with the blood of a goat, its strength restores.
- A magnet stored in brine from a suckerfish can extract gold that has fallen into wells.
- Some magnets can attract silver, diamonds, jasper, glass, and even meat.

It is interesting to look at this list now. The idea that a magnet opens locks has come true today. There are magnetic picks, but they only open electromagnetic locking devices, which simply did not exist in Gilbert's times.

A magnet can attract more than just iron. An ordinary magnet attracts all ferromagnetic substances. And substances related to paramagnets attract, but very weakly. However, a very strong electromagnet can attract aluminium or other paramagnets. We discussed all of this in detail in Section 4.1, during our journey, but why not recall the details?

In Gilbert's times, people tried to magnetically treat many diseases, in particular gynaecological ones. Interestingly, there are still people who wear magnetic bracelets or other devices on their bodies or try to improve water or gasoline by exposing them to a magnetic field.

What about trying to scientifically study or explain the properties of a magnet? The first attempt was made in 1269 by the Frenchman Pierre Pelerin de Maricourt. He wrote a book, *Epistola Petri Peregrini de Maricourt ad Sygerum de Foucaucourt, militem, de magnete* (Letter of Peter Peregrinus of Maricourt to Sygerus of Foucaucourt, Soldier, on the Magnet), which contains a lot of observations about the magnet accumulated before him and made by him personally. Peregrine for the first time speaks about the poles of magnets, about the attraction of opposite poles and repulsion of the same ones, about the manufacture of artificial magnets, about the penetration of magnetic forces through glass and water, about the compass.

By the way, he has the idea of a magnet in the constellation Ursa Major. Here is what he wrote:

> This stone hides in itself a semblance of the sky (below I will teach in detail how to clearly prove this by experience). And if there are two points in the sky that are more significant than others, sphere, and one of them is called the Arctic or North Pole, and the other — Antarctic or South Pole, then in this stone you must accurately determine two points, one — North, the other — South ... Take a wooden round vessel, like a dish or bowl, and put a stone in it, and then this vessel, together with the stone placed in it, place in another large vessel filled with water ... this stone, so placed, will rotate its small vessel until its north pole stops directly opposite the north the poles of the sky, and the south is directly opposite the south.

Here we already see an appeal to experience, and not to general reasoning.

It is clear that de Maricourt's reasoning is nothing more than an attempt to give a primitive explanation for the fact that the compass needle points either to the north or to a nearby strong magnet. Maybe there is a super-strong magnet somewhere far in the north direction, which the arrow points to when there is no other magnet nearby?

Without a compass, sailing in the open ocean is impossible, only coastal sailing near the coast. Of course, you can navigate by the Sun and stars, but the Sun can be under the horizon or hidden behind the clouds; the stars are also not visible during a storm or hurricane. Therefore, the life of sailors often depended on the presence of a compass.

By the way, the sailors found out that the compass does not point exactly to the north, but somewhere in the direction of the north. The angle of deflection is called magnetic declination and increases as you approach the pole. The voyage of Christopher Columbus to America made it possible to establish that the declination of the magnetic needle at the same parallel changes with longitude.

In 1544, the vicar of the cathedral of St Sebald at Nuremberg Georg Hartmann wrote to Duke Albrecht of Prussia that he measured the magnetic declination in Rome in 1510 to be 6° east, while in Nuremberg it was 10°. The fourth viceroy of Portuguese India, João de Castro, took 43 declination measurements off the west coast of India and in the Red Sea between 1538 and 1541.

The protagonist of the current section, Gilbert, was friends with captains Francis Drake and Henry Cavendish. These privateers, corsairs in royal service, informed Gilbert that in the Southern Hemisphere, the compass needle continues to point north and even showed their secret maps with magnetic declination in different places. It was very valuable information for them.

But let's go back to de Maricourt's book. As was usual in those days, it contained a lot of nonsense. For example, that a magnet is able to reconcile a husband and wife; this was a widespread opinion in the Middle Ages. At the end of the treatise, de Maricourt makes another antiscientific conclusion, calling for the use of the properties of magnets to build a perpetual motion machine.

Giambattista della Porta tries even more to test the findings by experience in his book *Natural Magic* (*Magiae naturalis sive de miraculis rerum naturalium*), published in Naples in 1558 and in an expanded version in 1589. Here is a quote from it:

> There is a widespread belief among sailors that onions and garlic destroy the effect of a magnet. The helmsman and those who deal with the compass have no right to eat the onion, because the needle of the compass will be thrown around from side to side.

But when I tried to check it out, I found that all this was wrong: not only breathing, but also burping after eating garlic, does not weaken the properties of magnetic ironstone; even if you smear the entire stone with garlic juice, it performs its service as well as if it were not touched at all. They won't think I'm trying to discredit the ancients' claim, but I didn't notice much of a difference. In addition, when I asked the sailors if they were really forbidden to eat garlic and onions because of this, they replied that all this was old wives' tales and nonsense, and that they would rather die than not eat onions and garlic.

The book was published when the author was 23 years old. According to him, he wrote it at the age of 15. Soon, at the age of 25, Giambattista della Porta organized the first physics academy in Naples, called the *Academia Secretorum Naturae*, but at the direction of the Pope it was banned in 1560.

Can della Porta be ranked among the first scientists? No. Although he was engaged in some things related to science, he should rather be attributed to magicians, alchemists, and healers, which in those days were often viewed as closely related activities.

You can find recipes for witchcraft ointments obtained from witches in his writings. Della Porta not only observed and described how the witch prepared for a night flight with their help, but also tried the ointment on himself, although he did not write about the results of the experiment. No wonder the Inquisition for some time banned his activities in Italy and he moved to England. Della Porta also developed the pseudoscience of physiognomy, which was popular for many years, but is almost forgotten in our time, and many other delusions of his time, such as palmistry and the search for the philosopher's stone.

But can his studies of magnetism and the properties of a magnet be considered a science? Yes, but only partially. In the book *Extraordinary Popular Delusions and the Madness of Crowds*, written by Charles McKay in the middle of the 19th century, there is a section on magnetism and "animal magnetism", which also mentions Giambattista della Porta. This is due to the so-called weapon-salve.

Supporters of the healing properties of the magnet decided that with its help it is possible to heal wounds caused by metal objects.

Then they moved on to the idea that the magnet would help in the treatment of any wounds that, in principle, could have been inflicted with a sword. It is enough to magnetize this sword. Then Paracelsus proposed to use a substance called weapon-salve to treat wounds from cold weapons that do not penetrate the heart, brain, or arteries.

The following was the recipe given by Paracelsus:

> Take of moss growing on the head of a thief who has been hanged and left in the air; of real mummy; of human blood, still warm — of each, one ounce; of human suet, two ounces; of linseed oil, turpentine, and Armenian bole — of each, two drachms. Mix all well in a mortar, and keep the salve in an oblong, narrow urn.

This salve was supposed to lubricate the weapon, previously dipped in the blood from the wound. In addition, the wound was processed and bandaged in a standard way for that time. Now I would say that this magical ritual served as a placebo for a purely psychological purpose. Della Porta worked on improving the weapon-salve and used it in his medical practice. And a recipe like "an ounce of moss growing on the head of a thief who has been hanged and left in the air" does not go well with the title of a scientist. So in this case, we see the beginnings of the application of the scientific method against the background of the approach familiar for those times.

But all this was before Gilbert's book *De Magnete, Magneticisque Corporibus, et de Magno Magnete Tellure* (On the Magnet and Magnetic Bodies, and on the Great Magnet the Earth; a new physiology, demonstrated by many arguments and experiments), published in 1600. He wrote it for 18 years. The book describes about 600 experiments conducted by Gilbert in order to find out the properties of magnets. Galileo, after reading it, declared Gilbert "great to such a degree that causes envy". Indeed, the author collected all the reliable data of that time on magnetism, adding to them his enormous contribution.

Gilbert once again confirmed that every magnet has two poles that cannot be separated. Poles of the same name repel, and unlike poles attract. A magnet can turn a piece of iron into a magnet magnetizing it. Submersion in water does not weaken the effect of the magnet, but hitting the magnet can do it. Moreover, each of these facts was based not on an assumption, but on the experiments made and repeated. Hilbert also checked and refuted numerous parables and superstitions about the magnet, created by ancient scientists.

Gilbert wrote:

> And as yet we have not set ourselves to overthrow by argument those errors and impotent reasonings of theirs, nor many other fables told about the loadstone, nor the superstitions of impostors and fabulists: for instance, Franciscus Rueus' doubt whether the loadstone were not an imposture of evil spirits: or that, placed underneath the head of an unconscious woman while asleep, it drives her away from the bed if an adulteress: or that the loadstone is of use to thieves by its fume and sheen, being a stone born, as it were, to aid theft: or that it opens bars and locks, as Serapio crazily writes: or that iron held up by a loadstone, when placed in the scales, added nothing to the weight of the loadstone, as though the gravity of the iron were absorbed by the force of the stone: or that, as Serapio and the Moors relate, in India there exist certain rocks of the sea abounding in loadstone, which draw out all the nails of the ships which are driven toward them, and so stop their sailing; which fable Olaus Magnus does not omit, saying that there are mountains in the north of such great powers of attraction, that ships are built with wooden pegs, lest the iron nails should be drawn from the timber as they passed by amongst the magnetick crags ... Or that by day it has a certain power of attracting iron, but by night the power is feeble, or rather null: Or that when weak and dulled the virtue is renewed by goats' blood, as Ruellius writes: Or that Goats' blood sets a loadstone free from the venom of a diamond, so that the lost power is revived when bathed in goats' blood by reason of the discord between that blood and the diamond: Or that it removed sorcery from women, and put to flight demons, as Arnaldus de Villanova dreams.

He wrote very harshly about book authors "who in all branches of learning are seen to hand on errors and occasionally add something false of their own".

In addition to these experiments with magnets, Gilbert figured out the force guiding the compass needle. Among other experiments, he described the experiment with a rounded loadstone and realized that this ball is essentially a model of the Earth. Gilbert called it *terella*, that is, small Earth (*Terra*). Experiments with *terella* demonstrated all the facts known at that time: magnetic declination, inclination and their change when moving. Thus, Gilbert was the first to find out that the Earth is a magnet.

So we can see something just by comparing the approach taken in three different books written at different times. Recall that they are talking about the study of a then unknown phenomenon — magnetism, which then no one called the fundamental interaction due to the fact that these words had not even been invented.

De Maricourt's approach differs little from, say, Herodotus' description of distant countries. The historian painstakingly writes down all the tales about the countries where people with dog heads or bald cone-eaters live. If you think about it, then Peregrine has slight differences from the descriptive purely natural-philosophical approach. He tries to give at least some explanation (a magnetic stone in the *Ursa Major*) and tries to come up with a practical application of the described phenomenon, even if it is a perpetual motion machine, which, as was later proved, is impossible.

Della Porta was already at least checking a part of the information received from others and setting up the simplest experiments. But Gilbert checked everything and had a clear orientation towards experimental verification and the experimental plan as a whole. He put forward hypotheses (the Earth is a huge magnet) and tested them on a specially designed installation (*terella*). Nothing was taken for granted. This was already a science, albeit still very young.

8.2.5. *Electricity*

Let me tell you how physicists came to the concept of electric charge, from which began a long journey of learning the secrets, first of electricity, and then of electromagnetism. Along the way, I will try to use this example to see how new concepts arise in physics and how they become clearer and more specific. Now the charge is a well-defined quantity that can be measured and that is related to other physical quantities.

In the beginning, it was just a word used to describe electrical phenomena. In addition, there was no understanding that electricity in nature is one and the charge obtained in different ways is the same charge. Recall that the first naturalists could produce a charge by friction, feel discharges from electric eels[e] and other fish that have the ability to generate electricity, and observe lightning and sparks. And to guess that the painful sensation after communicating with the fish is somehow connected with the ability of the cat's hair to attract light objects was not at all easy. Here it is worth taking another look at Fig. 2.4.

In the old days, friction was the main way to generate electrical charge for physical experiments. The engraving in Fig. 8.3 shows one variation of this process. It started with a ball of sulphur, depicted in Fig. 8.4, rotating on an axis. It was made in 1660 by the

Figure 8.3.   Obtaining an electric charge by friction. *Source*: Wikipedia.

---

[e]The Roman emperor Nero was treated for rheumatism by bathing in an electric eel bath.

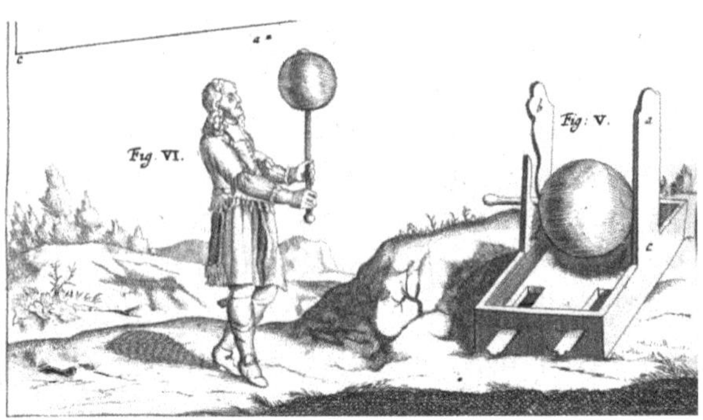

Figure 8.4.   Otto von Guericke and his ball of sulphur. Image is in public domain.

mayor of the German city of Magdeburg and part-time physicist Otto von Guericke. When you put your hand to it, the hand and ball are charged by friction. The charge could be used for experiments or to simply run a current through a chain of musketeers holding hands for the entertainment of the audience.

In 1733, the French naturalist Charles François de Cisternay du Fay proved that all the types of charge known to science can be reduced to two: resin electricity, obtained by rubbing resin, wax, or silk on wool, and glass electricity, obtained by rubbing glass or wool, for example, glass on silk. Note that before du Fay, some scientists, such as the British chemist and astronomer Stephen Gray, argued that the electrical properties of a body depend on other properties, such as colour. Dufay showed empirically that they were wrong.

Soon, the American Benjamin Franklin proved that this is what we now call negative and positive electric charges. So if we charge a body with resin electricity, then by adding a certain amount of glass electricity, we can get an uncharged body. The positive and negative charges balance each other out.

What was the difference between resin and glass electricity? An object electrified by any of them repelled objects electrified with the same kind of electricity and attracted objects electrified with the opposite kind. After Franklin's experiments, physicists talk about the attraction of charges of the opposite sign and the repulsion of charges of the same sign.

In Section 4.3, I described in more detail how the electrification of droplets by friction against air causes lightning. In the meantime, I add how scientists were able to make sure that the electric charge, no matter how it was received, has the same nature. The principle is well described by the duck test. If it looks like a duck, swims like a duck, and quacks like a duck, then it probably is a duck. If electricity obtained in different ways from different sources in experiments manifests itself in the same way, then this is the same electricity. For example, it is capable of emitting sparks that can ignite alcohol vapours, or attracts and repels the same objects. And even a cat can be made to emit sparks. Du Fay did this by stroking her on a silk pillow. He managed to produce sparks from a boy, electrified on a piece of resin.

But if there is electricity, then it can be stored somewhere. This "vessel for a charge" is called the Leyden jar. It was the first capacitor, independently invented in 1745 by the Dutchman Pieter van Musschenbroek and the German Ewald Georg von Kleist.

A jar is called a Leiden one because van Muschenbroek lived in the Dutch city of Leiden, where he taught at a local university. In addition, a resident of Leiden, a student of van Muschenbruck Kühneus, was the first man to experience a shock from the discharge of a device. One Leiden resident came up with a jar for the storage of electricity, which was discharged in the city of Leiden through a resident of this city. The question is, what else could this object be called?

It is a jar for the simple reason that the main part of its design is an ordinary glass jar, like those used by apothecaries at that time. The neck of the jar is closed with a wooden lid, through which a metal pin passes inside. In the minimal version of the device, a large nail is simply hammered with a head outward and a point inward. With the rest of the details, options are possible.

Mercury or water could be poured inside so that the pin touches the liquid. Or you can glue the inside of the jar by 2/3 of its height with tin foil connected to the pin with a metal chain. Outside, the jar was also covered with foil, but in the very first samples, they simply wrapped the palms of their hands around it.

When charged objects touched the pin, the charge through it and the chain was transferred to the inner foil or to the surface of mercury. A potential difference arose between it and the outer foil, increasing with the addition of new portions of a charge of the same sign. When the jar was discharged, sparks were visible, sometimes quite long.

Figure 8.5 shows a Leyden jar and a device for discharging. It was the first-born in the family of electric capacitors, i.e. devices for accumulating and storing charge. Over time, it had many relatives and descendants.

After the invention of the Leyden jar, experiments to test the unity of all types of electricity were reduced to the simple procedure. This capacitor was charged with electricity obtained from a certain type of source and with its help, a certain standard set of experiments was performed. All of them, regardless of the type of source, always showed the same results, which proved the unity of all types of electricity.

Figure 8.5. The Leiden jar and the device for its discharge and demonstration of electric sparks. Extracted from Magnet Academy at the National High Magnetic Field Laboratory.

It is clear that obtaining some types of charge, for example, "celestial electricity" from lightning, was a rather risky business. Franklin did it in America; a judge and amateur scientist Jacques de Roma did it in France. And in Russia, Georg Wilhelm Richmann, a member of the Academy of Sciences and Arts, was killed by a lightning strike in 1753 when trying to repeat their experiments.

Naturally, with the development of science, scientists turned from the ability to provide the same set of effects to the coincidence of their quantitative characteristics. After all, not everything that is similar turns out to be the same. And here the charges, obtained in different ways, showed themselves in the same way.

It is worth mentioning a simple but useful device called the gold-leaf electroscope, invented by the British clergyman and physicist Abraham Bennet in 1787. It is shown in Fig. 8.6. The electroscope demonstrates roughly the magnitude of the electric charge. The design of the first devices was very similar to the Leyden jar: a glass vessel, a lid, a vertical metal pin or a rod. A disc

Figure 8.6. The principle of operation of the electroscope.

or ball terminal is often attached to the top of the rod, where the charge to be tested is applied.

It used two parallel strips of thin flexible gold-leaf, or thin metal foil. Their upper ends were attached to the end of the rod. The charge was transferred from the terminal along the pin to the leaves, which diverged to opposite sides due to electrical repulsion. When the metal terminal was touched with a charged object, the gold leaves deviated to the sides formed the Greek letter Λ that is an inverted letter V.

The larger the charge, the more the angle between the leaves. And by that time, scientists knew how to measure angles. So it was not difficult to find out which charge from which object charged the electroscope more and which less. But one "more or less" is still not enough. I would like to have a normal device with a scale on which a deflected gold leaf would show the amount of charge, no matter in what units.

8.2.6. *Measurement of electric charge and its units*

The idea of charge separation came to the rescue. If we have two identical metal balls, then when they touch, their total charge is redistributed over them. If the balls are separated, then it is logical to assume that their charges will be the same, because there is no difference between the balls, which means there is no reason why the charge of one ball would be greater than the charge of another. This is an obvious manifestation of the principle of symmetry, which we will discuss in Section 8.3. Indeed, if you touch any of the two separated balls to the electroscope, the gold leaf will move apart at the same angle. Bingo! The device is almost ready.

So, we have two balls, let's call them ball 1 and ball 2. After contact, they have the same charges. Which ones? We do not know, and we have not yet invented the units for measuring the electric charge. We take out one more ball of the same kind and call it ball 3. We touch one of the charged balls with it, say ball 2, and then separate them. Each of balls 2 and 3 has half of the original charge of ball 2 before it comes into contact with the uncharged ball 3, or the half the charge of ball 1, with which we did nothing.

Let's call the charge of the ball 1 the conventional unit of charge. We touch the electroscope with it, mark the position of its leaves and write the number 1 near this mark. We discharge the electroscope, then charge it with a ball number 2, put the second mark and write 1/2 next to it. If we are not tired yet or do not start to celebrate the introduction of a new unit of measurement of a physical quantity, then we take the ball number 3, touch it with an uncharged ball (e.g. with ball 2 after grounding it), disconnect them, touch the electroscope with ball 3, and draw the 1/4 mark.

I think the principle is clear. Moreover, connecting a ball with a charge of 1/2 and a ball with a charge of 1/4, we get two balls with charges of $(1/2 + 1/4)/2 = 3/8$ of our conventional unit. If desired, we can calibrate the device in any detail.

But all marks will correspond to charges less than one on our conventional scale. How to continue it for larger values? We take a lot of identical balls, charge them, touch them one with the other many times repeatedly to come to the conclusion that their charges are identical. Then we discharge one of them to an electrometer calibrated by us. Let its leaves deviate to a position approximately corresponding to mark 1. At first, the deviation may be greater or less, but we can increase the charge of a single ball, electrifying it more or decrease it by adding a few uncharged balls to the charged ones and levelling their charges by contact. We will achieve the desired charge after another attempt. This means that we have several balls with charges that are almost equal to the charge of ball 1, from which our procedure began.

Now we can discharge one of them into an electroscope, in which the leaves are already in position with mark 1 and draw mark 2 near their new position. Then discharge one more ball and put a mark 3. In addition, we can calibrate any other electroscope using our balls with a reference charge.

Is there a catch in this procedure? Yes and it is an essential one. When we pour a litre of water into two identical vessels, standing at the same level, so that the liquid levels in them coincide, then half a litre will get into each of the vessels. Simply because the volume and weight of water during this operation is preserved, naturally, provided that we did not spill some of it.

But who said that the same can be said about the electric charge? When scientists began to conduct the first experiments with electricity, they developed a natural analogy between charge and liquid. According to it, when we charge a body, then a certain "electric liquid" flows into it; when we discharge it, it flows out. Its volume or weight is our charge. Why not assume that it is some kind of charge conservation law?

It is possible to speculate, but the hypothesis must be tested. So let's start with a simple test. Let us charge a pair of identical balls in contact so that after touching one of them, the electroscope we calibrated will point to the 1/2 mark. Then we add to it the charge from the second ball. The leaves will deflect to the position of mark 1 or close to it within the accuracy of our method. This will be the first experiment in a long chain of testing the law of conservation of charge. And it passed all the tests many times. Until now, none of the experiments has observed anything similar to a violation of the law of conservation of electric charge, no matter how carefully one looks for it.

Therefore, modern physics is confident that the full electric charge is conserved in any process. So the proposed method for calibrating the electroscope is quite correct. After this procedure, we get a device called an electrometer, capable of not only showing the presence of an electric charge, but also measuring it.

Now, armed with the law of conservation of charge and an electroscope with a scale calibrated in conventional units of charge, the physicist can start repeating Coulomb's experiments. To do this, this person needs to charge two small balls and measure the distance between them and the force that acts on one of them.

Finally, we got to the force and this step was especially important. Why? First, do not forget that we are discussing forces or interactions. And when we do not just talk about force, but measure it, this brings us significantly closer to the modern scientific method. Secondly, force is a mechanical quantity and is measured in units that have nothing to do with the units of charge. Measuring the force caused by electrical interaction allows us to link mechanical and electrical units.

Coulomb in his experiments measured the force of interaction of two electrically charged bodies, the distance between which was much greater than their size. It is very close to the force of interaction of two point charges. Coulomb proved that this force is directed along the line connecting them. Charges of the same sign are repelled. Charges of different signs are attracted. The magnitude of this force is proportional to the product of two charges and is inversely proportional to the square of the distance between them.

We are now ready to discuss the units of measure for electric charges. Charles-Augustin de Coulomb published the results of his experiments in the 80s of the 18th century. The metric system of units, the development of which is the modern international system of units or the SI system from the French *Système international (d'unités)*, was introduced in France only in the next decade, so that Coulomb in presenting his experiments used more ancient units, such as inches, feet, and pounds. However, when it comes to proportionality, the unit of measurement becomes irrelevant. But it is important for the law in its entirety.

So, there is Coulomb's law and there is a specific system of units to which this law must be adapted. And at this stage there is a fork in the path, a choice of two options. In the SI system, the unit of electric charge is called a coulomb (symbol: C) and it is named, naturally, in honour of de Coulomb. And, surprisingly, it is, to a certain extent, just a conventional unit.

After all, this is a derived unit obtained from the basic electrical unit, namely the unit of electric current — ampere (symbol: A) and the unit of time — seconds. And the ampere is determined from the magnetic, not the electrical forces. The Ampère force acts on currents due to their magnetic interaction. If electric currents of 1 A flow through two infinitely long straight parallel wires, with the distance 1 m between them, then the Ampère force is $2 \cdot 10^{-7}$ N per meter of length. If the currents flow in one direction, then they are attracted, if in the opposite, then repel. So the coulomb is a unit of electrical charge that came from a completely different interaction, albeit related. Be that as it may, we have the coulomb as a unit of charge in the SI system and usually forget about all the conventions of such a choice.

Let us formulate Coulomb's law in this form: the force of interaction of point charges is directed along the line connecting them and is equal to the product of their charges multiplied by a coefficient and divided by the square of the distance between them. And then we will determine the value of this coefficient from the experiments of Coulomb and his followers. In other words, we determine the force with which two charges of 1 C interact at a distance of 1 m.

Since a charge of 1 C is a large enough charge, the force is enormous. It is equal to $9 \cdot 10^9$ N or the force with which the Earth would attract an object lying on its surface with a mass of about 1 million tons. The dimension of the coefficient is easy to find, it is a newton multiplied by the metre squared and divided by the square of the coulomb. Only with this choice the force described by Coulomb's law is expressed in units of force, and not, say, volume.

But the SI system is not the only possible system of units, although it considers itself so and fights competitors with administrative measures. There is, for example, the CGS (centimetre–gram–second) system that is beloved by theoretical physicists, which is also called the Gaussian system of units. In it, when choosing a path at a fork, we turn in the other direction.

Namely, we do not introduce any conventional units of charge, but introduce units that naturally result from Coulomb's law. Two unit point charges of the CGS system, located at a unit distance of 1 cm, interact with a unit force of 1 dyne or 10^{-5} N. So the coefficient in the formula is equal to 1, and the unit of charge is obtained directly from Coulomb's law. It turns out to be $3 \cdot 10^9$ times smaller than the coulomb.

Which choice is better? If we recall similar choices in other cases, it becomes clear that there is simply no obviously preferable one. We measure the time intervals in years, days, hours, and seconds associated with the Earth's revolution around the Sun and its rotation around its axis. We measure the length in metres. This unit is related to the length of the Earth's equator. But the naturally chosen unit of length would be derived from the light second. In fact, the length of 1 m in the SI system now is determined through fractions of a light second, but the coefficient when converting these units is not very round. A metre is one-299,792,458th part of a light second.

What about mass? A kilogram was originally the mass of water with a volume of 1 litre, i.e. $0.001\,\mathrm{m}^3$ at certain values of temperature and pressure. It would be possible to introduce the unit of mass from the law of universal gravitation so that two unit masses at a unit distance would be attracted with a unit force. But we don't do that.[f]

So let's leave the pointless debate about which system is better. No one prevents us from measuring the voltage in the electrical socket in volts and using the CGS system in scientific calculations. And, accordingly, use different units of charge. All the same, if a calculation is correct, the results should coincide. The main thing is that we now have two suitable ways to determine the unit of electric charge. That's exactly two more than nothing. And which of the two possibilities we implement is no longer so important.

I described the whole chain of ideas and concepts that physicists had to go through in order to make the initially rather vague concept of an electric charge a specific one, to get the unit of measurement for it and to make the device that carries out measurements. As a result, a charge, say $1\,\mathrm{nC} = 10^{-9}\,\mathrm{C}$, is not a term defined in general terms in Wikipedia or a philosophical dictionary, but a measure of the amount of electricity in the electrometer when its indicator points to a certain place on the scale.

To achieve this we needed two very simple devices: an electrometer and an installation on which Coulomb conducted his experiments. And also the idea of the unity of an electric charge, tested in other experiments, the assumption of the conservation

---

[f]There is a system of units, often used in general relativity, in which mass, length, and time have the same dimensions. Time is measured in seconds, distance in light-seconds, approximately equal to 300,000 km, and a unit mass at a distance of one light-second provides an acceleration of gravity equal to a light-second per second squared. There is also the Planck system of units, in which all quantities are dimensionless.

of charge, Franklin's brilliant hypothesis about charges of different signs and the efforts of many scientists, who together go to an initially incomprehensible goal for them and are often unaware of each other's existence.

8.2.7. *Electric current*

However, for a modern person, electricity is associated not with a charge, but with an electric current. This is a natural phenomenon in nature. Lightning and the aurora are both forms of electric current, and the currents in the Earth's core creates its magnetic field. Nero did not know anything about electric current, but he felt it in different parts of his body every time he splashed in a bath with electric eels. In all cases, these currents flowed without wires. By the way, the phrase "the current flows" is connected with the old analogy between a charge and a mysterious electric fluid.

The first artificial current also did not flow through the wires. Having received at their disposal a source of charge produced by friction and a device for storing the charge — a Leyden jar, scientists were able to discharge the charge accumulated in them. This created an electric current flowing to the ground. But it just flowed along a chain of people holding hands, for example, through a platoon of musketeers carrying out the commander's order. And the assembled audience watched with interest their grimaces and listened to their strong language and expressions.

Abbot Jean-Antoine Nollet, who was also a physicist and a member of the Paris Academy of Sciences, became famous for such entertainments at the court of King Louis XV of France. His record is the current discharged through a circuit of 180 people. Note that at that time rubber boots had not yet been invented, so the length of the chain depended on the moisture content of the soil, the conductivity of the shoe soles of "current conductors" and various other factors.

In some cases, their hands were connected not directly with one another, but with different objects. It was easy to understand that metals conduct current, but wood does not. The first ideas about conductors and insulators were obtained in this way. The only measure of the current strength at that time was only the sensations of the person through whom this current flowed.

I am sure that everyone will agree that producing current by friction is as ineffective as producing fire by friction. But progress did not halt. Scientists have learned how to generate electric current using different types of batteries. More precisely, the sources of electric current in question have gradually evolved from a leg cut off from a frog to modern batteries and accumulators.

All of them are now called galvanic cells in honour of the Italian physician Luigi Galvani, who accidentally discovered that the leg of a dead frog twitches when touched at the same time by copper and iron objects, which are also in contact. Historically, it was a steel scalpel and a copper screw on a microscope, and young Luigi dissected the frog to entertain his young wife, who visited his laboratory. When the severed foot twitched, Signora Galvani's reaction was very violent. The name Galvani can be easily seen in the words *galvanometer*, *galvanoplasty*, and even the outdated but fashionable in the 18th century phrase "galvanization of corpses".

His research was continued by another Italian — Alessandro Giuseppe Antonio Anastasio Gerolamo Umberto Volta. His surname gave the name to the unit of voltage with the SI system and formed the basis for the words and combinations *voltmeter*, *voltaic pile*, and *voltaic arc*. The voltaic pile was the first source of a sufficiently high voltage that worked for a long time, and not in pulses, like the discharge of a charged capacitor.

The appearance of the first galvanic cells allowed physicists to begin the mass study of direct electric current and its properties. This opened the door to the world of practical use of electricity in technology and everyday life. Over time, through this door, already wide open, electrical appliances burst into our house: light bulbs, stoves, refrigerators, vacuum cleaners, hair dryers, irons, and then televisions with computers.

8.2.8. *Vacuum, atmospheric pressure, and their discoverers*

The vacuum or emptiness was the subject of heated debate in antiquity and the Middle Ages. Naturally, not scientific, but philosophical or natural-philosophical debate, i.e. not based on experience and facts. Even Aristotle came to the conclusion that "nature abhors a vacuum". This principle is briefly called *horror vacui*. The logical chain that led to this conclusion was given at the beginning of this chapter. Aristotle's conclusion was supported by his numerous medieval followers and the authority of the church.

However, in 1277, the theologians of the Sorbonne, headed by the Bishop of Paris E. Tampier, came to the conclusion that the denial of any possibility of the existence of a vacuum would limit the omnipotence of God. By a special decree, emptiness was assigned to the category of *causus divini*, that is, phenomena that do not exist in nature, but are possible for God.

Therefore, obtaining a vacuum was an important stage in the formation of science, which came into direct conflict with both natural philosophy and the religion supporting it. It is curious that this achievement is associated with the progress of not so much science as technology.

I begin the story from afar, from the depths of centuries. Remember the ancient Greek inventor, mathematician, and mechanic Ctesibius, the first head of the Alexandrian Museum? The works of Ctesibius himself have not reached us, but they are described by his student Geron ("Ηρων ὁ Ἀλεξανδρεύς), and his inventions are also known from the finds of preserved devices.

Ctesibius invented a siphon, a hydraulic organ, a water clock, a pneumatic crossbow, and a pump, which all cyclists still use today, albeit in a more modern design. This pump was also used to raise water from wells, and this has always been a very important matter. In addition to supplying water for drinking and irrigation, it was also used to pump water out of ancient mines, which would have simply flooded without drainage. Ctesibius made a fire pump based on the same device.

At that time, many mechanical devices were invented for the purpose of raising water, such as a water wheel and an Archimedean screw. Before that, the water was lifted on ropes, sometimes with the help of a winch, in buckets or other containers like leather bags. The device, called *sakia* or *sakieh*, raised the water in the days of Pharaoh Ramses. Much later, steam engines began to be used to raise water, followed by electric motors.

But we are interested not so much in technology as in the underlying physics. Therefore, let's take a closer look at the pumps that create a vacuum to suck in water and raise it. Already Ctesibia's pump was the prototype of the vacuum piston pump, which appeared almost 2000 years later. Let me draw your attention to the word *vacuum*.

It was the fear of emptiness that was used at that time to explain the operation of the pump that sucks water (but not the injection pumps that create overpressure). If the water did not rise up the pipe, then there would be a void so unloved by nature. Therefore, the water has no choice but to rise, thus filling the place that would otherwise be occupied by this nasty vacuum.

But life likes to make fun of lovers of speculative reasoning. In his *Conversations* published in 1638, Galileo noted, referring to the experience of Florentine plumbers, that there is a maximum possible height to which water can be raised by pumps of this type. And this height is always the same and is about 18 cubits (the combined length of the forearm and extended hand), i.e. about 10 m. The water has to be raised to a higher level in several stages: first to an available intermediate height, and then as many times as necessary. The conclusion made by Galileo: the power of the fear of emptiness is limited.

Further research after Galileo's death was continued by his disciple Evangelista Torricelli. He suggested that another student, and later Galileo's biographer, Vincenzo Viviani, replace water with the much heavier mercury.

In 1643, he filled a long glass tube sealed from one end with mercury, plugged the hole with his finger, turned the tube over, and immersed its open end in a basin of mercury. And then he removed his finger, allowing the mercury to flow from the vertical tube into the basin. It flowed out, but not all of it. A column of mercury remained in the tube, about 28 inches high, or 76 centimetres in modern units of measurement. And above it, through the transparent walls, the supposedly impossible emptiness could be seen.

Well, what did he see? The vacuum turned out to be transparent and illuminated objects could be seen through it and through the glass walls. It is curious that soon physicists, having seized upon experiments with such an exotic object as a vacuum, found that the vacuum does not pass sound. This was an important distinction between light and sound, which in those days were often regarded as something related.

Torricelli himself joined the experiments and a year later found that the height of the mercury column is insignificant, but changes on different days. Based on this observation, he introduced the concept of atmospheric pressure, which has become standard today. The atmosphere pressed against mercury, providing the force needed to lift it. Its pressure was equal to that created by the liquid in the device, called the mercury barometer.

Standard atmospheric pressure is 760 millimetres of mercury. In SI, it can be expressed as 101325 Pa. However, depending on the weather, this pressure changes and the barometer informs us about its current value. The device proved to be useful for predicting the weather and became indispensable on ships.

But who prevents you from replacing mercury with water and taking a longer pipe? The height of the water column at normal atmospheric pressure is 10.33 m. So a high enough building would be required for a water barometer. Gasparo Berti equipped a similar device on the facade of his house in Rome. Some scientists experimented with the vacuum obtained by this particular device over a water column. However, it was a rather nominal vacuum. The pressure in this part was far from zero and amounted to several tens of millimetres of mercury. Another amateur physicist, Mayor of Magdeburg Otto von Guericke, who I have already mentioned in connection with his device for electrifying bodies, managed to achieve a significantly thin air, almost vacuum.

During his stay in Regensburg in 1654, he learned about Torricelli's experiments. Around 1657, Guericke installed a magnificent water barometer in Magdeburg. It consisted of a long copper tube attached to the outer wall of Guericke's three-story house. Its lower end was immersed in a vessel with water, and the upper end was equipped with a tap and could be connected to an air pump. After evacuating the air, the water in the tube rose to a height of 19 cubits.

Using this device, Guericke confirmed that atmospheric pressure is constantly changing, which is why he characterized his barometer with the words *Semper vivum* (always alive). He noticed the relationship between the height of the water column in the tube and the state of the weather.

For greater effect, a float in the form of a human figure with a hand pointing to a sign with inscriptions corresponding to various weather conditions was placed on the surface of the water in a glass tube inserted at the top of the barometer. Gericke called him Wettermännchen (weatherman). In 1660, the device surprised the residents of Magdeburg by predicting a severe storm two hours before it began.

To demonstrate atmospheric pressure, von Guericke performed his most famous vacuum experiment in Regensburg in 1654. It used two copper hemispheres about 35 cm in diameter, hollow inside and pressed together so that they formed a hollow sphere. In the presence of Emperor Ferdinand II and the Reichstag, he pumped out the air from this sphere, after which the hemispheres could not be separated by the two teams of eight horses pulling them in opposite directions (see Fig. 8.7). In 1656, Guericke repeated the experiment in Magdeburg, which gave him the nickname "Magdeburg Hemispheres", and in 1663, he repeated it in Berlin, with 24 horses.

8.2.9. *Why exactly then?*

"Gilbert shall live till loadstones cease to draw / Or British fleets the boundless ocean awe." English poet Dryden wrote these lines a century after Gilbert died. These words help answer the question that naturally arises: what happened, why at the turn of

Figure 8.7.   *Magdeburg Hemispheres*, a drawing by Gaspar Schott.

the 16th and 17th centuries did the outstanding scientists Galileo, Gilbert, Kepler, Torricelli, and their less famous contemporaries work simultaneously?

Several factors influenced them, including the general rise of the sciences and arts in Europe during the Renaissance and the transformation of the first universities into centres not only of theology, but also of secular education and science. But we must not forget the need for scientific results for practice. You use artillery in battles? This means that a theory of the motion of the cannonballs shot out from the barrels of cannons and howitzers is needed. Would you like to see in detail what the enemy is doing? The spyglass helps in this matter. The compass is necessary in the fleet, where knowledge of magnetic declination is no less important than a map of currents. In everyday life, you may need glasses to see well.[8] And if knowledge is important, then people appear who are ready to pay for it. This means that professional scientists appear who earn their living by their work, and not by another craft.

---

[8]Glasses were invented earlier, at the end of the 13th century. At the same time, the profession of manufacturer of glasses and lens grinder appeared. Therefore, by the time the telescope and microscope were invented, lenses were readily available to scientists and artisans.

But from professionals, you can demand not only wise reasoning, but also something that really works, including new devices, knowledge, and techniques. You need instructions on the angle at which to fire the cannon at the maximum range. The best spotting scopes with high magnification and minimal image distortion were needed. Naturally, all this was immediately tested in practice. Anything that failed the test was ruthlessly discarded as useless. This laid the foundation for the scientific method, separating science from speculative natural philosophy.

The process has begun. In the 17th century, we already see many famous physicists. They are Isaac Newton, Robert Hooke, Christian Huygens, Blaise Pascal, Rene Descartes, Robert Boyle, Edm Marriott, and many others. Later, the scientific method penetrated into chemistry, biology, and other sciences.

It is now used in very different fields. For example, in medicine, the position of the evidence-based medicine has strengthened, when doctors must convincingly prove that the proposed method of treatment or medicine really helps more than harms. Experiments are used on groups of volunteers. Their results are carefully processed using methods of mathematical statistics.

To avoid conscious or unconscious influence of the experimenter, a double-blind method is used, when neither the patient nor the doctor knows whether the patient is receiving a medicine or a placebo, a pill made of chalk and sugar. The fact is that a placebo also helps, simply because of the peculiarities of the human psyche, but the medicine should help significantly more.

As an example of the use of the scientific method in biology, I will tell you how we found out what the body does with the food we eat. Everyone knew that it is digesting, but what is the essence of this vital process? Some scientists have argued that this is a physical process, similar to the grinding of grain, and food is simply crushed in the stomach.

Others believed that this is a chemical process and food is involved in some kind of reactions. Generations of ancient philosophers could have exchanged witty arguments, and purely speculative ones, for centuries in such a dispute. But other times have come and the scientific method demanded to know it empirically.

But how can that be done? In 1752, the French naturalist René-Antoine Ferchault de Réaumur put a piece of meat into a small piece of pipe, closed at both ends with a metal net, and fed it to a hawk, which, as befits these birds, after a while belched a foreign body from the oesophagus. The analysis showed that the meat was partially digested. And since no physical influence was possible, this became a serious argument in favour of chemical reactions in the stomach.

Then Reaumur fed sponge to a hawk and, after its return in the same way, squeezed out of it the gastric juice, which was successfully able to chemically affect the meat moistened with it. The issue was resolved once and for all. I note that Reaumur did not limit himself to the first experiment refuting the competing hypothesis, but also put another one confirming the correctness of the proper hypothesis.

8.3. Science and Philosophy

Traditionally, scientists who have proven their ability to engage in science are awarded the title Doctor of Philosophy. These are just memories of the times when science in Europe, then called natural philosophy, was considered part of philosophy. Then their paths parted. I wrote about how this happened in the previous chapter.

Now these are two independent directions of knowledge. Consider the differences between these two respected intellectual pursuits. They are simple: philosophy is not always based on facts and experience and it does not follow the procedure described in Section 8.1. It can fascinating, but it's not science. Nevertheless, some connection between science and philosophy has been preserved and I am happy to talk about it.

8.3.1. *Occam's razor*

What philosophical principles are used in science in general and in physics in particular? Let's start with the so-called Occam's razor and the principle of sufficient reason, which relates to methodology. They do not speak about nature and its structure, but about the principles of selecting new hypotheses and theories. The medieval English scholastic philosopher William of Ockham stated, *"Entia non sunt multiplicanda praeter necessitatem."* This means that entities should not be multiplied beyond necessity.

In science, it is understood that it is not necessary to introduce new concepts or laws in order to explain some new phenomenon, if this phenomenon can be exhaustively explained by the old laws. It would seem that this is a prohibitive principle. But centuries ago, scientists did not know anything about the atomic nucleus, thermonuclear reactions, particle spin, not to mention neutrinos, mesons, quarks, inflation of the Universe, or dark matter. Why did these new entities not only enter science, but also confidently settle down there? If you can easily answer this question, then you should probably skip right to the beginning of the next section.

Well, for those who do not have an answer or are not sure of its correctness, let us clarify that Occam's razor is usually understood as a general principle stating that if there are several alternative explanations of a phenomenon that explain it equally well, then the simplest of them should be considered true. From the point of view of the history of ideas, this principle is based on the principle of sufficient reason, introduced by Aristotle: it is possible to assert the existence of something (an object, phenomenon, connection, pattern, etc.) only if there is grounds for this. Therefore, comparing simple and complex explanations from the point of view of this principle, it is easy to see that if the simple explanation is complete and exhaustive, then there is simply not sufficient reason to introduce additional components into the reasoning.

Then how do new concepts get into science? It's all about details such as "equally well" or "complete and exhaustive". New concepts in science appear when old concepts, hypotheses, theories, paradigms cease to be equally good or complete and exhaustive compared to new ones. And it's up to the scientists themselves to decide.

Note that many apply Occam's razor principle so often, sometimes implicitly, that they take it for granted. And this is not so. In addition to the principle of sufficient reason, on which it is based, ancient philosophers offered many alternative and even opposite things.

For example, Democritus of Abdera ($\Delta\eta\mu\acute{o}\kappa\rho\iota\tau o\varsigma$) adhered to the isonomy principle, which came from jurisprudence, more precisely from the principle of equality of all before the law. Briefly, it can be formulated as a well-known postulate: "Everything that is not forbidden is allowed", or rather exists in nature, albeit somewhere in a remote corner of the world. This principle is also called the principle of lack of sufficient reason.

Quite simply: if you are able to imagine a mermaid, and your opponent cannot prove that mermaids do not exist, or rather, their existence is prohibited for some compelling reason, then they are definitely splashing in the water somewhere. Indeed, why are they worse than goats or jellyfish, which we can meet in nature and therefore know about their existence (here equality before the law has surfaced.).

So when he sees a well-known figure with a characteristic bag in his hand, Occam will understand that this is a neighbour taking out garbage, and Democritus will suspect that this is a masterfully disguised alien carrying food for a domestic intelligent reptilian. Naturally, the scientific method is based on Occam's razor, although at the stage of generating ideas and putting forward hypotheses, it is permissible to be slightly like Democritus.

This principle also applies to the process of choosing new concepts and warns us against a false start: you cannot transfer any hypothesis to the category of theory or theory to the rank of a paradigm if it does not explain nature much better than its competitors. Even in spite of its beauty. In an imaginary museum of the history of science, there is a closet where beautiful but incorrect theories are stored in jars with formalin. Even if the theory "explains a little better," this is just a reason to consider it a promising candidate and continue testing. Any new concept used for explanations must also pass this test.

In conclusion, I quote the words of Albert Einstein: "Everything should be simplified as long as possible, but no more than that." It can be considered another formulation of the same principle of Occam's razor.

8.3.2. *Causality principle*

Now is the time to talk about the principles that are related with the properties of the very nature that we are studying. There are not so many of them, and I already talked in Section 5.3 about the uncertainty principle. I'll add a few more, only those that can be safely ranked among the main ones. Let's start with the principle of causality. We know that everything moves in time, and in one direction. We remember the past and we cannot influence it, but we can influence the future. These are experimental facts, because observation is also an experiment.

Let's add two rather simple terms, cause and effect, and we get that the cause always precedes the effect. Let's cross this conclusion with the constancy of the speed of light in vacuum and the conclusions of the special theory of relativity and find out what the principle of causality looks like in physics. For this, it is necessary to introduce the concept of a light cone and conclude that the cause always lies in the cone of the absolute past of the effect, and the effect always lies in the cone of the absolute future of the cause. The idea of the light cone is described in detail in books, in particular in my book *How the Universe Works* and Wikipedia. It is very important for science. Let's show this with a simple example. In the general theory of relativity, there are formal solutions to its equations that admit closed world lines.

Travelling along such a line, you can return to the starting point of space and time, having visited your past along the way. Based on the principle of causality, physicists reject such solutions, considering them non-physical. An important argument in favour of the theory of inflation was the analysis of whether two different sources of the CMB could be causally related to something that happened during or after the Big Bang. That is, is it possible that something can affect both sources of the relic radiation at once?

However, in the history of physics there have been ideas that violate the principle of causality. So, in 1949, an article was published in which the electromagnetic interaction of two particles was considered, using the half-sum of a retarded and an advanced wave. A retarded wave is an electromagnetic wave familiar to everyone, which, when emitted, propagates in all directions. A leading wave is a retarded wave, but after the reversal of time. It comes from all sides and gathers at one point.

But the main thing is that it propagates "against time" and is observed before it is emitted. This solution in physics has always been considered as violating the principle of causality and, therefore, non-physical. The authors proposed a quantum approach using both types of waves. Maybe they were some kind of profane?

No, John Wheeler was a famous physicist, a student of Einstein, and Richard Phillips Feynman later received the Nobel Prize in Physics. The ideas of the article did not receive further development, but the Wheeler–Feynman electrodynamics is remembered precisely because of the attempt to do without the principle of causality. However, it can be viewed as an unusual, but rejected idea, which remained mainly in the history of physics.

8.3.3. *Copernican principle*

Now let's recall the so-called Copernican principle. The name is associated with the transition from the geocentric system of the world to the heliocentric one. At the same time, Nicolaus Copernicus implicitly put forward the postulate that the Earth does not occupy any special position in the Universe. He placed the Sun in the centre, but now we consider it an ordinary star, one of many. The principle, named after the great astronomer, assumes that the laws of nature are universal and operate in the same way everywhere.

It is sometimes also called the principle of mediocrity, because in the Universe, according to it, there is no outstanding place. I have already mentioned that scientists checked whether there is evidence of a difference in the laws of physics operating in the region of distant astronomical objects and those that we observe here and now, and did not find any hints on this.

In connection with the Copernican principle, two details can be recalled. The first is related to an attempt to generalize it by adding the sameness of all moments in time. But there was a highlighted moment in the life of the Universe, it was the Big Bang. Having accepted the extended principle, its proponents must suppose that there was no Big Bang.

When the relic radiation was not yet discovered and the answer to the question of whether the Big Bang was not as obvious as it is now, a group of supporters of such an extended version proposed the so-called stationary cosmological model. In it, the Universe expanded, but remained the same at all times due to the continuous formation of new matter from nothing, or, in a mild version, from some new field unknown to science, called the C-field. The letter C is for the creation. Now the supporters of this theory are extremely rare.

The second question is related to the homogeneity of the Universe. If you take Copernicus' principle too literally, you can face a lot of questions. After all, it is clear that different places in space are not exactly the same. Points in interstellar space, on the surface of the Earth and at the centre of the Sun are difficult to confuse. On even larger spatial scales, we observe the large-scale structure of the Universe.

In some places, there are clusters of galaxies; in others, there are voids, where galaxies are almost absent. It can be assumed that there is some particularly large scale on which the Universe becomes homogeneous. Then the Copernican principle should be applied with this refinement.

And if there is no such scale, then we can recall the idea of the hierarchy of structures. Until now, all objects have been combined into something larger. Quarks unite in hadrons, nucleons in nuclei, nuclei, and electrons in atoms, those in molecules, which form bodies. Planets and stars form star systems. Those form galaxies, their clusters, and superclusters. At the same time, it seemed to scientists more than once that the limit had been reached, both on the smallest scales and on large ones, but each time there were even smaller structures at one end and even bigger ones at the other. Maybe the Copernican principle should be formulated taking into account the possible hierarchy of structures.

Note that we have a naturally occurring maximum scale. This is the size of the so-called cosmological horizon, which I mentioned in Section 2.10. Let me remind you that the age of the Universe and the speed of light are finite, so there is a limiting distance that light can travel from the Big Bang to the present moment.

This is the radius of the cosmological horizon. We cannot see the objects located further than it, because the light from them simply did not have time to reach us. If the Universe were transparent all the time, then we would see at the very horizon the light emitted immediately at the moment of the Big Bang, reddened due to expansion and turned into very long radio waves. The size of the cosmological horizon is growing all the time. But its very existence changes a lot in the approach to many cosmological issues.

Scientists have speculated a lot about whether the Universe is finite or infinite. This is an interesting question, but if we understand that we cannot see anything beyond the cosmological horizon, then it is no longer so fundamental.

If we recall the hierarchy of structures, then we should not be interested in structures with scales larger than the horizon radius, because they are still impossible to detect. More precisely, this question may interest us, but we must be aware that we do not have the opportunity to verify the correctness of any answer to it.

8.3.4. *Symmetry*

Another question that naturally arises in connection with the Copernican principle is the question of the symmetry of the Universe and its violation. It is well known that the existence of any symmetry in a problem greatly facilitates its solution. The famous mathematician Amalie Emmy Noether showed that the law of conservation of momentum follows from the uniformity of space, the law of conservation of energy follows from uniformity in time, and the law of conservation of angular momentum follows from the isotropy of space.

And these homogeneities are nothing more than the symmetry of space-time with respect to shifts in space and time. Isotropy is symmetry about rotations in space. We have repeatedly checked these conservation laws and each time we made sure that they are fulfilled. Does this mean that the Universe is obviously symmetric?

Let us define the operation of mirror reflection or the equivalent operation of the simultaneous change of the sign of all three spatial coordinates. It is also called the P-transformation. In classical physics, it does not change mass, energy, angular momentum and electric charges and potentials, as well as the strength of the magnetic field. But it changes the signs of coordinates, velocities, accelerations, impulses, forces and strengths of the electric field. In a world mirroring ours, you must use a left-handed corkscrew in the formulation of their corkscrew rule.

The concept of the parity of elementary particles, associated with the symmetry with respect to the P-transformation, turned out to be very valuable in the physics of the microcosm. Each particle is described by a certain function called a wave function. It either remains unchanged or changes its sign after the P-transformation. In the first case, the parity of the particle is considered to be 1, in the second it is equal to -1.

Half of the 1963 Nobel Prize in Physics was awarded to the Hungarian-born American physicist Eugene Wigner (Wigner Jenő Pál) with the formulation "for his contributions to the theory of the atomic nucleus and the elementary particles, particularly through the discovery and application of fundamental symmetry principles". For a time, physicists were convinced that parity (more formally P-parity) is conserved. Then they found out that it really conserved, but only with strong and electromagnetic interactions.

American theoretical physicists of Chinese origin Chen Ning Yang and Tsung-Dao Lee suspected that two particles known at that time, having different parities, but the same set of other properties, are actually the same particle. And they assumed that in processes caused by weak interaction, P-invariance is violated. This would

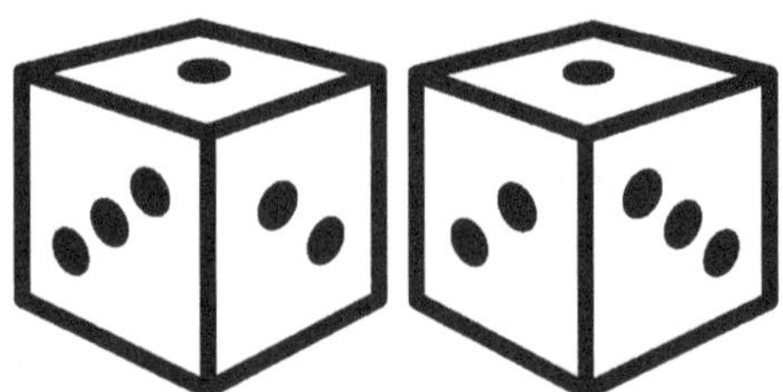

Figure 8.8. An ordinary dice, when reflected, does not transform into itself. The so-called "right" dice passes into the "left" one, and vice versa. The case was clearly not without weak interaction.

mean that the world is not symmetrical with respect to the mirroring and the "right side" in it differs from the "left". The differences between them are shown in Fig. 8.8 for the example of "right" and "left" dices.

To test their hypothesis, Chen Ning Yang and Tsung-Dao Lee came up with a not very complicated experiment that does not require an accelerator and is not associated with exotic particles. Chien-Shiung Wu agreed to conduct it according to the proposed scheme. She was also born in China and worked in the USA, but she was not a theorist, but an experimenter.

Radioactive cobalt-60 decays by beta decay, i.e. due to weak forces, according to the scheme $^{60}\text{Co} \to {}^{60}\text{Ni} + e^- + \bar{\nu}_e + 2\gamma$ with the emission of an electron, antineutrino and two gamma quanta. In Madame Wu's experiment, cobalt atoms were cooled to a very low temperature and placed in a magnetic field. As a result, their magnetic and angular moments are aligned along this field. This fixed the orientation of the atoms, naturally, with the exception of the weak influence of zero-point vibrations.

Experiment has shown that during beta decay, electrons flew out of them mainly in the direction opposite to the direction of the magnetic field. At the same time, 60% of gamma quanta were emitted in one direction (the upper hemisphere in Fig. 8.9), and 40% in the opposite direction.

So, our world is not mirror-symmetrical. If it were, electrons and gamma quanta would radiate up and down with equal probability. In Fig. 8.9 the magnetic field is directed upward. Electrons in the

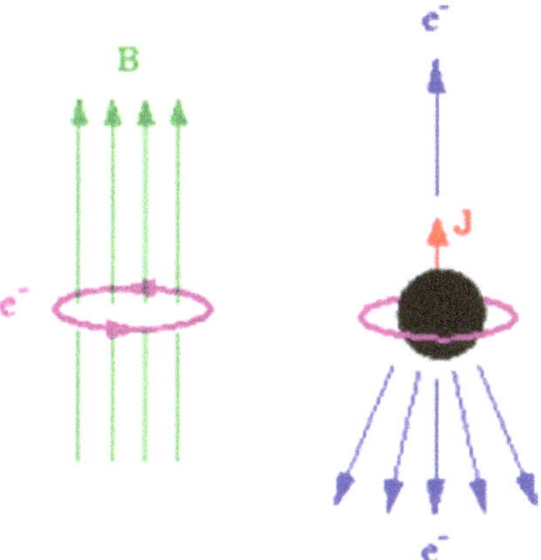

Figure 8.9. Scheme of the experiment to demonstrate the violation of parity in weak interactions.

real world are mostly emitted downward. If you reflect Fig. 8.9 in a horizontal mirror orthogonal to the direction of the field, then the top and bottom on it will change places, and the direction of the magnetic field will remain the same, because it is created by an electric current flowing through a ring or coil. The current reflected in the mirror flows along the ring in the same direction as the real one and creates the same magnetic field. But the electrons in the mirror world are mainly radiated upwards.

Chen Ning Yang and Tsung-Dao Lee received the Nobel Prize in Physics for 1957 "for their penetrating investigation of the so-called parity laws which has led to important discoveries regarding the elementary particles". By the way, they were right: two supposedly different elementary particles really turned out to be the same.

The news of parity non-conservation in weak interactions greatly upset physicists. To restore their peace of mind and glue the cracked picture of the world, Lev Landau suggested changing the signs of all electric charges simultaneously with reflection. Invariance during such an operation was called combined parity or CP symmetry.

However, in 1964, the Americans James Watson Cronin and Val Logsdon Fitch showed that combined parity is also violated, with the same weak interaction. For this discovery they won the 1980

Nobel Prize in Physics "for the discovery of violations of fundamental symmetry principles in the decay of neutral K-mesons."

Now we understand that the symmetry of the equations with respect to any transformation is not enough for the symmetry of their solutions. I already talked in Section 6.3 about spontaneous symmetry breaking and its consequence — the Higgs mechanism. If earlier in the theory of the microcosm, symmetry was in vogue, then later its violation. At the same time, both fashion leaders — Wigner and Nambu— received Nobel Prizes. However, the ideas of spontaneous symmetry breaking have a long history. They explain, for example, the existence of permanent magnets and other nuances of ferromagnetism.

You see what happens when a philosophical principle, formulated in a rather general way, in our case, the Copernican principle, falls into the hands of scientists. To begin with, a rather vague formulation is replaced by a much more specific concept of symmetry with respect to a certain class of transformations and thereby formalized.

Physicists, using a mathematical apparatus, transform the initial idea into new physical concepts, the same parity, and immediately check the conclusions experimentally. The difference in approaches typical of science and philosophy is obvious. Emmy Noether, the author of the theorem she proved in 1918 on the connection between symmetry and conservation laws, would never have connected Copernicus' principle with the conservation of momentum if she had reasoned like a philosopher.

Now let's think about symmetry when changing the direction of the fourth coordinate — time. This question is far from obvious. Time differs from spatial coordinates in that any object, including ourselves, must move along the time axis in a certain direction, regardless of its will. In the real physical world, there is always the so-called arrow of time, a term coined by Sir Arthur Stanley Eddington. It is directed from the past to the future.

We can distinguish the past from the future based on the principle of causality. The difference between the past and the future for a person is clearly visible from the fact that he or she remembers the past, but does not know the future. Therefore, the direction of the arrow of time associated with the principle of causality is sometimes called the psychological arrow of time. But, in my opinion, it is preferable to talk about the arrow of time associated with the principle of causality, without explicitly drawing on secondary concepts from psychology.

How does the arrow of time manifest itself in various branches of physics? Does the arrow of time always exist in them? In mechanics, the laws of motion of bodies are symmetrical relative to the change in the sign of time until friction or other dissipative forces appear. In thermodynamics, all processes are divided into reversible, for which the sign of time is not important, and irreversible, for which the direction of the arrow of time is determined by the increase in entropy.

Let's go to the microcosm. From sufficiently convincing theoretical considerations, it can be shown that, under fairly simple assumptions, the world should be invariant with respect to the CPT transformation, when the sign of time, all spatial coordinates and the signs of all electric charges change simultaneously. That is, a change in the direction of the flow of time is added to the previously named CP transformations. Since it has been experimentally proven that CP invariance is violated, this must mean violation of T invariance.

It is clear that in real life we have no way to reverse the flow of time and all this reasoning is nothing more than classical "thought experiments". And thought experiments, unlike real ones, are useful in presenting some material, but they cannot bring unexpected results. However, this does not prevent writing long treatises on the reversibility and irreversibility of time, introducing several different arrows of time. For example, some introduce the cosmological arrow of time — the direction of time in which the Universe expands — and then start long speculations about whether time will turn back the moment the Universe stops expanding and begins to contract. However, from modern astronomical data, it turns out that this is impossible and the Universe will expand forever.

8.3.5. *Anthropic principle*

Let's move on to the so-called anthropic principle and briefly describe its main provisions. It is related not so much to science as to philosophy. Many scientists have written more than once that intelligent life, represented in the Universe by at least humanity, was made possible by a series of incredibly fortunate circumstances. A small change in the interaction parameters or the masses of elementary particles would lead to the impossibility of intelligent life. Moreover, the whole picture of the Universe would be completely different.

Naturally, the question arose: why is the Universe so adapted for the emergence of intelligent life? The answer of theologians is not difficult to foresee. But physicists found it very difficult to offer their own version of the answer.

The following considerations came to the rescue: in order to discuss the issue of the convenience of our Universe for the creation of intelligent life, we need not only a Universe that satisfies this criterion, but also interlocutors who discuss it. In other words, if the Universe is poorly adapted for the emergence of intelligent life, intelligent beings will not appear in it and, accordingly, there will be no one to complain about its poor design.

The fact that you are reading this book is also connected with the fact that the Universe with all its interactions, constants, and other parameters made possible the existence of both author and readers. If we have many options for the structure of the Universe, then discussion of this issue is possible only in those that have someone to discuss it. To some extent, this echoes the ideas of the influence of the device on the measurement process, only the whole of humanity, or rather its existence, acts as a device.

I will draw a more specific picture of the application of the anthropic principle below describing the concept of the Multiverse. In the meantime, I will note that no one and nothing will prevent us from applying the idea underlying the anthropic principle, not to the entire Universe, but to its part. I'll start with an example. Astronomers using the Kepler space telescope and the Gaia (Global Astrometric Interferometer for Astrophysics) satellite have studied the change in the brightness of 369 stars similar to the Sun. It turned out that the overwhelming majority of them have much greater variability than our star.

Should we start to wonder? After all, the Copernican principle states that the Sun should not stand out in any way among stars similar to it, but here it clearly stands out. But considerations related to the anthropic principle can also be applied. Let us assume that the strong variability of the radiation of the central star prevents the formation of intelligent life on its planets. For example, it simply kills all living things or makes their existence so difficult that life does not reach the stage of reason. The whole evolution is aimed at solving the problem of survival. You have to fight not with other living beings, but with the whims of the star. If this assumption is correct (of which there is no certainty), then intelligent life can only be found near stars with low variability. Then the result will be quite expected and in no way violate the Copernican principle.

So the anthropic principle somewhat clarifies the Copernican principle, understood literally. All the stars are equal, but the Sun stands out because we, who are discussing it, are connected with this particular star. The same can be said about the Milky Way and other galaxies.

8.3.6. *Scientific criterion*

It's time to discuss the criteria for the scientific validity of various theories, teachings, and hypotheses. For a long time, this question was more concerned with philosophers developing the so-called philosophy of science than scientists themselves.

It is clear that scientists in the process of learning have acquired the ability to distinguish theories according to their degree of scientific knowledge, even without having a clearly formulated criterion. But the philosophers wanted to have an explicit formulation of the criterion. In addition, they clearly distinguished between the concepts of scientific and correctness. The hypothesis that the Moon consists of green cheese is clearly wrong, but it is quite scientific.

The modern criterion of scientific knowledge is based on the works of the Austrian philosopher Karl Raimund Popper and his followers — Thomas Kuhn, Paul Feyerabend, Imre Lakatos, and others. The main thing in it is the principle of falsifiability of the hypothesis, that is, under what conditions we are ready to recognize it as incorrect.

So, in the case of the aforementioned hypothesis that the Moon consists of green cheese, in order to refute it, astronauts need only find a piece of something on the Moon that is neither green cheese nor an object alien to the Moon, such as a fallen meteorite. But the hypothesis that the Moon is partially composed of green cheese is not so easy to refute.

If we find at least one piece of green cheese inside the Moon, then we will brilliantly confirm the hypothesis, but in order to refute it, we need to dig up all the bowels of the Moon and find no green cheese at all. However, we have a principal opportunity to refute this hypothesis, albeit a purely theoretical one. Those hypotheses that, in principle, cannot be refuted should be considered unscientific ones.

8.3.7. *Multiverse as an example of beautiful unscientific ideas*

Note that if an idea does not meet the scientific criterion, then it should not be immediately rejected with indignation. It's just not a scientific hypothesis; there are many of them in our life.

Even in science there are ideas that fall into this category, for example the idea of the Multiverse, which I will describe. Let's start with the name. The words Multiverse or Omniverse are used instead of the Universe. The famous Polish science fiction writer Stanislaw Lem used the term *Polyversum*, which replaces the Latin word *Universum*.

What is this idea? It arises when combining the ideas of inflation and spontaneous symmetry breaking and uses the anthropic principle. To begin with, at the start of inflation, the region that now corresponds to the cosmological horizon with a radius of about 40 billion light-years was substantially smaller than the atomic nucleus.

If the size of the Universe at that moment was significantly larger, then now in the Universe there are a huge number of parts that are not causally related to each other. If we take the size of the Universe before the beginning of inflation at 1 mm, then there would be at least 10^{69} parts. Any estimate will give us an incredible number of regions of the Universe that are not observed from any other ones, and it is impossible to know what is happening even in the nearest of the other parts during the entire existence of the Universe.

Suppose now that inflation is associated with the transition from a false vacuum to a true one, which, in turn, is associated with spontaneous symmetry breaking and the Higgs mechanism. Moreover, let's assume that in different parts of the Universe that are not causally connected, the results of spontaneous symmetry breaking may be different. That's why it is called spontaneous one and not predetermined in advance.

Parts of the Universe that have grown from these pieces would have various symmetry breakings. This means different physics, namely different interaction constants, different properties of elementary particles, in the extreme case — even a different number of dimensions of space due to a different number of compactified dimensions. Returning to the concept of Multiverse, we will get not only a gigantic number of parts of the Universe independent from each other, but a huge number of parts of the Universe with different physics.

It may seem that this is a flight of imagination. However, nature has made sure that we have a very simple and intuitive model of *Polyversum*. As I explained in Section 4.1, a piece of iron is made up of various regions called magnetic domains, within which all the magnetic moments of the atoms are lined up in parallel. In other domains, they are also lined up in parallel, but in different directions. The boundaries between domains — the so-called domain walls — have an additional energy density compared to domains. Without an external magnetic field, the direction of the magnetic moments inside the domains can be considered the result of spontaneous symmetry breaking.

Imagine that the Universe also consists of separate domains, within which the physical laws are the same. However, they are different in different domains. Domains are separated by domain walls. The size of the cosmological horizon is significantly smaller than the dimensions of domains, therefore the observed part of the Universe falls into one domain and the same physical laws with the same parameters operate in it. And somewhere far away, beyond the cosmological horizons, there are many other parts with different laws of nature. One can only regret that experimental physicists will never be able to use their instruments to study objects from other domains.

The idea of a Polyversum with a domain structure allows many different possible worlds to be realized practically independently of each other. In such a picture, the anthropic principle comes to the fore. If the idea that a unique Universe is well arranged for intelligent life looks strange, then it is much less surprising that we can find at least one piece of the Universe suitable for the emergence of intelligent life from a huge number of options with different physical laws. It is incredibly difficult to win a jackpot by buying one lottery ticket, but it is natural if you bought 10^{69} ones.

Agree that this is a beautiful idea based on modern scientific concepts. However, modern physics is based on the fact that nothing can move faster than the speed of light, so we do not have cannot test the hypothesis of the Multiverse without coming up with a way to quickly move over great distances, say through a wormhole.[h] The absence of falsifiability forces us to recognize it as unscientific. One cannot fail to note the intellectual beauty of this theory. But you cannot argue with the scientific criterion, and that's why I am talking about this wonderful idea here and not in parts of the book, which described the real structure of the world.

Getting to know the theory of the Multiverse, you involuntarily become like an ancient Greek listening or reading Aristotle's

---

[h]In existing models of such hypothetical tunnels, connecting different distant corners of space-time, any macroscopic object travelling through the wormhole will be torn apart by tidal forces into elementary particles. Moreover, if such holes were formed during the Big Bang, inflation moved them far beyond our cosmological horizon. This makes their detection more than questionable.

reasoning. It sounds beautiful and convincing, you can't check it, but you want to believe it.

We know that only less than a millennium has passed since scientists realized that they had to escape from the shaky building with the signboard "Natural Philosophy" and start building a solid building with the inscription "Science". But this is no reason not to visit old neighbours sometimes, especially since philosophy is by no means reduced to natural philosophy. And the evidence of this is the scientific criterion formulated by philosophers, which was discussed in this book.

SUMMARY

The bonus miles are over. Now we can safely say that our journey is complete. I hope it was useful and even interesting in some ways. We briefly ran through the entire nature of the world around us, on all its scales. Starting from the Universe as a whole, we got to elementary particles and even quarks, which make up some of these particles. We learned something about previously unknown interactions. About what is inside the Sun and the Earth, about the climate and the magnetic field of our planet, about the states of matter, exotic and not so, about the electronic shells of atoms and their nuclei. About rainbows, lightning and magnets, phase transitions, superconductivity and superfluidity. We got acquainted with the mysterious world of quanta, neutron stars, black holes and strange naked singularities.

For lovers of the humanitarian aspects of science, I added a bonus. I told about the scientific method, about the historically important stage of separating science from natural philosophy, and about philosophical principles and criteria in science.

WHAT ELSE TO READ?

It is clear that our excursion could not cover everything, and we examined rather superficially what we visited. If you want to continue acquaintance with the structure of the world, then you have the entire Internet for this, starting with Wikipedia and popular science magazines. Keywords taken from this book could greatly speed up the search process. But if you are interested in books, then I am ready to recommend a couple to you. This is my subjective choice, which does not pretend to be optimal or comprehensive.

I'll start with classic books written a long time ago. On astronomy, it is worth reading Isaac Asimov's book *The Universe: From Flat Earth to Quasar*. He also wrote a couple of books about physics. It is clear that these books, written more than half a century ago, cannot contain descriptions of the discoveries of this time, but I hope that the book you are reading fills this gap.

The foundations of the special theory of relativity are perfectly described with formulas in the book *Assumption and myth in physical theory* by Hermann Bondi. Cosmology, general relativity, black holes and related issues are described in our book *How the Universe Works: An Introduction into modern cosmology* by Serge Parnovsky and Aleksei Parnowski. I recommend the books *Wonders of Physics* by Lev Aslamazov and Andrey Varlamov and *The kaleidoscope of physics* by Attilio Rigamonti, Andrey Varlamov and Jacques Vilen, which perfectly complement what I wrote here. I have already mentioned some other books in the text.

Well, there is also a world of hundreds and thousands of books for every taste, level and diligence. It is worth casting a line into this sea from time to time in an attempt to fish out your goldfish. Believe me, they could be found there.

CPSIA information can be obtained
at www.ICGtesting.com
Printed in the USA
BVHW060216291222
654929BV00002B/8